Methods in Molecular Biology

Series Editor
John M. Walker
School of Life and Medical Sciences
University of Hertfordshire
Hatfield, Hertfordshire, UK

For further volumes:
http://www.springer.com/series/7651

For over 35 years, biological scientists have come to rely on the research protocols and methodologies in the critically acclaimed *Methods in Molecular Biology* series. The series was the first to introduce the step-by-step protocols approach that has become the standard in all biomedical protocol publishing. Each protocol is provided in readily-reproducible step-by-step fashion, opening with an introductory overview, a list of the materials and reagents needed to complete the experiment, and followed by a detailed procedure that is supported with a helpful notes section offering tips and tricks of the trade as well as troubleshooting advice. These hallmark features were introduced by series editor Dr. John Walker and constitute the key ingredient in each and every volume of the *Methods in Molecular Biology* series. Tested and trusted, comprehensive and reliable, all protocols from the series are indexed in PubMed.

Single Cell Transcriptomics

Methods and Protocols

Edited by

Raffaele A. Calogero

Molecular Biotechnology Center, University of Torino, Torino, Italy

Vladimir Benes

European Molecular Biology Laboratory, Heidelberg, Germany

Humana Press

Editors
Raffaele A. Calogero
Molecular Biotechnology Center
University of Torino
Torino, Italy

Vladimir Benes
European Molecular Biology Laboratory
Heidelberg, Germany

ISSN 1064-3745 ISSN 1940-6029 (electronic)
Methods in Molecular Biology
ISBN 978-1-0716-2758-7 ISBN 978-1-0716-2756-3 (eBook)
https://doi.org/10.1007/978-1-0716-2756-3

This Humana imprint is published by the registered company Springer Science+Business Media, LLC, part of Springer Nature.
The registered company address is: 1 New York Plaza, New York, NY 10004, U.S.A.

Preface

The first article on bona fide single-cell transcriptomics (scRNAseq) [1] is now 13 years old. Undisputedly, this article has paved the way to establishing a flourishing methodology focusing on analyses of transcripts' repertoire in individual cells. With the increasing interest in this type of analyses and realization of their potential, caveats and limitations of this technology not only in the laboratory but also in analysis of generated data have become apparent too. A "re-purposing" of the MMLV reverse transcriptase originally used for mapping of 5′-end [2] termed "SMART" (Switching Mechanism at 5'end of RNA Transcript) has started the scRNA-seq method in earnest [3]. Essentially, all full-length as well as 3′-end focused single-cell transcriptomic data are generated with the application of this approach.

However, fairly soon, it has also turned out that there are very few biological systems with freely available single cells directly amenable to the protocol. Therefore, it has been and still is necessary to develop protocols for fixation, dissociation, and separation of individual cells, steps particularly critical for success of any single-cell transcriptomic experiment. That has become even more relevant with the introduction of microfluidic droplet-based systems [4].

With individual contributions we wish to bring readers' attention to options available, including those enabling exploration of plant and bacterial single cells. Protocols included in this tome are by no means only possibilities; every day we are seeing new efforts to optimize current methods or provide entirely new protocols, and this trend will undoubtedly continue.

As happened with other high-throughput technologies, single-cell transcriptomics also require bioinformatics to make the data suitable to further interpretation. Single-cell RNA-seq bioinformatics is today a very active area and provides a wide variety of software to work with and interpret scRNA-seq data. This book was developed to provide hands on examples of both basic and advance bioinformatics methodologies. It includes a set of chapters describing, on one side, basic analytical procedures as single-cell data QC, dimensionality reduction, clustering, and cluster-specific features selection. On the other side, there are a set of chapters dedicated to the analysis of advance methodologies as RNA velocity, multi-modal data integration, and single-cell RNA editing.

Taken together, the overall structure of this book provides an up-to-date instrument to facilitate an informed selection of single-cell wet and bioinformatics protocols based on the researcher experiment requirements.

Torino, Italy — *Raffaele A. Calogero*
Heidelberg, Germany — *Vladimir Benes*

References

1. Tang F, Barbacioru C, Wang Y, Nordman E, Lee C, Xu N et al (2009) mRNA-Seq whole-transcriptome analysis of a single cell. Nat Methods 6(5):377–382. https://doi.org/10.1038/nmeth.1315

2. Chenchik A, Diachenko L, Moqadam F, Tarabykin V, Lukyanov S, Siebert PD (1996) Full-length cDNA cloning and determination of mRNA 5′ and 3′ ends by amplification of adaptor-ligated cDNA. Biotechniques 21(3):526–534. https://doi.org/10.2144/96213pf02

3. Picelli S, Björklund ÅK, Faridani OR, Sagasser S, Winberg G, Sandberg R (2013) Smart-seq2 for sensitive full-length transcriptome profiling in single cells. Nat Methods 10(11):1096–1098. https://doi.org/10.1038/nmeth.2639

4. Macosko EZ, Basu A, Satija R, Nemesh J, Shekhar K, Goldman M, Tirosh I, Bialas AR, Kamitaki N, Martersteck EM, Trombetta JJ, Weitz DA, Sanes JR, Shalek AK, Regev A, McCarroll SA (2015) Highly parallel genome-wide expression profiling of individual cells using nanoliter droplets. Cell 161(5):1202–1214. https://doi.org/10.1016/j.cell.2015.05.002

Contents

Contributors

MARCO ALDINUCCI • *Department of Computer Science, University of Turin, Turin, Italy*
LUCA ALESSANDRI • *Molecular Biotechnology Center, University of Torino, Turin, Italy; Molecular Biotechnology Center, University of Turin, Turin, Italy*
FEDERICA ANTICO • *Molecular Biotechnology Center, University of Torino, Torino, Italy*
MADDALENA ARIGONI • *Molecular Biotechnology Center, University of Torino, Torino, Italy*
MARCO BECCUTI • *Department of Computer Science, University of Torino, Turin, Italy; Department of Computer Science, University of Turin, Turin, Italy*
IVAN BEREST • *Structural and Computational Biology Unit, European Molecular Biology Laboratory, Heidelberg, Germany; Data Science, Pharma Research and Development, Roche Innovation Center Zurich, Zurich, Switzerland*
RAFFAELE A. CALOGERO • *Molecular Biotechnology Center, University of Torino, Torino, Italy*
IACOPO COLONNELLI • *Department of Computer Science, University of Turin, Turin, Italy*
SANDRO G. CONTALDO • *Department of Computer Science, University of Turin, Turin, Italy*
FRANCESCA CORDERO • *Department of Computer Science, University of Torino, Turin, Italy*
KATHARINA DANIELSKI • *10x Genomics B.V, Leiden, The Netherlands*
RILEY S. DRAKE • *Institute for Medical Engineering and Science (IMES), Massachusetts Institute of Technology, Cambridge, MA, USA; Department of Chemistry, Massachusetts Institute of Technology, Cambridge, MA, USA; Koch Institute for Integrative Cancer Research, Massachusetts Institute of Technology, Cambridge, MA, USA; The Ragon Institute of Massachusetts General Hospital, Massachusetts Institute of Technology and Harvard University, Cambridge, MA, USA; Broad Institute of MIT and Harvard, Cambridge, MA, USA*
ANDREW FARMER • *National Center for Genome Resources, Santa Fe, NM, USA*
ADRIANO FONZINO • *Department of Biosciences, Biotechnology and Biopharmaceutics, University of Bari "A. Moro", Bari, Italy*
MARTA GAI • *Molecular Biotechnology Center, University of Torino, Torino, Italy*
IAN GERRARD • *BD Biosciences GmbH, Heidelberg, Germany*
VINCENT HAHAUT • *Institute of Molecular and Clinical Ophthalmology Basel (IOB), Basel, Switzerland*
CHRISTINA HOMBERGER • *University of Würzburg, Würzburg, Germany*
RENATA JURKOWSKA • *School of Biosciences, Cardiff University, Cardiff, Wales, UK*
MARC LIBAULT • *Center for Plant Science Innovation, University of Nebraska-Lincoln, Beadle Center, Lincoln, NE, USA; Department of Agronomy and Horticulture, University of Nebraska-Lincoln, Beadle Center, Lincoln, NE, USA*
VADIR LOPEZ-SALMERON • *BD Biosciences GmbH, Heidelberg, Germany*
J. CHRISTOPHER LOVE • *Koch Institute for Integrative Cancer Research, Massachusetts Institute of Technology, Cambridge, MA, USA; The Ragon Institute of Massachusetts General Hospital, Massachusetts Institute of Technology and Harvard University, Cambridge, MA, USA; Broad Institute of MIT and Harvard, Cambridge, MA, USA; Department of Chemical Engineering, Massachusetts Institute of Technology, Cambridge, MA, USA*

PIETRO MANDREOLI • *Institute of Biomembranes, Bioenergetics and Molecular Biotechnologies, National Research Council (CNR), Bari, Italy; Department of Biosciences, University of Milan, Milan, Italy*

ANDREW NAVIA • *Institute for Medical Engineering and Science (IMES), Massachusetts Institute of Technology, Cambridge, MA, USA; Department of Chemistry, Massachusetts Institute of Technology, Cambridge, MA, USA; Koch Institute for Integrative Cancer Research, Massachusetts Institute of Technology, Cambridge, MA, USA; The Ragon Institute of Massachusetts General Hospital, Massachusetts Institute of Technology and Harvard University, Cambridge, MA, USA; Broad Institute of MIT and Harvard, Cambridge, MA, USA*

MARTINA OLIVERO • *Department of Oncology, University of Torino, Torino, Italy; Candiolo Cancer Institute-FPO, IRCCS, Candiolo, TO, Italy*

MALTE PAULSEN • *DanStem Institute, University Copenhagen, Copenhagen, Denmark*

GRAZIANO PESOLE • *Department of Biosciences, Biotechnology and Biopharmaceutics, University of Bari "A. Moro", Bari, Italy; Institute of Biomembranes, Bioenergetics and Molecular Biotechnologies, National Research Council, Bari, Italy*

ERNESTO PICARDI • *Department of Biosciences, Biotechnology and Biopharmaceutics, University of Bari "A. Moro", Bari, Italy; Institute of Biomembranes, Bioenergetics and Molecular Biotechnologies, National Research Council, Bari, Italy*

SIMONE PICELLI • *Institute of Molecular and Clinical Ophthalmology Basel (IOB), Basel, Switzerland*

MARIA LUISA RATTO • *Molecular Biotechnology Center, University of Torino, Turin, Italy*

DANIELA D. RUSSO • *Institute for Medical Engineering and Science (IMES), Massachusetts Institute of Technology, Cambridge, MA, USA; Department of Chemistry, Massachusetts Institute of Technology, Cambridge, MA, USA; Koch Institute for Integrative Cancer Research, Massachusetts Institute of Technology, Cambridge, MA, USA; The Ragon Institute of Massachusetts General Hospital, Massachusetts Institute of Technology and Harvard University, Cambridge, MA, USA; Broad Institute of MIT and Harvard, Cambridge, MA, USA*

ANTOINE-EMMANUEL SALIBA • *Helmholtz Institute for RNA-based Infection Research, Helmholtz Centre for Infection Research, Würzburg, Germany*

SUSAN SCHROEDER • *Department of Chemistry & Biochemistry, University of Oklahoma, Norman, OK, USA; Department of Microbiology & Plant Biology, University of Oklahoma, Norman, OK, USA*

ALEX K. SHALEK • *Institute for Medical Engineering and Science (IMES), Massachusetts Institute of Technology, Cambridge, MA, USA; Department of Chemistry, Massachusetts Institute of Technology, Cambridge, MA, USA; The Ragon Institute of Massachusetts General Hospital, Massachusetts Institute of Technology and Harvard University, Cambridge, MA, USA; Broad Institute of MIT and Harvard, Cambridge, MA, USA*

KELLY STREET • *Department of Data Science, Dana-Farber Cancer Institute, Boston, MA, USA; Department of Biostatistics, Harvard T.H. Chan School of Public Health, Boston, MA, USA*

MARCO ANTONIO TANGARO • *Institute of Biomembranes, Bioenergetics and Molecular Biotechnologies, National Research Council (CNR), Bari, Italy*

ANDREA TANGHERLONI • *Department of Human and Social Sciences, University of Bergamo, Bergamo, Italy*

SANDRA THIBIVILLIERS • *Center for Plant Science Innovation, University of Nebraska-Lincoln, Beadle Center, Lincoln, NE, USA*

SIMONE TIBERI • *Department of Molecular Life Sciences, University of Zurich, Zurich, Switzerland; Swiss Institute of Bioinformatics, University of Zurich, Zurich, Switzerland*
JANNES ULBRICH • *BD Biosciences GmbH, Heidelberg, Germany*
KOEN VAN DEN BERGE • *Department of Applied Mathematics, Computer Science and Statistics, Ghent University, Ghent, Belgium; Bioinformatics Institute Ghent, Ghent University, Ghent, Belgium; Department of Statistics, University of California, Berkeley, CA, USA*
MARTIN ARREOLA VILLANUEVA • *Institute for Medical Engineering and Science (IMES), Massachusetts Institute of Technology, Cambridge, MA, USA; Department of Chemistry, Massachusetts Institute of Technology, Cambridge, MA, USA; Koch Institute for Integrative Cancer Research, Massachusetts Institute of Technology, Cambridge, MA, USA; The Ragon Institute of Massachusetts General Hospital, Massachusetts Institute of Technology and Harvard University, Cambridge, MA, USA; Broad Institute of MIT and Harvard, Cambridge, MA, USA; Department of Biology, Massachusetts Institute of Technology, Cambridge, MA, USA*
MIKE VILME • *Institute for Medical Engineering and Science (IMES), Massachusetts Institute of Technology, Cambridge, MA, USA; Department of Chemistry, Massachusetts Institute of Technology, Cambridge, MA, USA; Koch Institute for Integrative Cancer Research, Massachusetts Institute of Technology, Cambridge, MA, USA; The Ragon Institute of Massachusetts General Hospital, Massachusetts Institute of Technology and Harvard University, Cambridge, MA, USA; Broad Institute of MIT and Harvard, Cambridge, MA, USA*
JÖRG VOGEL • *University of Würzburg, Würzburg, Germany; Helmholtz Institute for RNA-based Infection Research, Helmholtz Centre for Infection Research, Würzburg, Germany*
PHILIPP WEILER • *Institute of Computational Biology, Helmholtz Center Munich, Munich, Germany; Department of Mathematics, Technical University of Munich, Munich, Germany*
FEDERICO ZAMBELLI • *Institute of Biomembranes, Bioenergetics and Molecular Biotechnologies, National Research Council (CNR), Bari, Italy; Department of Biosciences, University of Milan, Milan, Italy*

Chapter 1

Guidance on Processing the 10x Genomics Single Cell Gene Expression Assay

Katharina Danielski

Abstract

The demand for technologies that allow the study of gene expression at single cell resolution continues to increase. One such assay was launched in 2016 by the US-based company 10x Genomics Inc. Utilizing the power of the single cell on a large scale (Zheng et al. Nat Commun 8:14049, 2017)—capturing thousands of cells at once—has shaped life sciences ever since and allowed researchers to discover new insights within their respective fields of study such as oncology, neurobiology, and immunology (among others). Obtaining high-data quality is the key to being able to make these meaningful discoveries, which in turn is directly linked to the quality of the initial cell (or nuclei) suspension that is used to load the 10x Genomics Chromium Single Cell Gene Expression assay. A successful workflow relies on a cell suspension which is fully dissociated, extremely clean, and of high viability. While the workflow itself has been detailed elsewhere (De Simone et al. Methods Mol Biol 1979:87–110, 2019), in this chapter we will focus on the importance of the quality of the initial cell suspension, as well as common mistakes that can occur while running a Single Cell Gene Expression assay. The descriptions of these tips and tricks refer to the current version of the 10x Genomics User Guide (Chromium Single Cell 3' Reagent Kits User Guide (v3.1 Chemistry Dual Index). https://support10xgenomicscom/single-cell-gene-expression/index/doc/user-guide-chromium-single-cell-3-reagent-kits-user-guide-v31-chemistry-dual-index) which can be downloaded from the Support section on the 10x Genomics website (10x Genomics website. https://www10xgenomicscom). These documents and user guides are continuously improved and updated; hence, it is important to regularly check the company's website for the most recent version.

Key words 10x Genomics, Chromium, Single Cell Gene Expression, Gel Bead-in-emulsion, Cell Ranger, Single cell RNA sequencing, Single cell transcriptomics

1 Introduction

The Chromium Single Cell Gene Expression solution [3] leverages a scalable microfluidic platform to allow for the direct measurement of 3′ digital gene expression at the single cell level. Profiling can be performed for 500–10,000 individual cells per sample, and the resulting data analysis can be used to quantify cell population heterogeneity, characterize cell types and states, and identify dynamic cellular transitions. A pool of ~3,500,000 10x Barcodes

Raffaele A. Calogero and Vladimir Benes (eds.), *Single Cell Transcriptomics: Methods and Protocols*, Methods in Molecular Biology, vol. 2584, https://doi.org/10.1007/978-1-0716-2756-3_1,

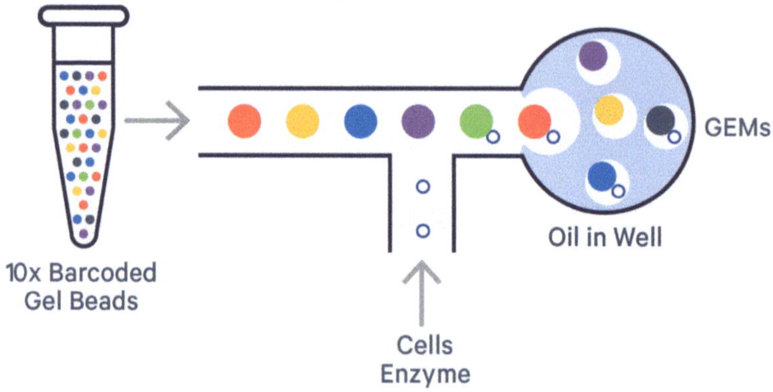

Fig. 1 Gel Beads-in-emulsion (GEM) generation. GEMs are generated by combining barcoded Single Cell 3′ v3.1 Gel Beads, Partitioning Oil, and a master mix containing cells, oligos, and enzymes onto Chromium Next GEM Chip

are sampled separately to index each cell's transcriptome. It is done by partitioning thousands of cells into nanoliter-scale Gel Beads-in-emulsion (GEMs). GEMs are generated by combining barcoded Single Cell 3′ v3.1 Gel Beads, Partitioning Oil, and a master mix containing cells, oligos, and enzymes onto Chromium Next GEM Chip G (Fig. 1). To achieve single cell resolution, cells are delivered at a limiting dilution, such that the majority (~90–99%) of generated GEMs contain no cell, while the remainder largely contain a single cell.

Immediately following GEM generation, the Gel Bead is dissolved, primers are released, and any co-partitioned cell is lysed. Primers containing an Illumina TruSeq Read 1 (read 1 sequencing primer), 16 nt 10x Barcode, 12 nt unique molecular identifier (UMI), and a 30 nt poly(dT) sequence are mixed with the cell lysate and a master mix containing reverse transcription (RT) reagents. Incubation of the GEMs produces barcoded, full-length cDNA from polyadenylated mRNA (Fig. 2). This assay was developed based on human and mouse cell types but has been used for a wide range of various species in the meantime. However, the lysis buffer is not capable of breaking down plant cell walls, which means those sample types require additional optimization upstream of the assay.

After incubation, GEMs are broken and pooled fractions are recovered. Silane magnetic beads are used to purify the first-strand cDNA from the post-GEM-RT reaction mixture, which includes leftover biochemical reagents and primers. Barcoded, full-length cDNA is amplified via PCR to generate sufficient mass for library construction. Enzymatic fragmentation and size selection are used to optimize the cDNA amplicon size. P5, P7, i7, and i5 sample indexes, and TruSeq Read 2 (read 2 primer sequence) are added via End Repair, A-tailing, Adaptor Ligation, and PCR (Fig. 3).

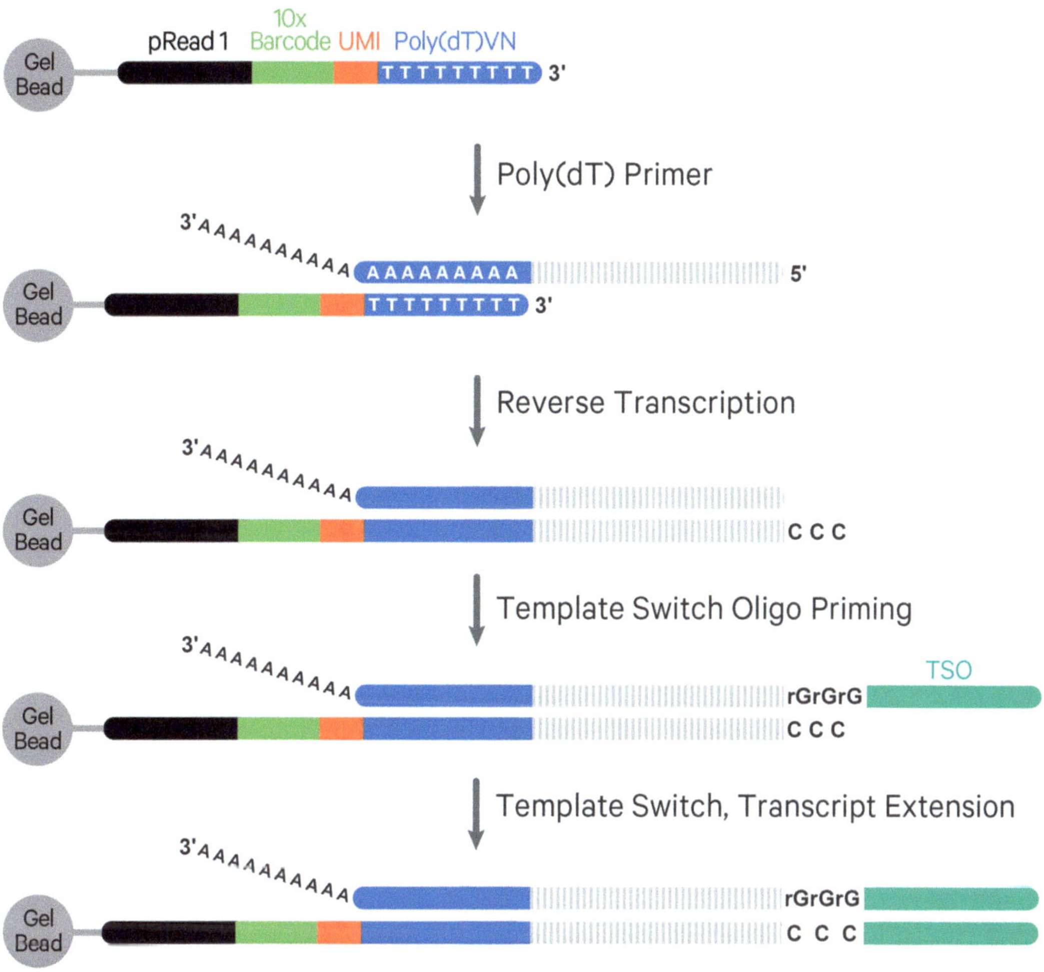

Fig. 2 cDNA generation. Primers containing an Illumina TruSeq Read 1, 16 nt 10x Barcode, 12 nt unique molecular identifier (UMI), and a 30 nt poly(dT) sequence are mixed with the cell lysate and a master mix containing reverse transcription (RT) reagents to produce barcoded, full-length cDNA from polyadenylated mRNA

A Chromium Single Cell 3' Gene Expression Dual Index library comprises standard Illumina paired-end constructs which begin and end with P5 and P7 (Fig. 4). The 16 bp 10x Barcode and 12 bp UMI are encoded in Read 1, while Read 2 is used to sequence the cDNA fragment. i7 and i5 index sequences are incorporated as the sample index reads. TruSeq Read 1 and TruSeq Read 2 are standard Illumina sequencing primer sites used in paired-end sequencing. The 10x Barcodes are used to associate individual reads back to the individual GEMs.

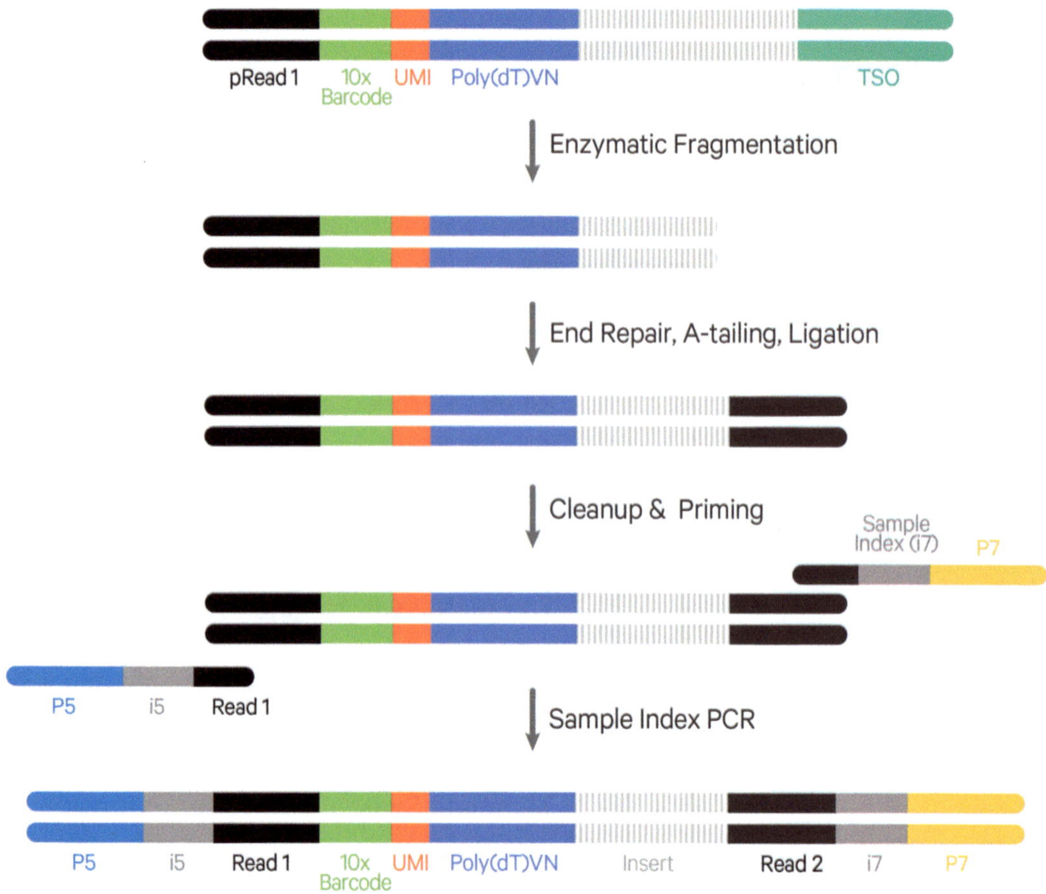

Fig. 3 Libraries generation. Enzymatic fragmentation and size selection are used to optimize the cDNA amplicon size. P5, P7, i7, and i5 sample indexes and TruSeq Read 2 are added via End Repair, A-tailing, Adaptor Ligation, and PCR

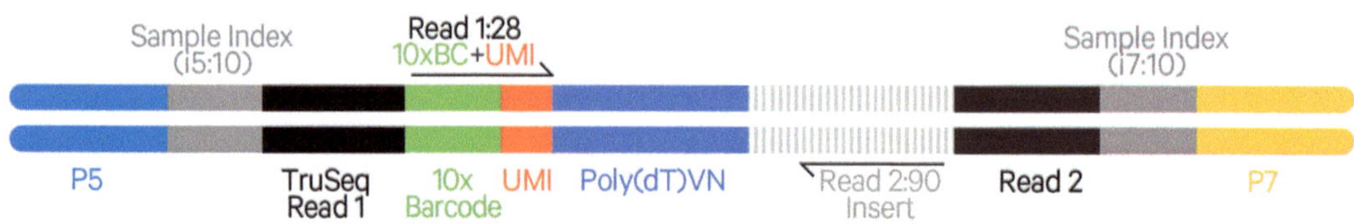

Fig. 4 Chromium Dual Index library. Chromium Single Cell 3′ Gene Expression Dual Index library comprises standard Illumina paired-end constructs which begin and end with P5 and P7

2 Materials

2.1 Plasticware

It is very important to use the recommended plasticware as indicated in the user guide [3], especially when working with GEMs. This includes the 0.2 ml 8-tube PCR strips and pipette tips. The recommended plastics have been tested and validated to be safe for the emulsions. In rare cases, other plasticware could break the emulsion before the reverse transcription and transcript barcoding can take place. This would lead to a loss of single cell characteristics for the sample, as the transcripts from the cells and the barcoded oligos would not be confined to an individual GEM droplet, but would float freely within the strip tube in a randomized manner. While the usage of strip tubes is limited to those that are specifically listed in the user guide, and it is strongly recommended to use the brand of pipette tips as indicated in the user guide as well, there are many users who work with alternative pipette vendors without issues, as long as the pipette tips are filtered and low binding (or low retention). Using nonrecommended pipette tips does bear some risks though, as some instances of broken emulsions are not always visible to the naked eye. The usage of DNA LoBind 1.5 ml or 2.0 ml microcentrifuge tubes is recommended to minimize the loss of sample and reagents due to adherence on the walls of the tubes. The same principle applies to the recommended low binding / low retention pipette tips.

2.2 Kits and Reagents

Regarding additional reagents that are required to run this assay, two in particular should be given careful consideration. First, the TE buffer used to resuspend the Template Switch Oligo (TSO) into a working stock solution needs to be of low EDTA concentration, i.e., only 0.1 mM EDTA. This reagent is going to be added to the reverse transcription master mix where the presence of too much EDTA could cause an inhibition of the RT enzyme, thereby causing the experiment to fail. It is recommended to purchase premade low-EDTA TE buffer to avoid any mistakes in final EDTA concentration.

The second reagent requiring special consideration is the SPRI-select Reagent Kit. While it is technically possible to use AMPure XP Beads instead, the use of the SPRI-select Reagent Kit is recommended due to the additional quality control performed by the manufacturer which negates the need to calibrate the beads manually in between different LOT numbers.

2.3 Equipment

It is not strictly necessary to use one of the thermal cyclers listed in the user guide [3]. Many other machines are also compatible with this assay. The important thing to keep in mind is that the thermal cycler needs to have a deep well module which fits 0.2 ml strip tubes and can cover 100 μl of liquid by the module wells. To ensure even heating of the entire 100 μl reaction volume, the reactions inside the strip tubes must not stand over the edge of the wells. Furthermore, the thermal cycler also needs to have a temperature-controlled lid which is capable of holding low temperatures as well. Please refer to the thermal cycler programs in the 10x Genomics User Guide to check on the required lid temperatures. Not conforming to the listed dedicated lid temperatures may compromise assay performance, particularly at the GEM-RT step, as higher lid temperatures could compromise the emulsions. Some thermal cyclers immediately ramp up to a standard lid temperature upon switching it on; in those instances, ensure that the machine has ample time to equilibrate to the right lid temperature before running the machine with the samples.

3 Methods

3.1 Cell Preparation

Preparation of a high-quality single cell suspension is the key to performing the Single Cell Gene Expression assay and yielding high-quality data from the experiment. This means that the single cell suspension needs to be fully dissociated, clean from various contaminants, and have high cell viability. Technically, this is not part of the single cell assay itself, but it is the most important preparation work for the assay. The extraction, dissociation, and purification of the cells and/or nuclei must be practiced and optimized before attempting to run the actual single cell assay. Some sample or tissue types are much more difficult to work with and will require substantial optimization up front. In the following paragraphs, we will dive into more specifics on how the quality of the single cell suspension should be controlled.

3.1.1 Cell Viability

The viability of the cell suspension should be as high as possible, ideally above 90%, but at least 70%. A high fraction of nonviable or dying cells may decrease cell recovery (*See* **Note 1**).

Cell viability can be checked in several ways. The most common method is Trypan blue staining which will stain dead cells dark blue, while leaving live cells unstained (*See* **Note 2**). Any cell viability below 70% is entering a gray zone for the resulting data quality. The lower the viability drops, the more ambient RNA will be included in the suspension. This ambient RNA could then be captured together with the viable cells, hence increasing background noise. If cell viability is far below 70%, a dead cell removal method could improve the overall cell viability as well as the data quality in the end [5] (*See* **Note 3**).

3.1.2 Cell Counting

When it comes to counting the number of cells in a suspension, manual methods (e.g., a hemocytometer or Neubauer chamber) or automated cell counters are appropriate. Each method has its advantages and disadvantages (*See* **Note 4**). Counting should always be done on the whole suspension, including both dead and alive cells (*See* **Note 5**). Cell counting can be done at various steps during the cell preparation protocol in order to have a better estimate of cell number so that they can be resuspended in an appropriate volume. However, it is important to have an accurate cell count of the final cell suspension that is going to be loaded into the single cell assay (*See* **Note 6**).

3.1.3 Fully Dissociated Cell Suspension

Always inspect the cell suspension under a microscope to ensure it is clean and fully dissociated (*See* **Note 7**).

3.1.4 Clean Cell Suspension

The single cell suspension must be free of aggregates, dust, debris, DNAse, RNAse, enzyme inhibitors, surfactants, and generally viscous reagents [6]. The reasons for these various exclusions are explained in detail in **Note 8**. However, in short, the presence of any of these factors in the single cell suspension could lead to a clog or wetting failure during the chip run on the Chromium instrument, or to a failure to barcode the mRNA from the cells.

3.1.5 How to Keep the Cells Alive and Happy

The default buffer used for cell washing and resuspension is typically magnesium- and calcium-free 1× PBS with 0.04% nonacetylated BSA. However, not all cell types do well in this type of buffer and may require an alternative to remain viable and aggregate-free in suspension (*See* **Note 9**).

The BSA content can also be increased up to 2% to help with cells or nuclei that are sticky and tend to form clumps. Serum up to 10% can be included in the washing and resuspension buffer if it helps the cells, but it is not necessary for the single cell assay itself (*See* **Note 10**). In addition to careful consideration of the chemical properties of buffers used for washing and resuspending cells, there are also physical properties that might need to be checked and optimized in order to obtain a viable single cell suspension: pipetting speed, wide-bore (wide-orifice) pipette tips, use of FACS sorter, temperature, and centrifugation (*See* **Note 11**).

3.1.6 Limited Cell Numbers

When it is necessary to work with very low cell numbers, it is likely that no pellet will be visible after spinning them down. While these situations are more challenging, many researchers face this scenario fairly often. There are a couple of points (listed in **Note 12**) to keep in mind when working with low cell input to get the most out of these samples [7].

3.1.7 Using Nuclei as Suspension Input

The Single Cell Gene Expression assay from 10x Genomics also allows the use of nuclei as input material, rather than whole cells [7, 8]. This can be useful when working with cell types that are too big to fit through the channels of the Next GEM chips. It can also be considered when the dissociation of the cells is too difficult or when tissue has been frozen and the extraction of viable whole cells from such circumstances is not possible (*See* **Note 13**).

3.2 GEM Generation and Barcoding

After ensuring that the single cell (or single nucleus) suspension is of high quality and viability, it can be used as input for the Single Cell Gene Expression assay. The first step is to create a master mix for the downstream reverse transcription (RT) and combine that with the cell suspension (*See* **Note 14**). Next, the final master mix will be loaded into the Next GEM Chip together with barcoded gel beads and partitioning oil (*See* **Note 15**). Running the chip on the Chromium instrument will generate an emulsion whereby one individual gel bead will be partitioned together with a single cell surrounded by the RT master mix. These GEMs (Gel Beads-in-emulsion) are then transferred to a strip tube, and the reverse transcription reaction will take place on a thermal cycler (*See* **Note 16**).

3.3 Post-GEM-RT Cleanup and cDNA Amplification

At this point, the reverse transcription reaction will have generated the first-strand cDNA from the captured mRNA transcripts, and the resulting RT products carry unique barcodes to their cells of origin. The emulsions can now be broken and the rest of the assay performed in the standard bulk fashion of Illumina library preparation. The barcoded cDNA will have been recovered from the broken emulsions and cleaned using a Dynabead cleanup step (*See* **Note 17**). Now the Read 1 adapter and the attached Template Switch Oligo can be used as primer sites to amplify all of the cDNA products that were captured in the RT step.

Depending on the number of cells targeted for recovery, the amount of PCR cycles used for amplification will need to be adapted (*See* **Note 18**). If nuclei were used as assay input instead of whole cells, it will be necessary to increase the number of PCR cycles by 1–2 cycles during cDNA amplification (*See* **Note 19**). After the amplification, the sample will be cleaned again. From this point forward, SPRI-select magnetic beads will be used for all cleaning steps. When checking cDNA traces on a Bioanalyzer, the user guide [3] recommends performing a 1:10 dilution for loading onto the Bioanalyzer chip (*See* **Note 20**).

3.4 Libraries Preparation and Sequencing

After cDNA amplification, the generated amplicons need to be optimized to be compatible with downstream sequencing (Fig. 3). Enzymatic fragmentation (*See* **Note 21**), which reduces the average size of the amplified fragments dramatically, is followed by an end repair and A-tailing step. Then, a Read 2 adapter can be ligated to the A-tailed ends of the fragments, which will allow attachment of the i5 and i7 sample indices, as well as the P5 and P7 adapter sequences in the final PCR step of the Single Cell Gene Expression assay (*See* **Note 22**). When checking the library traces on a Bioanalyzer, it is possible that a small amount of primer dimers will remain in the sample (*See* **Note 23**). To quantify the final library, a qPCR method, such as the KAPA Library Quantification Kit [9] is recommended (*See* **Note 24**).

3.5 Bioinformatic and Quality Control of the Data

10x Genomics has a license-free Linux pipeline called Cell Ranger that can be downloaded from the 10x Genomics support website [10] and used for data analysis of the Chromium Single Cell Gene Expression assay. The details of how this pipeline functions are beyond the scope of this work, but a description can be found at https://support.10xgenomics.com/. Of greater importance is the interpretation of the data after it has been processed by Cell Ranger [11–13]. The key output file for this purpose is the web_summary.html file, which contains the quality control metrics for the sample that was processed. There is a Technical Note available that helps with the interpretation of these metrics [14], but the most important ones are summarized in **Note 25**.

4 Notes

1. Premature cell lysis and cell damage release mRNA from dying cells into the microcentrifuge tubes containing the single cell suspension. This ambient mRNA is suspended among the viable single cells and will therefore also be loaded into the microfluidic chip during the master mix preparations. During the Chromium run, the ambient mRNA will be captured inside the emulsion droplets together with viable cells. These ambient transcripts will be barcoded with the same 10x Barcode as the transcripts originating from the viable cell, thereby contaminating the GEM and the data coming from this cell. The ambient mRNA from dead, dying, or lysing cells in the single cell suspension will cause background noise in single cell data and should be avoided at all costs.
2. Trypan blue can build up precipitates over time which are visible as dark blue flakes during counting. This could potentially lead to inaccurate results when using an automated cell counter. In this situation, it might be necessary to filter or spin down the Trypan blue tube before using it for cell staining, so that the solution remains free of precipitate. One alternative to Trypan blue staining is dual-fluorescence with acridine orange (AO) and propidium iodide (PI) viability staining [15]. While PI can only stain dead cells due to their damaged cell membranes, AO can permeate membranes and therefore stain both dead and live cells. In some circumstances, especially with samples containing heterogeneous cell populations, this method can be advantageous over the Trypan blue staining method.
3. Two examples of dead cell removal commonly used methods are as follows: (i) Live/dead fluorescence-activated cell sorting (FACS) of the cell suspension, by using a viability marker to determine plasma membrane integrity. (ii) Dead cell removal, by using the Miltenyi Biotec Dead Cell Removal Kit [16] to magnetically label cell debris, dead cells, and dying cells, which will then be removed by magnetic separation over a column.
4. Automated cell counters are faster and more streamlined than manual counting. Furthermore, (some) automated cell counters will allow the possibility of retaining images of the cell

suspensions, which can be very helpful for troubleshooting purposes. However, some machines could struggle to count accurately when: (i) the cells are extremely small and are falsely recognized as debris or dead cells, (ii) the focus is not set correctly, (iii) the loaded cell density is too dilute or too concentrated [17], (iv) there are many cell aggregates, and (v) there are big differences in cell sizes within heterogeneous cell suspension. These factors could lead to inaccurate cell counts, and some of these concerns also apply to manual counting, of course [18]. The disadvantages of manual cell counting are the extended time it takes to count the cells and the variability in counting outcomes between users based on the individual user experience.

5. Dead cells will still be present in the cell suspension that will be loaded into the chip on the Chromium instrument, and the downstream data analysis pipeline could potentially still recognize those cells. Ideally, cell counting should be performed at least in duplicate for higher accuracy, with the standard deviation of replicate counts being below 25% [19].

6. Filtering the cells with a cell strainer could lead to a loss in cell number and/or cell concentration, depending on which strainer is used. The cell suspension should also be recounted after FACS sorting as the cell numbers tend to be overestimated by FACS machines. It is important to count the suspension that is actually going to be used for the downstream steps. While some researchers like to sort cells into two different wells or tubes and then use one of them for counting and the other for loading the single cell assay, this practice could lead to discrepancies in cell numbers and should be avoided.

7. Cells that are still stuck together in multiplets or larger aggregates are not in single cell suspension. If included in the suspension loaded onto the Chromium instrument, these nondissociated cells will be encapsulated together inside the same GEM and will contribute to multiplets, which will be generated during the single cell assay.

8. *Aggregates, debris, and dust* The channels of the Next GEM chips that are processed on Chromium instruments have a fixed width. The maximum recommended cell diameter is 30 μm (in suspension). Loading cells with diameters up to 40 μm might be possible, but cell recovery rates may be compromised. The microfluidic channels themselves are the limiting factors here. While there is no lower cell size limit, cells that are larger than 40 μm in diameter could potentially clog the channel and lead to a loss of the sample. Furthermore, anything in the suspension that is not a single cell, e.g., cell aggregates, clumps, dust particles, or debris, could lead to the same failure type, if they are larger than the recommended cell diameter. The cell

suspension should be checked under a microscope to ensure there are no contaminants present. This check can also be done while performing the cell counting. If the suspension is not visibly clean, it should be filtered through a cell strainer with a maximum filter pore size of 40 μm to ensure that any particles larger than 40 μm are removed from the suspension. Further information on clogs is covered below, in the Subheading 3.2. Note that high levels of debris can also result in a wetting failure, in which emulsions will not form correctly and the single cell characteristics of the sample will be lost.

DNAses and RNAses The presence of these contaminants in the cell suspension is problematic as they will be loaded and partitioned inside the GEMs together with the cell master mix. Once the cells are lysed and reverse transcription has started, RNAses and DNAses can destroy the mRNA that would otherwise be captured, or destroy the barcoded cDNA intermediate product that would be generated during this step.

Enzyme inhibitors If enzyme inhibitors are part of the single cell suspension, they will be included in the GEMs produced by the Chromium instrument. The very first step of this assay is enzymatic, and if the reverse transcription is blocked by an enzyme inhibitor, the barcoded cDNA will not be generated. Media should not contain excessive amounts of EDTA (>0.1 mM), or magnesium (>3 mM) as those components will inhibit the reverse transcription reaction.

Surfactants and viscous reagents Another possible problem during a chip run is a "wetting failure," a phenomenon that occurs when the emulsion cannot be formed correctly. This topic is covered in more detail in Subheading 3.2. The presence of surfactants or any form of viscous reagents in the single cell suspension could lead to an instability of the emulsion formation. Failure to form GEMs means the cells will lyse in an uncontrolled manner and the primers from the gel beads will be released into the reaction mixture at random as well. In short, a wetting failure leads to the loss of single cell characteristics.

9. When cell viability and/or clumping problems are observed with the default 1× PBS + 0.04% BSA buffer, the buffers listed below can be tried instead. They have been verified to be compatible with 10x Genomics Chromium Single Cell Protocols with little or no loss of performance when profiling [7]: Dulbecco's Phosphate-Buffered Saline (DPBS), Hank's Balanced Salt Solution (HBSS), Eagle's Minimum Essential Medium (EMEM) + 10% FBS, Dulbecco's Modified Eagle

Medium (DMEM) + 10% FBS, Iscove's Modified Eagle Medium (IMEM) + 10% FBS, Roswell Park Memorial Institute (RPMI) + 10% FBS, Ham's F12 + 10% FBS,1:1 DMEM/ F12 + 10% FBS, M199.

10. It is important to note here that the list of alternative buffers and media above (*See* **Note 9**) is not exhaustive. Other media might be compatible, but they have not yet been tested. As long as the medium does not contain any of the disruptive ingredients listed in Subheading 3.1.4, they will pose a low risk. This is especially the case if the unknown buffer or medium has been used upstream during the cell dissociation process and has been sufficiently washed out before the final cell suspension is loaded into the actual Single Cell Gene Expression assay.

11. Two factors that are often overlooked at this point are as follows: (i) pipetting speed and (ii) the usage wide-bore (or wide-orifice) pipette tips. (i) Pipetting speed. The speed of pipetting while washing and resuspending cells is a subjective matter. One needs to keep in mind that cells are generally fragile, even the ones deemed to be robust. High-speed pipetting, as would be commonly used for a master mix, can be too harsh for cells. Pushing the cells at high speed through the narrow opening of the pipette tip can create shearing forces on the cells. This will then cause their cell membranes to rupture, releasing their mRNA contents into the microcentrifuge tube, hence contaminating the suspension with background noise through the presence of this ambient RNA. Slower pipetting speeds or the use of wide-bore pipette tips can help to reduce the shearing forces exerted on the cells. (ii) Wide-bore (wide-orifice) pipette tips. Some vendors offer wide-bore pipette tips for P1000 (and sometimes also P200) volumes. Using such ready-made pipette tips is the recommended method for washing and resuspending cells. Cutting normal pipette tips by hand to make wide bore tips would improve the shearing forces due to the wider opening of the tip, but this method causes other problems. By cutting the plastic, sharp edges can be created that would offset the reduced shearing forces. Furthermore, tiny pieces of plastic could break off and fall into the cell suspension, thereby potentially causing a clog during the Next GEM Chip run on the Chromium instrument. Additional physical properties that should be considered when optimizing the cell preparation are FACS, temperature, and centrifugation.

    In order to maintain cell viability and integrity when using a FACS sorter, it is not recommended to sort into dry tubes. Also, it is important to keep in mind that using FACS is not completely risk-free as some sensitive cell types can become leaky due to the sorting pressure. By default, cell suspensions are kept on ice during cell preparation protocols. However, not all cells do well when kept on ice as the cold temperature could cause them to stress, clump, or die. Some fragile cells might be better maintained at room temperature. The speed and timing of centrifugation might need to be adjusted for the particular cell types that are being studied. The default speeds and times listed in various protocols have usually been optimized for very specific cell types. When working with cell types that are larger or smaller, both the centrifugation speed and duration should be modified as necessary. In general, smaller cells might benefit from faster speeds, while larger or fragile cells might be better suited to slower speeds but longer centrifugation times.

12. *Washing and resuspension buffers.* Use the correct buffer for washing and resuspending the cells. If the cells are coming from a FACS machine, consider sorting them directly into the right buffer and then keeping them in the same type of buffer throughout the downstream processing steps. This way no buffer changes are needed, thus minimizing handling of the sample and loss of cells.

Straining If it is necessary to strain the cells, do so while the cells are still dilute and before spinning them down for the first washing step.

Counting Count the cells before the first washing step to estimate the appropriate buffer volume for resuspension.

Centrifugation Optimize the centrifugation speed and duration for your cell type. Always use low-binding centrifugation tubes to avoid unnecessary adhesion of cells to the tube walls. Depending on the centrifuge and rotor type, use of either 2.0 ml round-bottom or 1.5 ml V-shaped bottom microcentrifugation tubes can help with pellet placement. V-shaped tubes in a swing-bucket rotor centrifuge will place the pellet in the tip of the tube, allowing for very precise localization of the cell pellet, even if it is not visible to the eye.

Washing Remove the supernatant very carefully, recovering as much as possible, but retain the supernatant until after the washed and resuspended cell suspension has been checked. When the cell numbers are low and the centrifugation conditions are not entirely optimized yet, it is possible to accidentally take up the cell pellet while washing. Retaining the supernatant makes it possible to go back to it if cells are found to be missing during the final cell counting. Sometimes even careful pipetting of the supernatant is going to lead to the accidental removal of cells. If this continues to happen, consider leaving a bit of the supernatant behind (~50 μl) and resuspending the cells within the remainder of the supernatant.

This method will clean the suspension as well as the complete removal of the supernatant, but in some rare cases it might be the best option.

13. When the nuclei originate from a cell suspension, it is important to ensure that the stock cell suspension has high viability, ideally above 90%. Even though the cells will eventually be lysed, initial cell viability is directly related to the quality of the nuclei themselves. If the viability of the cell suspension is low to begin with, nuclei quality will suffer as well. In order to find the optimum duration for lysis, perform a lysis time course. Check cell viability at the beginning of the lysis as a reference point and then continue to check on cell viability regularly throughout the timeline. Ideally, there should be between 1 and 5% live cells remaining after a successful lysis. This ensures that the cells have not been overly lysed, which could happen if cell viability after lysis is 0%. Another quality control check for nuclei is to assess their membrane integrity under a microscope at 60× magnification. This should be done with and without staining solution. The membrane is more easily defined when it is unstained, and staining also makes the nuclei look swollen and sometimes grainy, depending on the nuclei type. First, look at the nuclei at low resolution to gauge the level of clumping. This can be done on a cell counter while counting. If there is extensive clumping, this could be a sign of potential over-lysis. After confirming that the nuclei suspension is clean, focus in at 60× magnification and look at a good number of nuclei to gauge membrane integrity. It is important not to cherry pick only the best-looking nuclei membranes. Counting nuclei suspensions can be tricky, especially if there is a lot of debris present from cell or tissue dissociation. In that case, using a fluorescent method such as ethidium homodimer-1 or other fluorescent dyes is advised.
14. It is recommended to make the master mix fresh and only after the final sample number has been confirmed. During sample preparation, it may become necessary to exclude a sample from the assay for various reasons (e.g., when cell viability is too low). In this case, preparing the master mix in advance would waste reagents since volumes would be calculated to accommodate a higher number of samples than will actually be processed. Before resuspending the Template Switch Oligo (TSO) with low-EDTA TE buffer, confirm the location of the TSO pellet. During transportation, it may have come loose and could be stuck in the cap of the tube. Always spin down the tube before opening it. The remaining volume of resuspended TSO must be stored at −80 °C to ensure the proper efficiency of the primer. Placing the TSO tube inside the box with the Gel Beads is a convenient solution that prevents the tube from

getting lost in the −80 °C freezer. When preparing the master mix, ensure that the correct volumes of the cell suspension and nuclease-free water are used. There are two options for determining these volumes: (i) Following the user guide. To help with the concentration calculations of both cell suspension and water, a pipetting table is included in the Single Cell Gene Expression user guide [3]. If the cell concentration of the stock solution and the number of cells targeted for recovery are known, the table can be used to determine the exact volumes of water and cell suspension required, as the cell recovery efficiency of ~60% is already built into that table. Important: Do not mix the cells with water! The table is an overview of the volumes required. Once you have generated the master mix and aliquoted it into the strip tubes, add the appropriate amount of water to the master mix and combine thoroughly. Into this master mix and water combination, add the appropriate volume of cell suspension. Use a P200 pipette tip to pipette up and down 5 times, with both the aspiration and dispersion steps each taking 3 s to complete. Using a multichannel pipette at this point is recommended. Once the cells are fully combined with the master mix, use the same pipette tip to transfer the suspension into the Next GEM Chip. Using the same pipette tip to mix and transfer avoids sample loss that could occur by unnecessarily switching to fresh pipette tips. (ii) Deviating from the user guide [3]. Some researchers prefer to calculate their own volumes for cell suspension and water. Instead of referring to the pipetting table before each run, they always load a fixed volume of cell suspension. This way, they also have a fixed volume of water which can be directly added into the master mix before it is aliquoted into the strip tubes. The desired cell concentration is calculated beforehand and adjusted within the fixed volume of cell suspension.

15. When loading a Next GEM Chip, there are several aspects that need to be kept in mind. The chip is an open system with microfluidic channels. In order for the Chromium instrument to process the liquids on the chip correctly and successfully form single cell emulsions, it is important to be aware of the potential reasons that a run might fail.

Clogs A clog is a situation where the emulsion was able to form correctly, but one of the channels became blocked and the remainder of the reagents was not able to flow through. The emulsion will look normal in appearance, but it will have less volume than it should. An example of this can be seen in pipette tip C in Fig. 5.

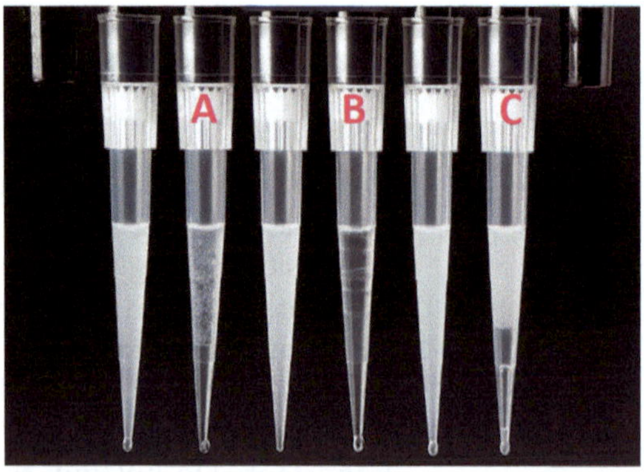

Fig. 5 Critical issues in the emulsion chip processing. (**a**, **b**) Images of incorrect emulsion formation that cause wetting failure; (**c**) example of clog which prevents reagents to flow through the chip

To avoid clogs, it is recommended to strain the cell suspension with a cell strainer that has a maximum filter size of 40 μm. Using a cell strainer with larger filter pores could potentially still include particles that are too big for the chip channels. However, the most overlooked risk factor is probably dust. Dust is always present in labs, and no matter how clean the cell suspension is, if a dust particle ends up in one of the channels of the Next GEM Chip, it can still cause a clog. To minimize this risk, cover every component that will eventually be loaded into the chip in some way, unless it is actively being used. This includes the chip itself, the individual reagent tubes from the Single Cell Gene Expression kit, the microcentrifuge tube used for preparing the master mix, the microcentrifuge tubes containing the cell suspensions, and the strip tubes used for the master mix and cell suspension aliquots.

Wetting failures A wetting failure means that the emulsion was not able to be formed correctly. In this case, cells will lyse and the Gel Beads will dissolve freely, without being confined to an individual emulsion droplet. This means that the sample will lose its single cell characteristics. Examples of wetting failures can be seen in pipette tips A and B in Fig. 5. Please note that there are varying levels of wetting failures and some are easier to spot than others. Always use a dark background (as shown in the figure below) when checking on the emulsion quality after the chip run. The reasons for wetting failures are quite varied [20], but the most common issues are temperature, loading orientation, pipetting speed, priming, pipetting style, air bubbles, and sample quality.

Temperature The Chromium instrument needs to be operated within a specific temperature range of 18–28 °C. The Gel Beads also need at least 30 min to warm up to room temperature (i.e., the same range as required for the Chromium instrument) before use. Placing and running the machine next to a south-facing window is not recommended as the temperatures will be too high for proper emulsion formation. The same logic applies to cold conditions—do not place and run the Chromium instrument within a cold storage room or in a nonheated room in winter. Operating the instrument outside of these conditions could cause a wetting failure because the emulsions consist of a delicate balance of oil, water, and (strictly controlled) viscous reagents such as Gel Beads.

Loading orientation The wells of the Next GEM Chip are labeled with the order in which they need to be loaded: the cell suspension combined with the RT master mix is loaded into Row 1, the gel beads are loaded into Row 2, and the partitioning oil is loaded into Row 3. Since the chip is an open system, as soon as liquids are loaded into the wells, they will be pulled through the channels by capillary forces. In order for the emulsions to be formed correctly, the reagents inside these channels need to be at the correct place at the correct time. Loading the wells in the correct order ensures this happens. Reagents loaded in reverse order will cause a wetting failure and loss of the sample.

Pipetting speed This is often overlooked in the user guide, but the appropriate speed for pipetting reagents into the chip is discussed in that document. Pipetting needs to be slow, such that dispensation of the reagents into the wells should take 5 s. This will ensure that the liquids are not pushed into the wells and into the channels. Faster pipetting could force the reagents through the channels and their junctions where they would otherwise not flow naturally. The result would be the same as if the reagents had been loaded out of order. *Priming*. There is always a 30 s waiting period between the loading of each row of reagents. This allows the capillary forces to prime the chip by pulling the reagents through the channels where they need to go in order for the emulsions to be formed correctly. *Pipetting style*. Always pipette into the center of the wells (Fig. 6) and slowly raise the pipette tip while dispensing the reagents within the required 5 s time frame.

Special care should be taken with the partitioning oil as this will be a small reagent volume pipetted into a large well. The partitioning oil needs to be in direct contact with the center section of the wells in Row 3, as this is where the gel beads and cells will come together to form the GEMs.

Air bubbles In general, avoid air bubbles as much as possible when loading the chip. However, there are some forms of bubbles which are far more damaging than others. If an air bubble gets stuck at the bottom of the well, it could cause the reagents to flow incorrectly through the channels, thereby creating a wetting failure. After loading all of the wells, carefully tilt the chip to check that no air bubbles are trapped at the bottoms of the wells. If an air bubble is

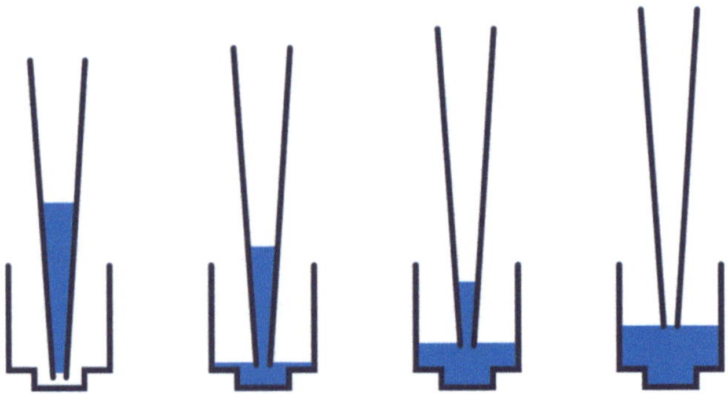

Fig. 6 Correct pipetting mode. Pipetting into the center of the well is strongly recommended

visible, carefully nudge it with a P10 pipette tip to dislodge the bubble. Once the bubble has floated to the top of the well, it is safe to proceed with the chip run. If there are bubbles at the top of the well and they cannot be removed with a P10 pipette tip, they can remain in the wells for the chip run, as they are small enough to not interfere with the assay. However, to be on the safe side, a needle can be used to get rid of even these small bubbles.

Sample quality Wetting failures can also be caused by poor sample quality in general. If suboptimal samples are used (e.g., suspensions that contain a lot of debris or dead cells), the risk of obtaining a wetting failure may increase.

Even when taking all of these precautions, it is still possible for a clog or wetting failure to occur. These failures can be observed at a low, single-digit percentage of the samples being processed. Fortunately, these problems will be apparent immediately after the Chromium instrument has finished its run and the emulsions need to be transferred to the strip tube for further processing on the thermal cycler. Monitor the Chromium instrument during the first minute of each run to ensure that there are no error messages. Next, set a timer to be sure to be back at the Chromium instrument just as the run is finishing. Do not let the chip sit before transferring the GEMs. If you notice a problematic sample that is showing signs of a clog or wetting failure like one of the examples from Fig. 5, there may still be a chance to save the sample. If the sample is precious and the experiment cannot be repeated, prepare a new chip and a new master mix and run the cell suspension again. In most cases, the repeat run will succeed and the assay move forward with all samples. If the issue was a wetting failure and the repeat of the sample is showing this same problem again, the sample will need to be discarded. In that case, a problem with sample quality or something in the cell suspension buffer is causing the emulsion to fail. Trying to repeat the run with the same buffer composition will not succeed. Secure all samples for proper storage and then return the 10x Genomics reagents to −20 °C or −80 °C.

Sometimes, the exact number of cells in a final cell suspension is unknown because the cell numbers are so low that it is too much of a risk to lose cells through the counting process. In this case, researchers might choose to skip the addition of the water to the master mix and use the full volume of the cell suspension instead. While the cell number will remain unknown, this option aims to capture as many cells as possible. This is a valid approach, however, there are risks involved. Although the odds of a run failure are very low, if one occurs, it would be a considerable loss for a sample that was extremely rare and precious. If working with such precious samples where the total cell suspension volume is too low to repeat if it should fail, one possibility is to run only half of the suspension, no matter the desired number of cells. This means capturing fewer cells, but if something goes wrong during the run and the sample is lost through a clog or a wetting failure, there is still the opportunity to repeat the sample and save it. Researchers will have to weigh the options and decide for themselves the best way to proceed for their given experiment.

16. As soon as a run has been completed on a Chromium instrument, the emulsions need to be recovered from the Next GEM Chip and transferred to a strip tube immediately. The pipetting speed for aspirating the emulsions is specified at (at least) 20 s. The dispensation into the strip tubes should be done at the same speed, for the duration of (at least) 20 s. This is done to avoid breaking the emulsion and also to allow the liquid to leave the pipette tip. The emulsion is slightly viscous, and dispensing it too quickly will leave emulsion behind inside the pipette tip, causing unnecessary sample loss. Using a multichannel pipette for this step is highly recommended. However, if a pipette tip falls off during this step, do not attempt to pipette the emulsions up and down to recover the lost sample. First, finish pipetting the other channels and then use a single channel pipette to recover the sample from the fallen tip. Proceed immediately to the reverse transcription section of the protocol. The emulsion can be stored on ice for up to 1 h if there is a second chip being processed on the Chromium instrument and both sets of samples will be run together on the thermal cycler. Please note that the assay has been developed and validated with a limited number of cell types. While this is a safe practice and can be used for running multiple chips back-to-back, it is possible that this delay in starting reverse transcription might not be suitable for certain cell or sample types. When loading the thermal cycler, place empty tubes on the corners of the block to avoid sample tubes being crushed by the lid, especially when running less than four samples.

Furthermore, ensure that the heated lid option is set at the correct temperature; running the heated lid at high temperatures (e.g., 100 °C) will break the GEMs prematurely. The thermal cycler program can be adjusted with an additional hold step at 53 °C at the beginning of the run. This would ensure that the lid and sample block are already at the correct temperature for the incubation of the GEMs when they are loaded onto the block. If you are incorporating this additional holding step, do not forget to advance the program to the actual RT steps. Forgetting to advance the program to the RT process is far more damaging to the sample than not including the optional preheated holding step. The finished RT reaction can be held at 4 °C overnight in the thermal cycler if scheduling makes it convenient to do so.

17. The emulsions are broken with the help of the pink Recovery Agent. No amplification rounds have been performed at this stage so each cDNA molecule is only represented once. Be careful not to lose material during the cleanup steps as there are no duplicates for any lost transcripts and that data cannot be recovered. After the Recovery Agent has been added, the upper aqueous phase will contain the barcoded cDNA product and the bottom pink phase will need to be discarded. In the user guide [3], the upper phase is shown with a clear, see-through appearance. Under certain circumstances, this upper phase could look differently [21] and may appear milky instead of clear if the cell suspension contains a high concentration of BSA or FCS; appear slightly pink or red if the cells were resuspended in a culture media of such color; and appear cloudy if the GEMs had been frozen at −20 °C after reverse transcription. After the incubation and breaking of the emulsion, be sure to pipette slowly when aspirating the pink Recovery Agent from the bottom of the strip tubes. If everything went well and there were no issues with the run, there will still be a bit of pink fluid left over at the bottom of the strip tubes after removing the indicated volume of Recovery Agent. This is done by design. If while pipetting, pink fluid is running low at the bottom of the well before reaching the top of the aspiration, simply stop earlier than the full volume of the pipette's setting. Continuing to pipette after the pink Recovery Agent is depleted would mean pipetting away and discarding sample. Avoid this at all costs. The Recovery Agent is viscous, so pipette slowly to allow the reagent time to settle down into the bottom of the tube. Once the Recovery Agent is removed, Dynabeads are used for sample cleanup. Before vortexing the Dynabeads, mix the whole volume of the Dynabeads reagent by pipetting up and down, as it is possible for Dynabeads to pellet at the bottom of the tube. Vortexing alone might not always be

efficient enough to dislodge that magnetic bead pellet. Make sure to vortex the tube for at least 30 s and then pipette the fresh magnetic beads directly to your cleanup master mix. The magnetic beads are quite heavy and will settle very quickly. If settling results in an incorrect concentration of magnetic beads in the cleanup master mix, this will lead to less efficient cDNA recovery. When adding the cleanup master mix to the strip tubes, do not attempt to close the lids on the strip tubes. The strip tube is designed for 0.2 ml fill volumes, but they will contain 300 μl at this point in the protocol. The full volume will fit, provided that the lid is kept open. The bottom phase will appear pink rather than white as indicated in the user guide [3]. While mixing, take care to mix the full depth of the tube, including the pink phase at the bottom (i.e., actively pipette into the bottom of the tube and then release the suspension from the top of the tube to maximize the mixing of the cDNA with the magnetic beads). It is normal to mix the pink phase alongside the upper brown phase, but that pink phase will settle down again immediately. A white interphase can sometimes be present between the bottom pink phase and the upper brown phase. This is completely normal and nothing to worry about. Following the incubation period, in which the cDNA is bound to the magnetic beads, place the strip tube on the Magnetic Separator. The magnetic beads will be pulled to the back of the strip tube into a small, tight brown pellet consisting of the magnetic beads with bound cDNA. The initial supernatant recovery calls for the removal of everything from the tube except the magnetic beads. This includes the pink phase and (if present) the white interphase at the bottom of the tube. After that, simply follow the user guide protocol for standard cleanup. Always make sure that the magnetic bead pellet is wet in some way, either with ethanol or the elution solution at the end of the protocol. The indicated air-drying times should not be exceeded, but rather strictly maintained to avoid overdrying of the bead pellet and a corresponding loss of cDNA product. Please note two important aspects of the Magnetic Separator from 10x Genomics:

(i) Magnetic pins may come loose. The tiny round pins at the back of the magnet could move to the outside or even fall out completely because they are not fixed nor glued in place. This is of particular concern if using a magnetic wall for storage. It is not an ideal storage location—when pulling the magnet off the wall, the magnetic pins at the back could be removed from the magnet. Even if the pins are simply moved a bit (instead of falling out entirely), this would lead to a bigger gap between the magnets and the magnetic beads carrying the sample, therefore leading to a worse recovery of the cDNA. If the magnet has been stored on a magnetic wall, make sure to check on the magnetic pins at the back and push them back into place if necessary.

(ii) Magnets are asymmetrical. There are two markings edged into the Magnetic Separator at the top and bottom of the frame. These indicate the High and Low positions of the magnets, which have been designed to be asymmetrical in the holder. If using the recommended Eppendorf strip tubes, the High and Low positions can be used as guidance, as indicated in the user guide. However, the recommended Microamp strip tubes do not reach the magnetic pins if the tubes are placed into the Low position of the Magnetic Separator. Due to their design, the Microamp strip tubes sit higher inside the magnet, so only the High position of the magnet should be used, and the High/Low indications in the user guide should be ignored. Last but not least, the magnet can always be switched between the High or Low position if the magnetic beads are not pelleting in an optimal way. Use of the High and Low positions as indicated in the user guide is not a firm rule.

18. Low cell numbers require more amplification, while large cell numbers will need less amplification. This means that amplification cycle numbers will vary between sample batches. Therefore, always confirm the PCR programming, and in particular the cycling conditions, when starting the thermal cycler to be sure that the amplification cycle number is appropriate for each set of samples.

19. Nuclei contain only a tiny fraction of the total RNA from the whole cell, so more amplification is needed here than with intact cells.

20. This dilution factor might be too high for some samples during quality control, especially when working with tiny cells or a small number of total cells. Instead, consider doing a 1:3 or 1:4 dilution which may prevent the need to repeat the Bioanalyzer run if the 1:10 dilution turns out to dilute the sample too strongly, beyond the point where traces can be correctly identified. Alternatively, a Qubit run could be used first to check on the concentration, providing information needed to load the Bioanalyzer chip within its correct range of detection. Furthermore, the cleaned, amplified cDNA is in a stable state at this stage and can be stored at −20 °C for up to 4 weeks. Some researchers may prefer to hold here and batch a couple of sample runs together on a Bioanalyzer chip or Tapestation screen tape before checking on the traces.

21. Enzymatic fragmentation is controlled only through temperature and duration, which means everything needs to be

precooled on ice, including the microcentrifuge tube used for the master mix and the strip tube with the samples. Furthermore, it is important to note here that the fragmentation program on the thermal cycler is programmed to start at 4 °C, meaning that the block will be cooled down to 4 °C when loading the samples. However, do not forget to press Start or Skip (depending on the thermal cycler model) to actually advance the fragmentation program through the 4 °C hold stage. Both the beginning and the end of the program hold at 4 °C and upon returning to the machine and seeing the 4 °C readout, it is easy to assume that the run has been completed, when instead it never progressed through the fragmentation. During this phase of the protocol, there are a few double-sided SPRI cleanup steps, in which fragments that are too small and fragments that are too big will be removed from the sample. In a standard SPRI cleanup, the large cDNA fragments are bound to the magnetic beads. The ratio between the sample and the added magnetic beads determines the fragment size that can be captured by the beads. This means that the smaller fragments are left behind in the supernatant which is usually discarded in typical one-sided SPRI cleanup. For a double-sided SPRI cleanup, the supernatant is kept at first, and the magnetic beads are discarded instead. This step removes fragments which are too large, as they are bound to the magnetic beads. Then, magnetic beads are added at a different ratio, and the supernatant is again discarded. This allows for sample cleanup within a very specific fragment size range. The most important point to remember for a double-sided SPRI cleanup is to not throw away the supernatant by force of habit.

22. When setting up the Sample Index PCR, make sure to choose the correct Sample Index plate as there are quite a few different ones available these days, and each one is very specific to both the assay and the library type that can be generated. Also, always write down which index was used for which sample; this information will be needed downstream during the data analysis. The number of PCR cycles needed for the Sample Index PCR depends on the cDNA mass that was calculated during the Bioanalyzer check of the amplified cDNA intermediate product. Keep in mind that the assay continued with only a quarter of the total cDNA mass, so the number of PCR cycles needs to be appropriate for 25% of the total cDNA that was measured. Furthermore, when considering PCR cycle number, one should always follow this simple guideline: amplify as much as is needed, but no more than is necessary.

23. If this is the case, evaluate whether it might be less expensive to increase the sequencing depth to compensate for this, versus

re-cleaning the library or re-constructing a new library from the leftover cDNA intermediate before sequencing the sample.

24. The reason being that this kit uses the P5 and P7 adapter sequences as primer sites, ensuring quantification is based on a complete and correctly built library. If the P5 and/or P7 adapters are missing from the library, the sample cannot be sequenced because it will be unable to bind to the Illumina flow cell. In addition to the KAPA kit, qPCR kits from alternative vendors are also available that function according to the same principle. Oftentimes, researchers skip the qPCR check due to cost and time factors. In these instances, the Bioanalyzer data is used to check on the fragment size, and Qubit is used to quantify the library instead. This is the most commonly used approach, and Qubit is accurate for this purpose. However, please note that Qubit will not be able to provide information on whether the P5 and/or P7 adapters might be missing, while the Bioanalyzer traces might be too ambiguous for detection of missing sequence. Without quantification via qPCR, there will be no opportunity to notice that something might be wrong with the library until after sequencing.

25. *Estimated Number of Cells.* This number should be close to the targeted cell recovery number. If this number is far removed from the target, it could indicate an issue with cell handling (e.g., inaccurate cell counting), a problem that occurred during the run on the Chromium instrument, or that the cell type was significantly more fragile to work with than anticipated.

Fraction Reads in Cells This metric indicates the percentage of reads that are associated with a valid cell barcode and are confidently mapped to the transcriptome. This metric should be as high as possible, but ideally above 70%. A high percentage here means that the data are clean from background noise and that the sample contained healthy and viable cells. If the cell handling during the sample preparation upstream of the Single Cell Gene Expression assay had been too harsh, the cells could have died or ruptured, causing the release of their mRNA contents into the microcentrifuge tube. This ambient RNA will not be called as a cell by the Cell Ranger pipeline, therefore increasing the background noise and contaminating the data from partitions containing individual cells.

Barcode Rank Plot This graph plots all detected barcodes versus the total number of UMIs that have been counted for each one of those respective barcodes, ranking the barcodes from the ones with the highest UMI count to the ones with the lowest (Fig. 7). Interpretation of this plot is probably the most difficult one as an algorithm will usually have a hard time interpreting the shape of a curve, while the human eye can easily see patterns that the algorithm cannot calculate properly. The Barcode Rank Plot should

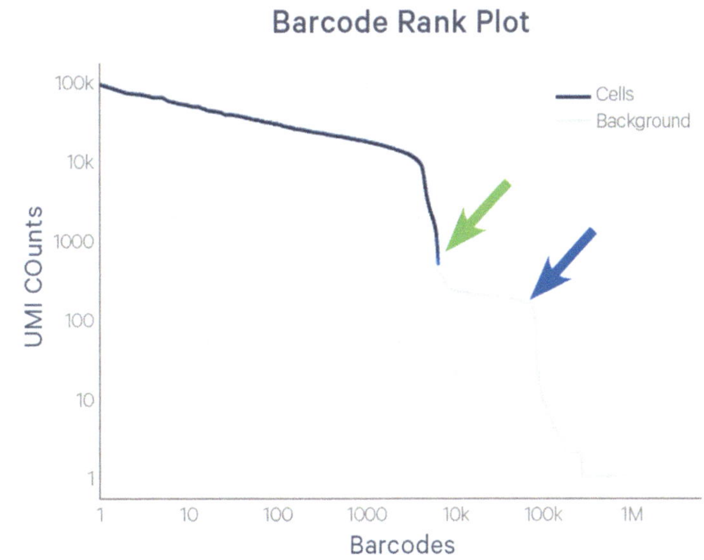

Fig. 7 Barcode Rank Plot in heterogeneous cell populations. Graph represents all detected barcode (x-axis) versus UMI counts (y-axis). Green arrow indicates the first typical sharp drop after the UMI-rich plateau (blue line), and blue arrow indicates a second drop in the form of a cliff (typical sign of a smaller cell population) in the background noise (gray line)

ideally have a plateau where the barcodes with the highest UMIs are located as this will contain the main bulk of the cells of the sample. From that UMI-rich plateau, we expect to find a sharp drop in the form of a cliff (green arrow, Fig. 7) into the background noise, i.e., the barcodes that contain very little or barely any UMIs. This would indicate the best separation between the actual single cell data and the noise. There will always be background noise in the data, but this should be kept to a minimum in order to avoid the ambient RNA contaminating the data from partitions containing real cells. When working with heterogeneous cell populations, there could be two cliffs before the plot drops off into the background noise. This is indicative of a smaller cell population that is presented within that lower second cliff. If the plot does not have a clearly defined plateau and cliff structure, this could be indicative of a wetting failure or a cell lysis event that caused the sample to lose its single cell characteristics.

Mean Reads Per Cell This is the sequencing depth of the sample, defined by the user when setting up the sequencing run. There is a minimum recommended number of reads per cell; however, this is strictly dependent on the cell type being processed. Large cells might require much higher sequencing depth than the minimum recommended amount.

Median Genes Per Cell and Median UMI Counts Per Cell These two values indicate data yield. Low numbers are related to the sequencing read depth in situations where the sequencing was too shallow for a given sample. However, if sequencing depth was not intentionally set this way, low numbers could be associated with lower cell viability, but they could also be cell type–dependent, e.g., when the cell is very small and/or not expressing that many transcripts. The numbers for these metrics should ideally be similar to previously processed samples of the same cell type.

Q30 Bases This is a summary of the Q-score ≥30 coming from the Illumina sequencer. These metrics should be as high as possible to ensure clean sequencing data; however, they are sequencing platform–dependent [22]. A drop in these metrics could indicate a potential problem with the sequencing run itself.

Sequencing Saturation This is the percentage of known reads, i.e., it is an indication of how deeply the sequencing has gone into the complexity of the library. This percentage depends on the cell type being processed as well as the targeted sequencing depth. Large cells are a lot more complex than small cells and therefore require more sequencing to reach the same level of saturation. The appropriate saturation will vary based on experimental needs. For some situations, a saturation level of 30% is sufficient for addressing the research objectives. On the other hand, sequencing very deep into the library—up to 90% saturation—may be preferred, in order to capture as much data from the cells as possible, allowing the identification of small differences between cell subtypes.

Reads Mapped Confidently to Transcriptome This is the percentage of reads that were able to be mapped to a unique gene in the reference transcriptome. These are the only reads which are used by Cell Ranger for UMI counting and therefore downstream secondary data analysis and gene expression calculations. How high this percentage needs to be, depends on the sample type, the species, and the quality of the reference file. Shortening the read length of Read 2 beyond the recommended length will negatively affect this metric.

Cell Cluster Interpretation Cell Clustering plots are the output from various clustering algorithms. The more similar the gene expression profile of the cells, the more likely they are going to be clustered together. Looking at the Cell Clustering plots requires a bit of background knowledge of the sample that was being processed, including the following: Was this a homogeneous or heterogeneous sample? What is the expectation for finding many cell types that are very different from each other, thereby forming many different clusters? If the Cell Cluster plots show one large, undefined mass of a cluster with a very homogeneous distribution of the cells and/or UMI values, this could indicate a problem with the sample, such as a Wetting Failure or cell lysis prior to partitioning.

Gene Expression Table Interpretation The Gene Expression Table is the output of a differential gene expression analysis that is performed automatically by Cell Ranger for all clusters that have been detected in the previous step. The same types of questions from the Cell Cluster interpretation apply here: Was the sample type a homogeneous or heterogeneous mixture of various cell types? What is the expectation for finding large or small differences between the cell clusters?

Furthermore, identifying mitochondrial or apoptotic markers in this table is a sign that the cells were not in their best shape and had suffered in terms of viability.

References

1. Zheng GX, Terry JM, Belgrader P, Ryvkin P, Bent ZW, Wilson R, Ziraldo SB, Wheeler TD, McDermott GP, Zhu J, Gregory MT, Shuga J, Montesclaros L, Underwood JG, Masquelier DA, Nishimura SY, Schnall-Levin M, Wyatt PW, Hindson CM, Bharadwaj R, Wong A, Ness KD, Beppu LW, Deeg HJ, McFarland C, Loeb KR, Valente WJ, Ericson NG, Stevens EA, Radich JP, Mikkelsen TS, Hindson BJ, Bielas JH (2017) Massively parallel digital transcriptional profiling of single cells. Nat Commun 8:14049. https://doi.org/10.1038/ncomms14049
2. De Simone M, Rossetti G, Pagani M (2019) Chromium 10x single-cell 3' mRNA sequencing of tumor-infiltrating lymphocytes. Methods Mol Biol 1979:87–110. https://doi.org/10.1007/978-1-4939-9240-9_7
3. Chromium Single Cell 3' Reagent kits user guide (v3.1 Chemistry Dual Index). https://www.10xgenomics.com/support/single-cell-gene-expression/documentation/steps/library-prep/chromium-single-cell-3-reagent-kits-user-guide-v-3-1-chemistry
4. 10x Genomics website. https://www.10xgenomics.com
5. Removal of dead cells from single cell suspensions improves performance for 10x Genomics® Single Cell Applications (2019) https://www.10xgenomics.com/support/single-cell-gene-expression/documentation/steps/sample-prep/removal-of-dead-cells-from-single-cell-suspensions-improves-performance-for-10-x-genomics-r-single-cell-applications
6. What buffers can be used for washing and cell resuspension? (2021) https://kb.10xgenomics.com/hc/en-us
7. Single Cell protocols - cell preparation guide (2021) https://support.10xgenomics.com/single-cell-gene-expression/sample-prep/doc/demonstrated-protocol-single-cell-protocols-cell-preparation-guide
8. How can I assess the quality of my nuclei for Single Cell ATAC or Single Cell Multiome ATAC+GEX Sequencing? (2021) https://kb.10xgenomics.com/hc/en-us
9. KAPA library quantification kits (2021) https://sequencing.roche.com/en/products solutions/products/sample-preparation/library-quantification/kapa-library-quantification-kits.html,
10. 10x Software Downloads (2021) https://support.10xgenomics.com/single-cell-gene-expression/software/downloads/latest
11. Alessandri L, Cordero F, Beccuti M, Licheri N, Arigoni M, Olivero M, Di Renzo MF, Sapino A, Calogero R (2021) Sparsely-connected autoencoder (SCA) for single cell RNAseq data mining. NPJ Syst Biol Appl 7(1):1. https://doi.org/10.1038/s41540-020-00162-6
12. Alessandri L, Ratto ML, Contaldo SG, Beccuti M, Cordero F, Arigoni M, Calogero RA (2021) Sparsely connected autoencoders: a multi-purpose tool for single cell omics analysis. Int J Mol Sci 22(23). https://doi.org/10.3390/ijms222312755
13. Alessandri L, Cordero F, Beccuti M, Arigoni M, Olivero M, Romano G, Rabellino S, Licheri N, De Libero G, Pace L,

Calogero RA (2019) rCASC: reproducible classification analysis of single-cell sequencing data. Gigascience 8(9). https://doi.org/10.1093/gigascience/giz105

14. Interpreting cell ranger web summary files for Single Cell Gene Expression Assays (2021) https://support.10xgenomics.com/single-cell-gene-expression/sequencing/doc/technical-note-interpreting-cell-ranger-web-summary-files-for-single-cell-gene-expression-assays
15. Chan LL, Smith T, Kumph KA, Kuksin D, Kessel S, Dery O, Cribbes S, Lai N, Qiu J (2016) A high-throughput AO/PI-based cell concentration and viability detection method using the Celigo image cytometry. Cytotechnology 68(5):2015–2025. https://doi.org/10.1007/s10616-016-0015-x
16. Dead Cell Removal Kit (2021) https://www.miltenyibiotec.com/DE-en/products/macs-sample-preparation/removal-reagents.html
17. Importance of cell stock concentration for accurate target cell recovery (2019) https://support.10xgenomics.com/single-cell-gene-expression/index/doc/technical-note-importance-of-cell-stock-concentration-for-accurate-target-cell-recovery
18. Guidelines for accurate target cell counts using 10x Genomics® Single Cell Solutions (2021) https://support.10xgenomics.com/single-cell-gene-expression/index/doc/technical-note-guidelines-for-accurate-target-cell-counts-using-10x-genomics-single-cell-solutions
19. Chromium single cell applications - guidelines for optimal sample preparation (2021) https://support.10xgenomics.com/single-cell-gene-expression/sample-prep/doc/technical-note-chromium-single-cell-applications-guidelines-for-optimal-sample-preparation
20. Isolation of nuclei for single cell RNA sequencing (2021) https://support.10xgenomics.com/single-cell-gene-expression/sample-prep/doc/demonstrated-protocol-isolation-of-nuclei-for-single-cell-rna-sequencing
21. Why is the aqueous layer cloudy after breaking GEMs? (2021) https://kb.10xgenomics.com/hc/en-us
22. Sequencing metrics & base composition of single cell 3′ v3.1 Dual Index Libraries (2021) https://support.10xgenomics.com/single-cell-gene-expression/sequencing/doc/technical-note-sequencing-metrics-and-base-composition-of-single-cell-3-v31-dual-index-libraries

Chapter 2

BD Rhapsody™ Single-Cell Analysis System Workflow: From Sample to Multimodal Single-Cell Sequencing Data

Jannes Ulbrich, Vadir Lopez-Salmeron, and Ian Gerrard

Abstract

Advancements in single-cell sequencing have revolutionized our understanding of complex biological systems such as the immune system and allowed us to overcome limitations in various disciplines of life science research such as oncology, developmental biology, or neurobiology (Perkel, Nature 595. https://www.nature.com/articles/d41586-021-01994-w, 2021).

The BD Rhapsody™ Single-Cell Analysis System enables us to capture multimodal information from thousands of single cells in parallel ("Multiomics") covering mRNA expression levels, protein expression levels, the immune repertoire for T-cell receptors (TCR) and B-cell receptors (BCR), and the identification of antigen-specific T cells and B cells using dCODE Dextramer® (RiO) from Immudex. The system utilizes microwell-based cartridges that allow to capture a broad range of single cells and an imaging device for sample quality control and workflow quality control (including viability and multiplets). The power of Multiomics relies on simultaneously measuring several aspects of single cells, including gene expression and protein abundance, using next generation sequencing (NGS) as a single readout.

Here we describe the complete BD Rhapsody™ Single-Cell Analysis System from the sample preparation including different options for the antibody and/or dCODE Dextramer® staining through to the data analysis.

For updated protocols, guides, and technical bulletins, please visit the BD Scomix page: https://scomix.bd.com/hc/en-us or the BDB webpage: https://www.bdbiosciences.com/en-eu.

Key words BD Biosciences, BD Rhapsody™ Single-Cell Analysis System, Multimodal Single-Cell Sequencing, Single-Cell Multiomics, Microwells, BD® AbSeq

1 Introduction

The BD Rhapsody™ Single-Cell Analysis System allows high-throughput capture of multiomic information from thousands of single cells (or isolated cell nuclei) using a simple cartridge workflow and a multitier barcoding system. The occurrence of multiplets (microwells with more than one cell captured) increases as more cells are loaded into the BD Rhapsody™ Cartridge, following Poisson distribution calculations (Table 1).

Raffaele A. Calogero and Vladimir Benes (eds.), *Single Cell Transcriptomics: Methods and Protocols*, Methods in Molecular Biology, vol. 2584, https://doi.org/10.1007/978-1-0716-2756-3_2,

Table 1
Theoretical percentage of multiplets increases as more cells are loaded into the BD Rhapsody™ Cartridge

# of cells loaded in BD Rhapsody Cartridge	% total theoretical multiplets
1000	0.2
5000	1.2
10,000	2.4
20,000	4.7
40,000	9.4

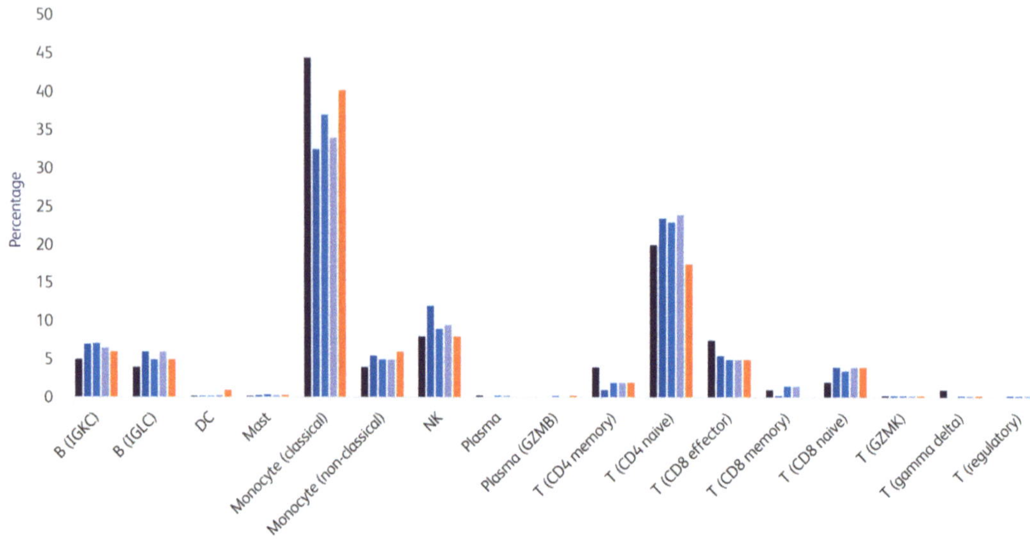

Fig. 1 Five different cell inputs (PBMCs) were put into the BD Rhapsody™ System and amplified with the immune response panel. Similar proportions of the cell types recovered from each of the inputs suggests strong reproducibility even with variable cell inputs. IGKC IgG kappa chain, IGLC IgG lambda chain, GZMB Granzyme B, GZMK Granzyme K

We recommend loading of up to 40,000 cells into a cartridge, but there are also examples of loading of 60,000 cells per cartridge successfully in multiple studies [12]. Also, this single-cell capture solution enables the processing of samples with very low cell counts, down to 100 single cells ([8]; Fig. 1).

The resulting captured information can be used to generate various types of multimodal single-cell next generation sequencing (NGS) libraries to provide high-dimensional resolution of single cells (Fig. 2).

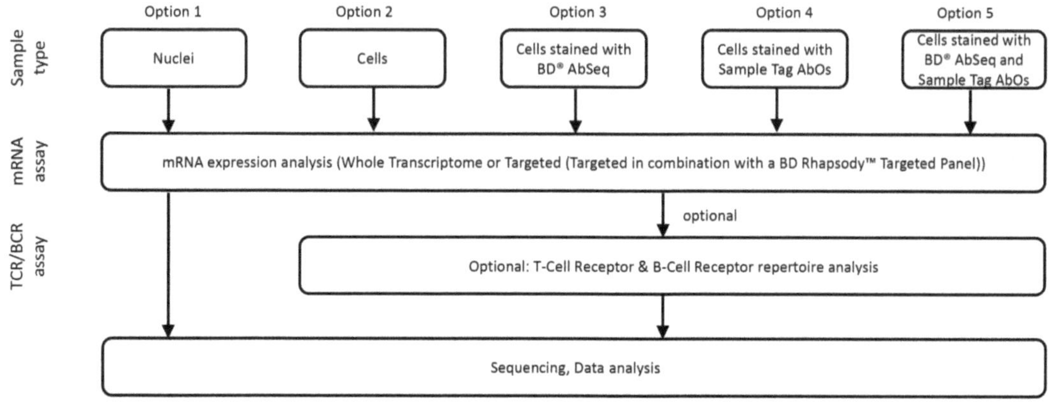

Fig. 2 BD Rhapsody™ Single-Cell Analysis System enables processing of cell suspensions or cell nuclei suspensions. Cell can be labeled with BD® AbSeq antibodies and/or Sample Multiplexing antibodies (Oligonucleotide-conjugated antibodies (AbOs)). Cells can be further labeled with dCODE Dextramers® (Immudex) for the identification of antigen specific T cells or B cells. Two mRNA-Seq assays are available, a Whole Transcriptome Amplification (WTA) assay and a targeted assay. The T-cell and the B-cell receptor repertoire can be analyzed in parallel with the TCR/BCR (VDJ) assay

2 Materials

2.1 The BD Rhapsody™ Single-Cell Analysis System

The BD Rhapsody™ Single-Cell Analysis System includes the BD Rhapsody™ Scanner and BD Rhapsody™ Express System. The BD Rhapsody™ Express System enables single-cell capture and barcoding of hundreds to thousands of single cells for analysis of transcriptomic and proteomic information, using proprietary, gentle, robust microwell-based single-cell partitioning technology [2]. The BD Rhapsody™ Scanner is designed to visualize all steps in the single-cell capture workflow and provide detailed analysis metrics at every step, enabling the user to make key decisions throughout the workflow.

2.2 BD Rhapsody™ Express Single-Cell Analysis System

The BD Rhapsody™ Cartridge (Fig. 3, Right) contains the microwell array used to capture single cells from a cell suspension. The cartridge requires use of the BD Rhapsody™ Express Single-Cell Analysis System. The mechanical station is used to load reagents, cells, and cell capture beads into the cartridge for bead capture and retrieval of the barcoded single-cell information. Each batch of BD Rhapsody™ Cartridges undergoes extensive quality control testing: functional testing (cell retention rate, bead loading efficiency, bead retrieval efficiency), physical attribute testing (height and width of partitions, minimal well defects, optimum optical quality, minimum number of fluorescing cell like particles), and fluidics testing (optimum buffer plug flow profile, leakage tests).

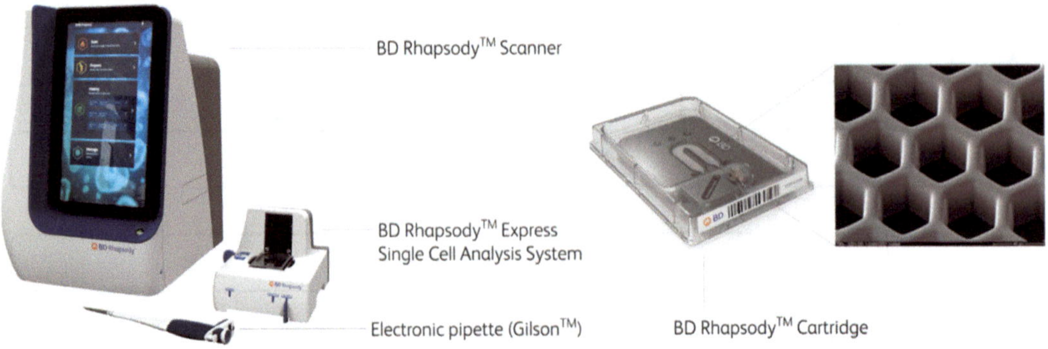

Fig. 3 BD Rhapsody™ Single-Cell Analysis System. Left: BD Rhapsody™ Single-Cell Analysis System, Right: BD Rhapsody™ Cartridge: The cartridge contains a sample inlet for the sample injection and a U-shaped microwell array with >220,000 microwells. The microwells are hexagon-shaped as shown in the scanning electron microscope image

2.3 BD Rhapsody™ Scanner

The BD Rhapsody™ Scanner (Fig. 3, Left) can be used to provide quality control measures at different stages of the workflow, such as cell capture rate and cell multiplet rate, by direct imaging. Scanner metrics can confirm quality of input cell sample and success of each step of the cartridge workflow, assess viability of cells until cells are lysed, and provide a reliable estimate of the number of cells retrieved by sequencing (*see* **Note 1**).

3 Methods

3.1 Sample Quality Control (Cell Viability Assessment and Cell Counting)

The BD Rhapsody™ Single-Cell Analysis System offers a sample and cartridge workflow quality control instrument, the BD Rhapsody™ Scanner, as mentioned above.

Before the cell suspension is loaded into the BD Rhapsody™ Cartridge, the BD Rhapsody™ Scanner is used to count cells and assess the cell viability by brightfield and dual band fluorescence imaging. The system then calculates volumes of cells and sample buffer needed to prepare single-cell suspensions for the BD Rhapsody™ workflow.

The suspended cells (620 μl) in cold sample buffer are labeled with Calcein AM (CAM) and Draq7™. After the labeling procedure, 10 μl of the sample is loaded onto a hemocytometer after which this is inserted and analyzed by the BD Rhapsody™ Scanner. The scan results include the concentration of the cells in the sample (cells/μl), the percentage of viable cells, and the expected multiplet rate with the scheduled number of cells to be captured in the following cartridge workflow (Fig. 4).

Fig. 4 BD Rhapsody™ Samples Calculator Screenshot

BD Rhapsody™ Cartridges are intended for single use. However, used cartridges in combination with the BD Rhapsody™ Scanner can be re-used to optimize the cell capture process for very challenging samples, e.g., cells outside of the recommended size range (cell diameter: 5–20 μm in suspension). The cartridges do not contain microfluidics, so issues associated to clogging are not observed even when handling larger cells; however, the capture rates are reduced when cells larger than 20 μm are processed. In order to compensate for the lower capture rates, simply more cells can be loaded in the cartridge and the cell numbers can be optimized with used cartridges (as prior mentioned) (*see* **Note 2**).

3.2 BD Rhapsody™ Cartridge Workflow Quality Control

The BD Rhapsody™ Scanner enables controlling each critical workflow step in the BD Rhapsody™ cartridge by imaging the microwell array. After the cartridge workflow, the scanner generates a full workflow report including the following metrics (Tables 2 and 3). You can find a detailed description of the microwell workflow below.

These variables are essential for single-cell sequencing approaches. Knowing the number of cells captured during the microwell workflow is a good basis for the sequencing depth to be applied. The determined number of multiplets can be used in the data analysis as a comparison value for the multiplet filtering (Tables 2 and 3).

3.3 BD Rhapsody™ Single-Cell Analysis System Workflow

Sample Preparation

Preparing Single-Cell Suspensions

- Cell handling: Minimize cell handling steps to reduce cell loss and decline in cell viability.
- Keep cells on ice when not handling them. Optimize centrifugation conditions according to cell number and cell phenotype to see a cell pellet.
- For high cell recovery, know the position of the cell pellet in the tube after centrifugation.
- If you detect doublets or clumps of cells in your sample, filter the cell suspension through an appropriately sized cell strainer multiple times to remove clumps and doublets.
- If no cell pellet is visible after the centrifugation, we recommend leave 50 μL of supernatant after each centrifugation step.

3.4 Cell Counting

- Filter cells to remove clumps and debris to ensure accurate cell counting. Debris in suspensions of small cells can lead to overestimated cell counts. Visually check filtered cell suspensions for debris.
- Avoid pipetting low volumes of cells (<2 μl).

Table 2
BD Rhapsody™ Scanner image analysis metrics

Step	Scanner metric	Definition	Passing threshold
Cell Load	Number of wells with viable cells at Cell Load[a]	Number of wells containing viable (CAM-positive) cells	–
Cell Load	Number of viable cells captured in wells at Cell Load	Number of viable (CAM-positive) cells captured in a well	–
Cell Load	Cell multiplet rate at Cell Load (%)	Number of wells containing multiple viable (CAM-positive) cells versus the total number of wells containing one or more viable cells	–
Bead Wash	Number of wells with viable cells and a bead	Total number of wells with one bead and ≥1 viable (CAM-positive) cells	Value
Bead Wash	Number of viable cells captured in wells with a bead	Total number of viable (CAM-positive) cells captured in a well with a bead	–
Bead Wash	Cell multiplet Rate (%)[a]	Number of wells containing one bead and multiple viable (CAM-positive) cells versus the total number of wells containing one bead and one or more viable cells	Value
Bead Wash	Bead loading efficiency (%)	Percent wells with one bead	≥80%
Bead Wash	Cell retention rate (%)	Number of viable (CAM positive) cells captured in wells prior to lysis versus number of viable cells captured in wells prior to bead loading	≥60%
Retrieval	Bead retrieval efficiency (%)	Percent of wells with beads removed	≥90%

[a]The estimated multiplet rate is calculated by Poisson distribution using the number of cells loaded into the cartridge. The cell multiplet rate reported by the BD Rhapsody™ Scanner is determined by imaging. If the cell multiplet rate ≫ estimated multiplet rate, this might indicate cell clumping

- Cell counting with the BD Rhapsody™ Scanner is most accurate when the cell concentration is in the range of ~200–800 cells/μl. If the cell concentration is >1000 cells/μl, it is recommended to dilute the cell suspension in cold sample buffer to ~200–800 cells/μl.
- Do not rely on FACS-based counts, because cell concentration might be overestimated by this method. Always re-count cells after FACS (Fluorescence Activated Cell Sorting).

The preparation of high-quality samples is key to obtain superior single-cell data [5].

Table 3
BD Rhapsody™ Scanner image analysis metrics and their usage

Scanner metric	Use
Number of wells with viable cells at Cell Load	Provides a preliminary estimate of the number of wells with viable cells captured[a]
Number of viable cells captured in wells at Cell Load	Provides a preliminary estimate of the total number of viable cells captured in the cartridge[b]
Cell multiplet rate at Cell Load (%)	Provides a measure of cell clumping
Number of wells with viable cells and a bead	Provides an estimate of the number of wells with viable cells captured with beads[c]
Number of viable cells captured in wells with a bead	Estimates the total number of viable cells captured on a cell capture bead at cell lysis[c]
Cell multiplet rate (%)	Provides an estimate of the occurrence of multiple cells captured by the same bead in a well
Bead loading efficiency (%)	Indicates if the cartridge is significantly underloaded with beads
Cell retention rate (%)	Indicates if a significant number of cells initially loaded into wells are lost or died during the workflow prior to lysis
Bead retrieval efficiency (%)	Indicates if the number of beads retrieved is significantly lower than expected

[a]The number of wells with viable cells is less than the total number of viable cells captured in wells because of wells containing cell multiplets

[b]Cell capture efficiency at Cell Load (%) = Number of viable cells captured in well at Cell Load/Total number of viable cells imaged * 100

[c]The metric, the number of wells with viable cells and a bead, is less than the metric, the number of viable cells captured in wells with a bead, due to wells containing cell multiplets. Multiple cells captured with one bead will appear as one cell in sequencing data. Because of this, the appropriate metric to use for an estimation of the number of cells that could be recovered in sequencing is the metric, number of wells with viable cells and a bead

It is advisable, to prepare cells as close in time to cartridge loading as possible. Some cell dissociation reagents, such as trypsin, might damage cell surface markers and decrease Sample Tag and BD® AbSeq AbO (antibody-oligonucleotide) sensitivity [11].

Use cell dissociation reagents suitable for cell surface staining and assess the marker detection with another technology like Flow Cytometry before proceeding. We optimized BD Horizon™ Tumor and Tissue Dissociation Reagent (TTDR) for effective dissociation of tissue and tumors for single-cell studies while maintaining cell surface epitopes for many cell markers (Fig. 5).

During the individual wash steps of the sample preparation protocol (please see section below), cells might be lost in the range of 25–50%. This needs to be taken into consideration especially for low-abundance samples (<100,000 cells).

Cell viability <50% might impact experimental results. If possible, use cells of as high viability as possible (≥90%).

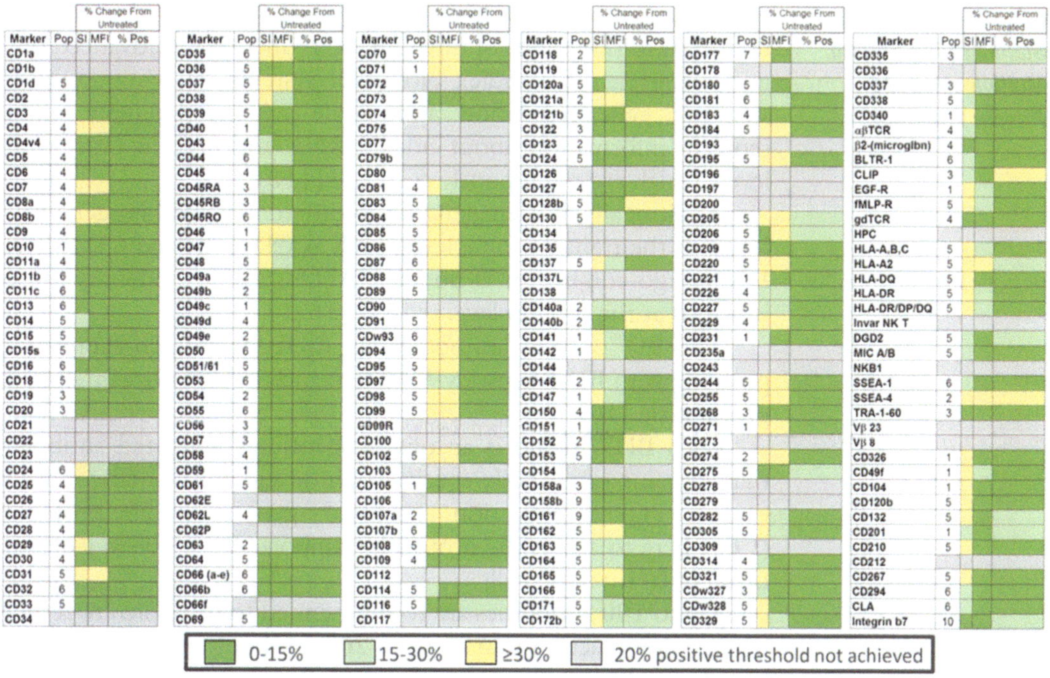

Fig. 5 Epitope preservation after TTDR treatment. Surface marker expression was assessed across the BD catalog in a mixed sample of two different cancer cell lines (RWPE-1 and MDA-MB-231) and PBMC from a donor pool (stimulated and unstimulated with PHA for 18 h). A total of 10 different populations were defined in this mixed sample: 1-RWPE, 2-MDA-MB-231, 3-unstimulated non-T cells (CD3−), 4-unstimulated T cells (CD3+), 5-unstimulated monocytes, 6-unstimulated granulocytes, 7-stimulated non-T cells (CD3−), 8-stimulated T cells (CD3+), 9-stimulated monocytes, and 10-stimulated granulocytes. For each marker, a single population (T cells, non-T cells, etc.) was chosen for the analysis. This target population is indicated by a number (1–10) that references these population assignment numbers (column heading is "Pop"). The mixed cell sample was then exposed to TTDR for 30 min or left untreated. Markers that demonstrated less than 20% expression are represented as grey boxes; 47 total markers. Untreated and treated (exposed to enzyme) samples were compared using Stain Index (SI), Median Fluorescence Intensity (MFI) of positive signal, and the % of the population that was clearly positive for a given marker (% Pos). Dark green indicates a change of less than 15%, light green indicates a change of 15–30%, and yellow indicates a change of 30% or more compared to the untreated control

BD Rhapsody™ Single-Cell Analysis System contains a sample and workflow quality control system that allows to assess the sample quality before the single-cell capture workflow, the BD Rhapsody™ Scanner as described above.

In order to process the cells as soon as possible after the sample preparation steps, the BD Rhapsody™ Cartridge should be set up for sample processing before sample preparation. If cell preparation takes ≥4 h, begin preparing cells before BD Rhapsody™ Cartridge preparation.

If the samples were not filtered before counting cells, pass the final dilution of the prepared single sample or pooled sample through a Falcon® Tube with Cell Strainer Cap (Corning™) before loading the sample into the cartridge.

3.5 Antibody Staining (Cell Sorting, Multiplexing, and Protein Profiling)

High-throughput sequencing has allowed researchers to examine hundreds to thousands of mRNA targets and greatly advanced our understanding of complex biological systems. However, understanding gene regulation and single-cell heterogeneity often requires information about both mRNA and protein expression levels which are often poorly correlated [13].

Additionally, it is challenging to differentiate certain gene isoforms at the transcriptome level, such as CD45RA and CD45RO. DNA-oligo-conjugated antibodies (AbOs) enable to identify those isoforms at the protein level reliably [15].

The BD® AbSeq antibodies are used for antigen-expression profiling with BD Rhapsody™ System and the downstream library preparation [10].

Each BD® AbSeq is an oligonucleotide-conjugated antibody that contains an antibody-specific barcode and poly(A)-tail for bead capture, PCR amplification, and library generation. The protocol (Click here to access the protocols) supports BD® AbSeq labeling of 20,000–1 million cells.

Up to 100 antibodies are recommended to be pooled together per staining reaction, but close to 200 antibodies have been combined already in different studies [14].

Staining cells with properly adjusted antibody-concentrations for single-cell sequencing workflows is equally important as in other single-cell technologies such as Flow Cytometry. Optimized AbO concentrations can help to prevent high background noise and an excess in reads for highly expressed markers while maintaining a high dynamic range in the protein expression quantification [1].

The concentration of each BD® AbSeq antibody is optimized for staining 20,000 to 1 million PBMCs so that individual BD® AbSeq antibodies can be easily pooled to an antibody panel without the need for further optimization. For other cell types, we recommend a Flow proxy protocol using poly(t)-conjugated fluorochromes (e.g., 5′ TTTTTTTTTTTTTTTTTT-AF647 3′) to optimize the concentration of the antibodies. The protocol can be downloaded here: (Click here to access the protocol).

The BD Rhapsody™ Single-Cell Multiplexing Kit utilizes an innovative antibody-oligo technology to provide higher sample throughput for single-cell library preparation [9].

Every antibody-oligo in a Multiplexing kit, referred to as a Sample Tag, has a unique sample barcode conjugated to a human or mouse antibody. Up to 12 samples can be labeled and pooled prior to single-cell capture with the BD Rhapsody™ Single-Cell Analysis system.

For most single-cell sequencing studies, sorting cells of interest, for example with Flow Cytometry Cell Sorting is a valuable tool to bypass sequencing unwanted cell populations. For this, the samples can be stained in parallel with fluorochrome-labeled antibodies, sample tags, and BD® AbSeqs (*see* **Note 3**) as described by us in a tech note: "Assessing obesity-induced inflammation in mice through single cell multiomics analysis" (Click here to access the tech note).

Please note that BD® AbSeq data can be used to gate for cells of interest when it is not possible to efficiently remove unwanted cell populations in the sorting process as described in this Innate Lymphoid Cell project: (Please click here to access the presentation).

Sufficient post-labeling washes are important for reducing noise that comes from residual unbound AbOs being captured onto 3′ capture beads during single-cell capture. However, some cell loss occurs with each additional wash, so an appropriate washing protocol has to be considered for every study.

We recommend washing cells by adding 3 ml of BD® Stain Buffer to labeled cells and pipet-mix for dilution followed by a centrifuge step to pellet the cells (*see* **Note 4**). The cells are then re-suspended in BD® Stain Buffer and the procedure is repeated two to three times to get rid of unbound AbOs.

After labeling the cells with the respective antibodies, they can be sorted directly into cold BD® Sample Buffer and can be further processed with the BD Rhapsody™ Single-Cell Analysis System [4].

3.6 Identification of Antigen-Specific T Cells and B Cells

dCODE Dextramer® (RiO) (Immudex) enable the identification of antigen specificity of T and B cells (*see* **Note 5**).

3.7 Preparation of a Nuclei Suspension

Obtaining single-cell suspensions from solid tissues requires dissociation methods that may damage cell integrity, particularly cells harboring intricate structures such as neurons. This problem can be compounded when using solid tissues preserved by freezing, which can further interfere with cell integrity before downstream single-cell capture methods are employed. However, these issues can be mitigated by using nuclei rather than whole cells for single-cell sequencing.

It is recommended to follow an established nuclei isolation protocol appropriate for the cell and tissue type of interest.

Isolated nuclei can be captured and amplified on the BD Rhapsody™ Single-Cell Analysis System with a modified protocol (Link to the protocol). A high viability of the stock cell suspension is key for obtaining good-quality data from isolated nuclei.

Throughout the BD Rhapsody™ cartridge workflow, ice cold sample buffer complemented with RNase inhibitor is used instead of sample buffer.

Cell nuclei can be counted with the BD Rhapsody™ Scanner, but the nuclei are stained with DyeCycle Green rather than Calcein/DRAQ7™. The BD Rhapsody™ Scanner will read 100% viability since all nuclei will be positive for the dye (*see* **Note 6**).

3.8 BD Rhapsody™ Cartridge Workflow

Preparation of the BD Rhapsody™ *Express*

For a step-by-step description on how to set up and operate the BD Rhapsody™ Express system (*see* **Note 7**).

3.9 Preparation of the BD Rhapsody™ Cartridge

The BD Rhapsody™ Cartridge needs to be primed before loading the cell sample. For this, the cartridge is be loaded with 100% (absolute) ethyl alcohol, Cartridge Wash Buffer 1, and Cartridge Wash Buffer 2. The cartridge can be stored at room temperature (15–25 °C) for ≤4 h.

Please note that the BD Rhapsody System includes two pre-programmed electronic pipettes to ensure optimal flow rates in each step of the workflow.

In the next step of the workflow, a volume of 615 μl of a cell suspension is loaded into the BD Rhapsody™ Cartridge. This enables processing of even dilute cell suspensions without the needing to concentrate them into very small volumes.

The cartridge contains a microwell-array (>220,000 microwells) that enables single-cell capture and molecular indexing of the analytes on magnetic barcoded beads (Fig. 6). The cells fall into the wells by gravity so that each cell occupies a single well according to Poisson statistics. This allows a highly efficient but gentle single-cell capturing with up to 80% cell capture rate and allows processing even highly fragile cells such as neutrophils with high efficacy [6]. The microwells are specifically treated to prevent cells from adhering to surfaces.

After the single cells have been captured in the microwells, barcoded magnetic cell capture beads are loaded close to saturation, so that each cell is paired with a bead. The geometry and diameter of the microwell and the beads ensure single-bead occupancy.

Upon cell lysis, the analytes of each cell can be captured to the beads by hybridization. The cell capture beads are subsequently retrieved from the microwells by magnets and pooled into a single 5 ml LoBind® Tube to allow simplified downstream next generation sequencing (NGS) library preparation (Fig. 6).

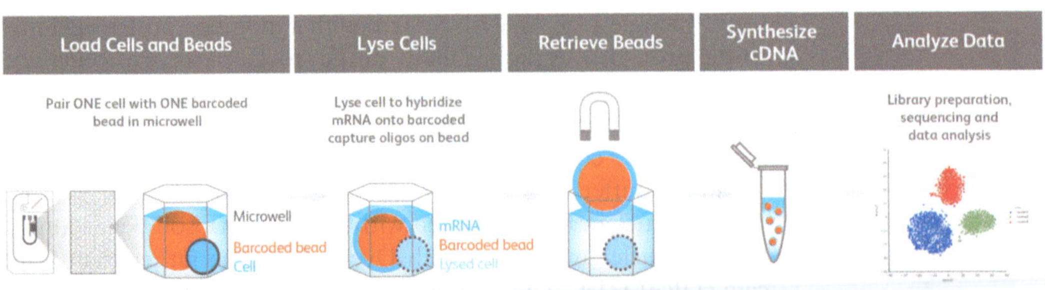

Fig. 6 The BD Rhapsody™ Single-Cell Analysis System workflow

3.10 Imaging Cells in a Cartridge

The BD Rhapsody™ Single-Cell Analysis System contains a sample quality control system and workflow quality control system, the BD Rhapsody™ Scanner. The imaging instrument can analyze each step in the cartridge workflow, and it summarizes the results in a full report.

For capturing bright field (530 nm) images of the microwells during the BD Rhapsody™ Cartridge Workflow, the system is equipped with LEDs as a light source and two additional fluorescence channels for reliable cell counting (Calcein-AM) and cell viability (DRAQ7™) assessment (*see* **Note 8**).

For a step-by-step description of the of the cell counting and viability assessment with the BD Rhapsody™ Scanner (*see* **Note 9**).

3.11 Cell Capture Beads

The magnetic cell capture beads contain two barcoding systems to enable capturing of single-cell information and correction for biased mRNA- and protein-expression levels introduced by the PCR amplification steps during the NGS library preparation.

First, each single DNA-oligo bound to a cell capture bead carries the same cell-label (CL) sequence to allow reliable allocation of information to a single cell. Second, adjacent to the CL sequence, a unique molecular identifier (UMI) sequence enables to eliminate the PCR bias from the library preparation. Each single DNA-oligo bound to a cell capture bead carries a unique UMI enabling reliable counting of analyte molecules in a multimodal dataset [3] in each single cell (Fig. 7).

The number of CL sequences is in great excess as compared to the number of cells analyzed per cartridge, such that each cell is paired with a bead that has a unique CL. Similarly, the diversity of UMIs that coat each bead allows almost all transcripts of the same gene from a cell to receive a different UMI. During cDNA synthesis on a dry block heater, the reverse transcriptase enzyme extends a cDNA complementary to the original mRNA transcript from dT, allowing the CL and UMI to be paired with each cDNA molecule. The resulting reaction generates a barcoded transcriptome of thousands of single cells that can be archived.

mRNAs, BD® AbSeqs, and the dCODE Dextramers® (RIO) are captured by poly-(A) capture oligos to the bead as described above. For retrieving the VDJ information from the T-cell receptor and B-cell receptor mRNAs, an additional set of capture oligos is present on the beads.

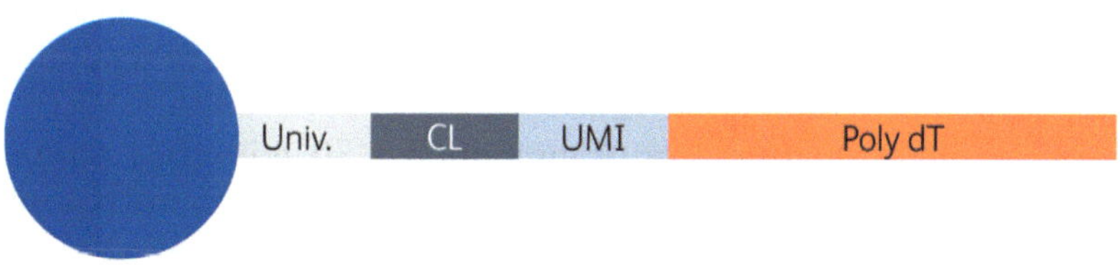

Fig. 7 Cell capture bead structure. Univ. universal primer binding site, CL cell label, UMI unique molecular identifier, Poly dT poly-dT sequence (25dTs)

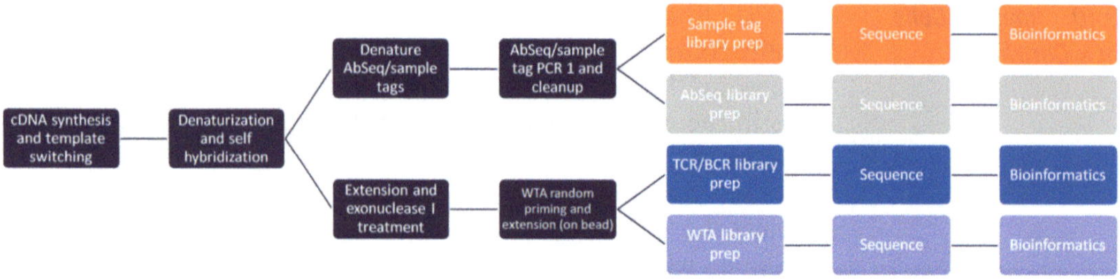

Fig. 8 WTA RNA/ BD® AbSeq/SMK/TCR/BCR library preparation strategy

3.12 NGS Library Preparation

To study the single-cell gene expression profiles, cDNA is amplified and converted into sequencing libraries. Multiple amplification schemes can be employed depending on user application after the reverse transcription allowing full flexibility for downstream assays.

The covalently bound cDNAs and AbO sequences on the cell capture beads enable high-throughput mRNA expression analysis (WTA or Targeted), protein analysis alongside T-cell receptor (TCR) and B-cell receptor (BCR) profiling of individual cells captured on the BD Rhapsody™ Single-Cell Analysis System. Here, we will focus on the protocol for preparing multimodal libraries including WTA RNA/BD® AbSeq/SMK/TCR/BCR library (Fig. 8).

The following section describes the library preparation for a WTA/BD® AbSeq/SMK/TCR/BCR library. For a detailed protocols for the preparation of different library types (*see* **Note 10**).

3.13 Cell Capture Bead Preparation for cDNA Synthesis and Reverse Transcription

Before the cDNA synthesis step, the capture beads have to be properly washed with bead wash buffer to prevent low performance of the reverse transcriptase in the next step. The buffer change is enabled through magnetic stands for pelleting the beads in the individual steps of the protocol. The beads are clearly visible at the side of the magnets as a dark brown pellet.

It is strongly advised to use of low retention pipet tips to handle the beads and to keep the B-cell capture beads cold during the washing steps to maintain high RNA integrity.

Furthermore, it is important to thoroughly remove the lysis buffer or lysis buffer bubbles from the tube containing the capture beads before adding the wash buffer. Otherwise, precipitates may form between the lysis buffer and bead wash buffer (added in the next step) that may cause the reverse transcription reaction to fail or impair its efficacy.

The synthesis of the cDNAs on the capture beads is conducted on a SmartBlock™ Thermoblock 1.5 mL (or equivalent) at 42 °C for 30 min and 1200 rpm.

Start reverse transcription ≤30 min after washing retrieved beads with bead wash buffer to ensure high mRNA integrity.

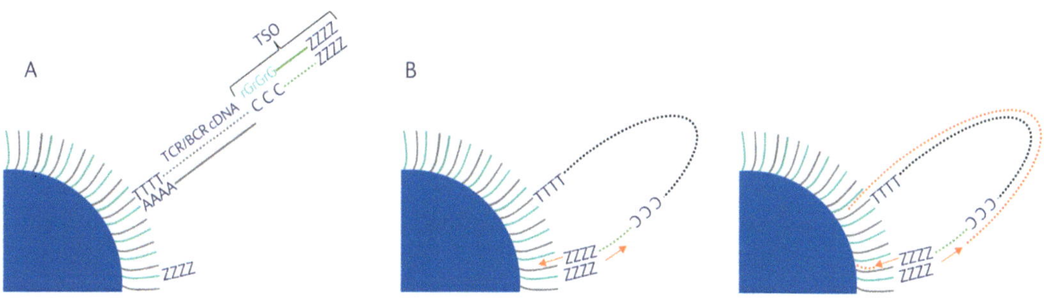

Fig. 9 Epitope preservation cDNA and template switching, hybridization, and extension. (**a**) cDNA and template switching: Capture of TCR/BCR transcript via a poly dT tail on a bead oligonucleotide. A unique sequence is added to the 3′end of the TCR/BCR cDNA transcript using a template-switch oligo (TSO). (**b**) Hybridization and extension: The unique sequence now added to the 3′end of the TCR/BCR cDNA hybridizes to a different bead oligo that has a complementary sequence. Addition of a DNA polymerase and other reagents allows. (a) copying new barcodes to the 3′end of the cDNA; (b) copying of the TCR/BCR cDNA

3.14 Template Switch Oligo

For the analysis of the TCR and BCR repertoire, the mRNAs encoding for T- and B-cell receptors are elongated at their 5′end by a sequence which is complementary to a dedicated TCR/BCR capture sequence on the cell capture beads using a Template Switch Oligo (TSO) (Fig. 9). This unique sequence then hybridizes to a set of capture oligos on the beads dedicated for the TCR/BCR cDNAs (*see* **Note 11**). The template switch reaction is started by simply adding the TSO mix to the reaction as described in the protocol (*see* **Note 12**). After the template switching reaction is finished, the beads can be stored up to 4 days at 2–8 °C after the buffer has been replaced with elution buffer.

3.15 Denaturation and Self-Hybridization

After the complementary sequence is added through the TSA reaction, the 5′-ends need to loop back to the capture bead oligos (Fig. 9). Additionally, the BD® AbSeq and sample tag information is transferred to the supernatant. The buffer is replaced with the elution buffer using a magnetic stand followed by mixing the suspension with a pipet.

To denature, tube samples are incubated at 95 °C in a thermomixer (no shaking) for 5 min.

The tube is briefly centrifuged, after which the beads are separated with a magnetic stand.

Please note, it is important to transfer the complete supernatant in a new Lowbind® Tube as it contains the BD® AbSeq and/or sample tag molecules (if the cells were incubated with either or both of them during the sample preparation). To minimize BD® AbSeq/sample tag contamination, ensure that all liquid is removed from the tube. Keep the supernatant tube at 4 °C until ready to proceed (please see below).

When the BD® AbSeq/sample tag information is separated, the TSO-modified cDNAs hybridize in the next step to the dedicated TCR/BCR capture oligos on the beads after which the TCR/BCR extension step is performed, as described in the protocol (*see* **Note 13**).

3.16 TCR/BCR Extension

After the 5′-ends of the mRNAs have hybridized with capture oligos, the single-stranded regions are filled up with a Klenow fragment to generate full-length cDNAs with a CL and UMI attached to their 5′-ends so that the 5′ VDJ information from the T-cell receptors and B-cell receptors can be obtained together with the 3′-gene expression information (Fig. 9).

3.17 Exonuclease I Treatment

For the Exonuclease I treatment, the buffer of the capture bead suspension is replaced with Exonuclease I mix with a magnetic stand. The suspension is incubated on the thermomixer at 1200 rpm and 37 °C for 30 min after which the Exonuclease I is inactivated by an 80 °C incubation step.

After the buffer has been replaced by bead resuspension buffer, the Exonuclease I-treated beads can be stored at 2–8 °C for ≤3 months or shipped to a sequencing service provider for the library preparation and the sequencing, as described below (*see* **Note 14**).

1. Bead storage: Cell capture beads can be stored for up to 12 weeks at 2–8 °C as demonstrated in the following test. Pooled beads (two subsamples) of 1000 cells obtained from a lymph node containing sparse metastatic breast cancer cells were profiled with the BD Rhapsody™ Onco-BC targeted panel. Samples were profiled at 0, 6, and 12 weeks after bead storage at 4 °C. Strong correlations of gene expression between subsamples indicated minimal batch effect and high stability of archived cDNA (Fig. 11).
2. Capture bead subsampling: The capture beads with the covalently bound cDNAs can be subsampled to prepare multiple NGS libraries from one set of beads as shown in Fig. 12. This allows to run multiple assays with one sample, such as a Whole Transcriptome Amplification (WTA) assay and a targeted RNAseq assay.
3. Streamlined shipment of cell capture beads

   The magnetic cell capture beads can be shipped at 2–8 °C to a sequencing provider if the library preparation and/or the sequencing should be performed in another lab. This setup is further facilitated by the sub-kit concept of the BD Rhapsody™ kits allowing a clean separation between the individual steps of the workflow.

3.18 Random Priming and Extension (RPE) on Cell Capture Beads with cDNA

In the next step, random priming products are generated to prepare the sequencing library. First, random primers are hybridized to the cDNA on the cell capture beads, followed by extension with an enzyme in order to amplify the WTA information. We strongly advice to use the entire sample of beads (*see* **Note 15** and **16**). Still subsampling of the beads is possible with other assays, for example, when combining gene expression analysis (WTA or Targeted) + BD® AbSeq.

The WTA-RPE products are cleaned up by a single-sided AMPure cleanup, which removes primer dimers and other small molecular weight by-products. The final product is purified single-stranded DNA which gets amplified through PCR, so that multiple copies of each random-primed molecule are available for the downstream steps.

After cleaning the PCR products (AMPure beads), the WTA-RPE PCR libraries can be stored at −20 °C for up to 6 months or at 4 °C for up to 6 weeks.

RPE PCR products have to be quantified and quality-controlled with a Qubit™ Fluorometer using the Qubit™ dsDNA HS Assay and the Agilent 2100 Bioanalyzer using the Agilent High Sensitivity DNA Kit.

The expected concentration from the Qubit Fluorometer is ~0.5–10 ng/μl, the Bioanalyzer trace should show a broad peak from ~150 to 2000 bp. Use the concentration from 200 to 600 bp to calculate volumes for the downstream amplification steps.

3.19 BD® AbSeq/ Sample Tag and TCR/ BCR Amplification

Both, the BD® AbSeq/Sample Tag information and the TCR/BCR information are amplified by PCRs using the supernatant from the denaturation and self-hybridization step (for the BD® AbSeq/Sample Tag information) and the capture beads (for the TCR/BCR information) after which the PCR products are purified using AMPure beads. The Sample Tag and TCR/BCR information is amplified a second time by using the amplicons from the first PCR as templates. After the first PCR, the BD® AbSeq PCR products and the Sample Tag PCR products need to be quantified to ensure an optimal ratio in the sequencing library.

3.20 Quantifying BD® AbSeq/Sample Tag PCR Products

The aliquots from the BD® AbSeq and Sample Tag PCR reactions are quantified by measuring the yield of the largest peak of the BD® AbSeq/Sample Tag PCR product (~150 bp) with the Agilent 2100 Bioanalyzer (High Sensitivity Kit assay). Based on the yield of the largest peak, dilute an aliquot of BD® AbSeq/Sample Tag PCR product to 0.1–1.1 ng/μL with Nuclease-Free water for the downstream steps.

3.21 Performing Random Priming and Extension (RPE) on TCR/BCR Amplicons from the Second PCR

TCR/BCR sequences undergo a third amplification by random priming followed by a purification step.

3.22 Performing TCR/BCR and BD® AbSeq/Sample Tag Index PCR

This section describes how to generate TCR/BCR and BD® AbSeq/Sample Tag libraries compatible with the Illumina sequencing platform, by adding full-length Illumina sequencing adapters and indices through PCR.

Please note that the number of PCR cycles for the BD® AbSeq and Sample Tag reactions has to be adjusted to the concentration of index PCR input. After the index PCR reactions, primer dimers from the TCR/BCR and BD® AbSeq/Sample Tag Index PCR products are removed by a single-sided AMPure cleanup.

After the cleanup, estimate the concentration (BD® Abseq, Sample Tag, TCR/BCR) by quantifying the final sequencing library with a Qubit™ Fluorometer using the Qubit dsDNA HS Kit to obtain an approximate concentration of PCR products to dilute for quantification on an Agilent 2100 Bioanalyzer system using the Agilent High Sensitivity D1000. The WTA index PCR product is purified with AMPure beads as well (single-sided cleanup).

The final product of the purification reactions is purified double-stranded DNA molecules with full-length Illumina adapter sequences.

3.23 Sequencing

The concentration of each sample is determined by quantifying the final sequencing library with a Qubit™ Fluorometer using the Qubit™ dsDNA HS Kit to obtain an approximate concentration of PCR products to dilute for quantification on an Agilent 2100 Bioanalyzer. The expected concentration of the libraries is >1.5 ng/μl.

Also, the average fragment size of the WTA and TCR/BCR libraries within the size range of 200–1000 bp are being measured by using the Agilent Bioanalyzer with the High Sensitivity Kit for 50–7000 bp, 5–1000 pg/μl (*see* **Note 17**).

3.24 Sequencing Flow Cell Loading and PhiX Concentrations

Calculate the molar concentration of WTA, BD® AbSeq, and TCR/BCR libraries using Qubit™ quantitation concentration (ng/μl) and average Bioanalyzer size (200–1000 bp). For TCR/BCR libraries, the expected Qubit™ concentration should be >1.5 ng/μl. Use the calculated molar concentrations to pool libraries. PhiX is a control library (Illumina®) that allows to increase the complexity of the NGS library.

Table 4
BD Rhapsody™ sequencing recommendations

Gene panel	Read requirement for data analysis	Adjusted reads for sequencing pooled TCR/BCR libraries on Illumina platforms[a]
BD Rhapsody™ WTA	~2000–20,000 reads/cell[b]	N/A
Sample Tag: samples from same type of cell (combining different donor PBMCs)	120 reads/cell	N/A
Sample Tag: samples from different type of cells (combining cell lines with PBMCs)	600 reads/cell	N/A
BD® AbSeq	200–1000 reads/cell/BD® AbSeq[b]	N/A
BD Rhapsody™ TCR	~5000 reads/T cell	~25,000 reads/T cell[a]
BD Rhapsody™ BCR	~5000 reads/B cell	~25,000 reads/T cell[a]

[a]Note: Because of the long amplicon sizes, TCR and BCR libraries will not cluster as efficiently as WTA libraries on Illumina platforms. To account for this, TCR and BCR libraries should be pooled with WTA libraries as if aiming for 5× the number of reads that are actually desired. See the following examples

[a]The amount of sequencing needed for BD® AbSeq libraries will vary depending on application, BD® AbSeq panel plexy, and cell type. We have observed that using 40,000 sequencing reads per cell for 40-plex BD® AbSeq libraries prepared from resting PBMCs achieves an RSEC sequencing depth of ~2

Table 5
Required parameters for BD® AbSeq/Sample Tag libraries

Parameter	Requirement
Platform	Illumina: 150 cycle kit
Paired-end-reads	Minimum of 51 × 75 paired read length
PhiX	Required (15%)
Analysis	See the BD® Single-Cell Multiomics Bioinformatics Handbook

3.25 BD® AbSeq/ Sample Tag Libraries

For a NextSeq High or Mid Output and MiniSeq High or Mid Output runs, load the flow cell at a concentration between 1.8 and 2.2 pM with 15% PhiX for a sequencing run (Tables 4 and 5).

3.26 WTA and TCR/ BCR Libraries (With or Without BD® AbSeq/ Sample Tag)

For a NextSeq High or Mid Output and MiniSeq High or Mid Output runs, load the flow cell at a concentration between 1.4 and 1.8 pM with 3% PhiX*. For other sequencers, follow Illumina recommendations for loading concentration and use 3% PhiX (Table 6).

Table 6
Required parameters for WTA and TCR/BCR libraries

Parameter	Requirement
Platform	Illumina: 300 cycle kit
Paired-end-reads	Minimum of 85 × 215[a] paired read length
PhiX	Required (3%)
Analysis	See the BD® Single-Cell Multiomics Bioinformatics Handbook (Doc ID: 54169)

[a]R2 length of 215 is recommended for optimal assembly. If necessary 150 × 150 read lengths can be used

*Note: if using less than 10,000 reads/cell for the WTA library, increase PhiX percentage to 5–10% to account for lower library diversity.

Set up sequencing run on Illumina® BaseSpace. Enter the pooled libraries as one sample if libraries were made with the same Library Forward primer but with different i7 indices (Table 6).

3.27 Bioinformatics

The BD Rhapsody™ assays are used to create sequencing libraries from single-cell multiomic experiments. The analysis pipeline works with paired-end FASTQ reads 1 (R1) and read 2 (R2) files generated from Illumina® sequencers. The minimum read length required is 51 bp for R1 and 71 bp for R2 and for the TCR/BCR libraries the minimum read length is 85 bp for R1 and 215 bp for R2. R1 reads contain information on the cell label and molecular identifier, and R2 reads contain information on the bioproduct.

After sequencing, the targeted analysis pipeline takes the FASTQ files, an mRNA reference file, and a BD® AbSeq reference file (if the latter is required) for alignment and annotation. The Whole Transcriptome Analysis (WTA) pipeline takes the FASTQ files, a reference genome, a supplemental reference (if needed), a transcriptome annotation file, and a BD® AbSeq reference file (if the latter is required).

For more detailed information about the data analysis pipeline, please refer to the BD® Single-Cell Multiomics Bioinformatics Handbook (*see* **Note 18**).

3.28 Filter by Read Quality

Filtering criteria Read 1 artifacts are first removed from Read 2. Then, read pairs with low sequencing quality are removed. This step reduces the influence of poor sequencing quality from the metrics that are specific to the BD Rhapsody™ assays. Read 1 artifacts are removed from read 2 with multiple filter algorithms.

3.29 Annotate R1 Reads

The quality filtered R1 reads are analyzed to identify the cell label sequences (CL), unique molecular identifier (UMI) sequence, and if applicable, poly(T) tail.

Reads are first checked for perfect matches in the CL sequences at the expected locations, where reads with perfect matches are kept. The remaining reads are subjected to another round of filtering to recover reads with base substitutions, insertions, and deletions caused by sequencing errors, PCR errors, or errors in oligonucleotide synthesis. A valid R1 read requires identified CLs, a UMI sequence with non-N bases, and if applicable, a poly(T) tail.

3.30 Annotate R2 Reads

A valid R2 requires the read to be uniquely mapped to a bioproduct in the reference. For a targeted assay, it must also have the correct PCR2 primer sequence at the start and an alignment match sufficient in length (*see* **Note 19**).

After the read alignment, valid R1 and R2 reads are combined for further processing.

3.31 Annotate Molecules

Reads with the same cell label, same UMI sequence, and same bioproduct are collapsed into a single raw molecule. The number of reads associated with each raw molecule is reported as the raw adjusted sequencing depth.

In the next step, artifact molecules are being removed using two UMI adjustment algorithms: recursive substitution error correction (RSEC) and distribution-based error correction (DBEC).

PCRs and sequencing often generate errors. If the error occurs within the UMI sequence, the R1/R2 read pair is called a unique molecule but is, in fact, an artifact.

Artifact molecules contribute to an overestimated molecule count of a gene in a cell. As sequencing depth increases, the number of raw molecules rises and never plateaus due to these artificial molecules. To remove the effect of UMI errors on molecule counting, BD Biosciences has developed a set of UMI adjustment algorithms. UMI errors that are single base substitution errors are identified and adjusted to the parent UMI barcode using RSEC.

For targeted sequencing analysis, other UMI errors derived from library preparation steps or sequencing base deletions are later adjusted using DBEC. Note that targeted sequencing analysis uses RSEC and DBEC, while WTA sequencing analysis uses RSEC only for mRNA libraries and RSEC and DBEC for BD® AbSeq libraries.

3.32 Determine Putative Cells

In theory, the number of unique cell labels detected by the bioinformatics pipeline should be similar to the number of cells captured and amplified by the BD Rhapsody™ workflow.

However, various processes throughout the workflow can introduce noise that contributes to excessive cell labels generated during sequencing analysis, including the following:

- Hybridizing polyadenylated poly(A)-oligonucleotides to beads residing in neighboring wells when the cell lysis step is too long
- Underloading beads in BD Rhapsody™ Cartridges resulting in cells without beads and the RNA from the cells diffusing to adjacent wells
- Experiencing low-level contamination during oligonucleotide and bead synthesis
- Generating errors during the PCR amplification steps of the workflow

To distinguish cell labels associated with putative cells from those associated with noise, a multi-step algorithm was designed for filtering cell labels.

3.33 Determine the Sample of Origin (Sample Multiplexing Only)

Up to 12 samples of cell suspensions can be loaded into a BD Rhapsody™ Cartridge using a BD® Single-Cell Multiplexing Kit. Each sample is labeled with a separate Sample Tag from the kit. When you start the BD Rhapsody™ Analysis pipeline, you can select the sample multiplex option. You can associate a name with a Sample Tag before the pipeline starts, and the specified sample names will be used in the output files. To account for every Sample Tag, each Sample Tag sequence in the kit is considered during pipeline analysis, whether the Sample Tags are used in the experiment or specified with a sample name. The pipeline automatically adds the Sample Tag sequences to the FASTA reference file. Reads that align to a Sample Tag sequence and associate with a putative cell are used to identify the sample for that cell.

3.34 Generate Expression Matrices (Reporting RSEC and DBEC Metrics)

RSEC-adjusted molecule counts and associated reads of each bioproduct for each putative cell and DBEC-adjusted molecule counts and associated reads are presented in either .csv or .st format.

3.35 Annotate BAM

The BAM file output by Bowtie2 or STAR is further annotated to summarize the results of the BD Rhapsody™ Analysis pipeline.

3.36 TCR and BCR Analysis (If Applicable)

In combination with other BD Rhapsody™ assays, optional protocols and products enable the generation of sequencing libraries specific to T- and B-cell receptors (TCR and BCR).

When enabled, the BD Rhapsody™ analysis pipeline can use the reads from these libraries to assemble contigs corresponding to rearranged TCR and BCR chain mRNA. These contigs are then analyzed to identify full-length single-cell level VDJ gene segments including complementary determining regions (CDRs), read and molecule counts, and per cell type chain-pairing. TCR and BCR analysis is supported for human and mouse.

Major TCR and BCR pipeline steps

- Identify reads derived from TCR or BCR mRNA
- Assemble reads into contigs
- Annotate contigs with VDJ gene segment information
- Select dominant contigs, chain family, cell type
- Error correction and contig trimming
- TCR and BCR output file

3.37 Sequencing Analysis Output Files

The data analysis pipeline at Seven Bridges Genomics generates multiple files for the analysis:

Metrics summary:	Report containing sequencing, molecules, and cell metrics.
BAM and BAM index:	BAM is an alignment file in binary format that is generated by the aligner. The aligner aligns R2 reads to the reference file and outputs tags related to alignment quality. This BAM file is sorted according to the alignment coordinates of R2 reads on each chromosome. The BAM Index is the index file associated with the coordinate-sorted BAM file.
Data tables	Reads per bioproduct per cell and molecules per bioproduct per cell, based on RSEC or DBEC. Unfiltered tables containing all cell labels of ≥10 reads.
Expression data	The expression sparse matrix, a table of counts in sparse format. Compressed file containing all cell labels of ≥10 reads.
Cell label filtering	Visualization of cell label filtering results.
Second derivative	Visualization of cell label filtering results.
Putative Cells	Algorithm that found the putative cell: basic or refined.
Protein Aggregates Exp.	Putative cells identified as protein aggregates.
Bioproduct Statistics	Metrics from RSEC and DBEC unique molecular identifier adjustment algorithms on a per-bioproduct basis.
Sample Tag metrics	Metrics from the sample determination algorithm.
Sample Tag calls	Assigned Sample Tag for each putative cell.
Per sample folder	Data tables metric summary, and expression matrix for a particular sample. Note: For putative cells that could not be assigned a specific Sample Tag, a Multiplet and Undetermined.zip file is also output.

4 Notes

1. This information gives users the power to change course and troubleshoot experiments, if necessary, before expensive downstream sequencing. These metrics allow users to track performance of workflows day to day or between various study sites.
2. Please note that cells loaded on a used cartridge must not be further processed for preparing sequencing libraries.
3. Labeling with Sample Tags or BD® AbSeqs is recommended in BD® Stain Buffer (FBS) (Cat. No. 554656).
4. Use polystyrene flow tubes and centrifuge cells using a benchtop centrifuge with swing out bucket rotor to minimize cell loss. Sort more cells than needed for cartridge loading. Sort cells into 5 mL polystyrene tube. Use the same 5 mL polystyrene tube that is used for sorting for cell labeling.
5. For further information and protocols, please visit: https://www.immudex.com/products/basic-research/dcode-dextramer-ngsmulti-omics/dcode-dextramer-rio/
6. Therefore, when processing cell nuclei, the viability measure is meaningless, but the counts generated are accurate (internal data).
7. https://www.youtube.com/watch?v=fVfB8ckmrfw
8. Fluorescence-imaging specifications: Excitation wavelengths: Blue: 482 nm, Red: 635 nm. Emission wavelengths: Green: 513–563 nm, Red: 662.5–707.5 nm.
9. https://www.youtube.com/watch?v=n7pkdZmohvU
10. https://scomix.bd.com/hc/en-us/categories/360000838932-Resource-Library
11. TSO mix must be prepared approximately within 2 min before the reverse transcription is finished.
12. https://www.bdbiosciences.com/content/dam/bdb/marketing-documents/vdj-tcr-bcr-protocols/Protocol%20Rhapsody%20VDJ%2BWTA%2BST%2BAbSeq%20RUO.pdf
13. https://www.bdbiosciences.com/content/dam/bdb/marketing-documents/vdj-tcr-bcr-protocols/Protocol%20Rhapsody%20VDJ%2BWTA%2BST%2BAbSeq%20RUO.pdf
14. After the reverse transcription and the Exo I inactivation, the cDNAs remain bound to the cell capture beads (Fig. 10) enabling enhanced flexibility in the downstream workflow steps.
15. Subsampling of beads is not recommended for TCR/BCR FL + WTA + BD® AbSeq + Sample Tag combination assays.

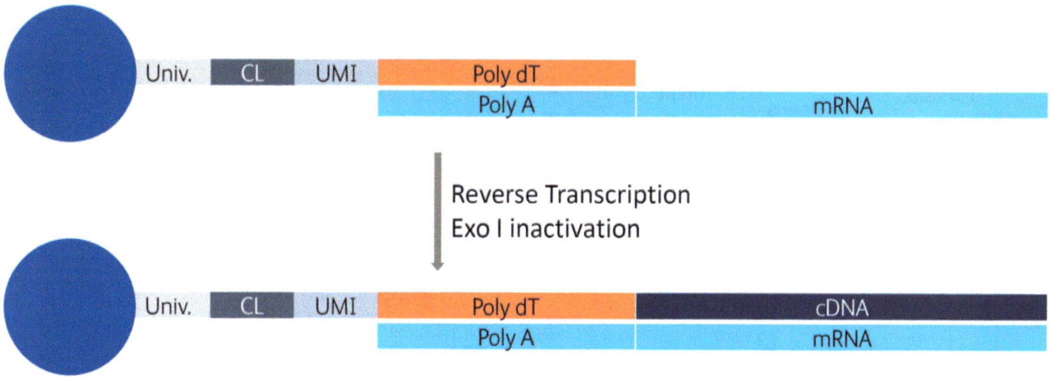

Fig. 10 Reverse transcription of captures mRNAs follow by Exonuclease (Exo) I inactivation leads to cDNAs covalently bound to the cell capture beads

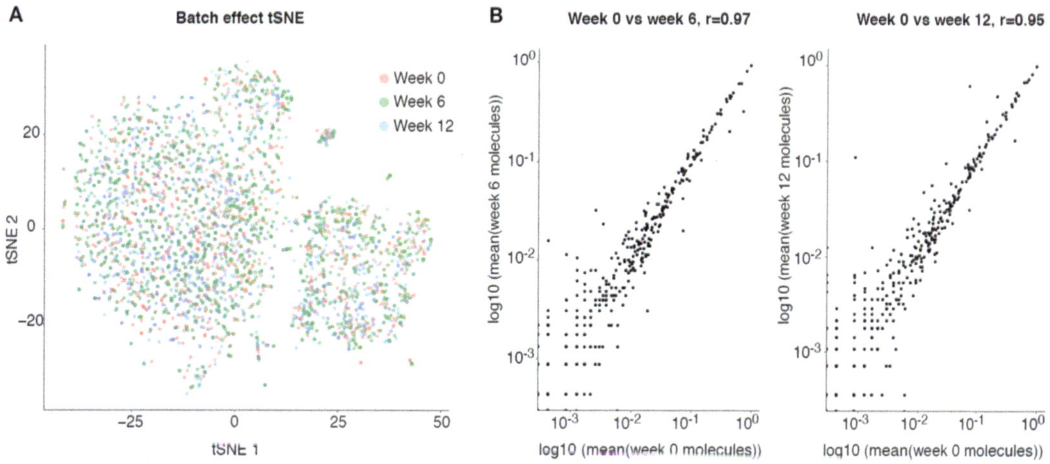

Fig. 11 Cell capture beads are stably stored with intact cDNA from one preparation and show little batch effect over time (**a**). Correlation of gene expression in batches prepared 12 weeks apart shows stability of archived sample (**b**)

16. After the reaction, the supernatant contains the WTA-RPE products which needs to be transferred to a new LoBind® tube, while the cell capture beads will be used for TCR/BCR and BD® AbSeq/Sample Tag target specific amplification.
17. Sequencing depth (WTA library) can vary depending on whether the sample contains high- or low-content RNA cells. For resting PBMCs, we recommend 10,000 reads per cell for shallow sequencing. Genes per cell and UMI per cell detected are generally lower but can be useful for cell type identification: 50,000 reads per cell for moderate sequencing and 100,000 reads per cell for highly saturated deep sequencing are used to identify the majority of UMIs in the library.

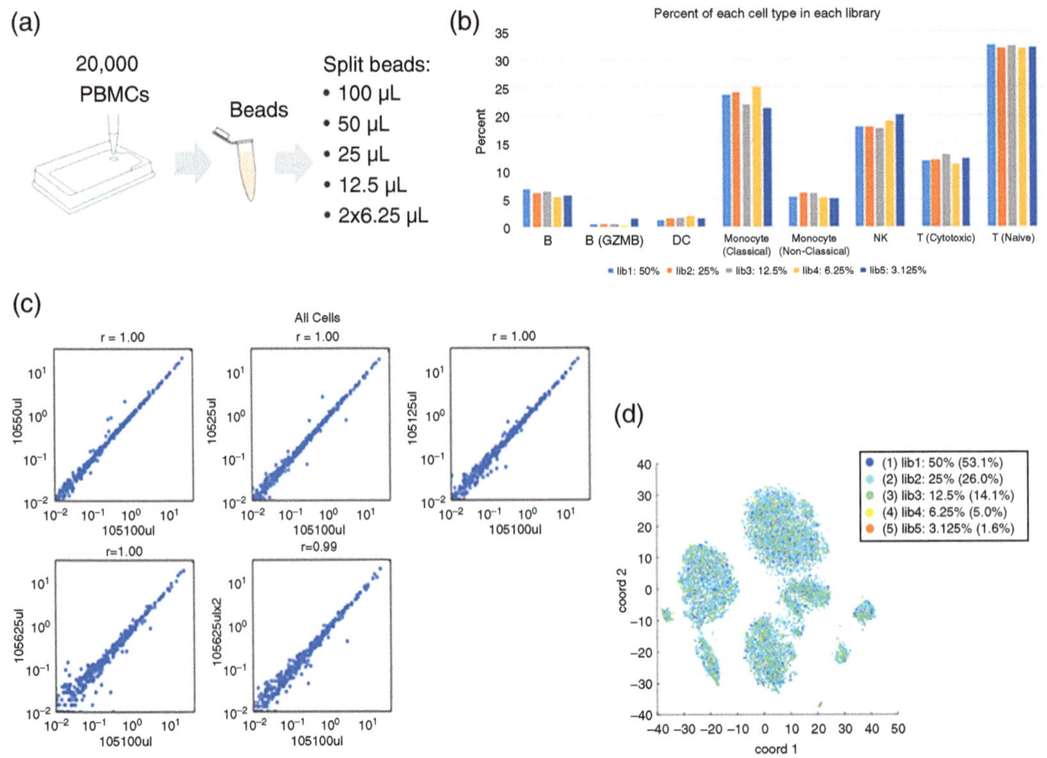

Fig. 12 20,000 PBMCs were loaded into a BD Rhapsody™ Cartridge. Different bead subsamples were taken through library prep using the human immune response panel (**a**). Cell numbers correspond nicely with the expected cell recovery (**b**), and similar proportions of each cell type are recovered from each bead subsample (**c**, **d**)

18. https://scomix.bd.com/hc/article_attachments/4423045606285/23-21713_02__scM_Bioinformatics_Handbook-ruo.pdf
19. Criteria for a valid R2 read: For mapping filtered reads to the transcriptome (WTA) or the reference panel sequences, the pipeline uses Bowtie2 for targeted panels and STAR for WTA. For a targeted panel, a R2 read is a valid alignment if all of the following criteria are met: (i) The R2 alignment begins within the first five nucleotides for mRNA, first 15 nucleotides for BD® AbSeq, and first 25 nucleotides for Sample Tags. This criterion ensures that the R2 read originates from an actual PCR priming event. (ii) The length of the alignment match (can be a match or mismatch) in the CIGAR (Compact Idiosyncratic Gapped Alignment Report) string which enables the identification of base alignments, insertions, and deletions with respect to the reference sequence. (iii) The read does not align to phiX174. WTA assays: An R2 is a valid gene alignment if all of these criteria are met: (i) The read aligns uniquely to a bioproduct in the reference. (ii) The read does not align to phiX174.

Glossary

BAM	An alignment file in binary format. A binary SAM file.
Bioproduct	Identifiers for biologically derived products such as mRNA and protein. Examples of identifiers are gene name for mRNA or AbSeq identifier for Abseq.
Bioproduct Type	Type of bioproducts such as mRNA or Abseq.
CIGAR	Compact Idiosyncratic Gapped Alignment Report. A sequence of base lengths to indicate base alignments, insertions, and deletions with respect to the reference sequence.
CLS	Cell label sequence.
DBEC	Distribution-based error correction.
FASTA	Text-based format that contains one or more DNA or RNA sequences.
FASTQ	A file in standardized, text-based format that contains the output of read bases and per-base quality values from a sequencer.
L	Common sequence
Molecule	A unique combination of a cell label, UMI sequence, and a bioproduct. Without UMI adjustment methods, it is called raw molecule. With RSEC UMI adjustment, it is called RSEC-adjusted molecule. With additional DBEC UMI adjustment, it is called DBEC-adjusted molecule.
PhiX	Control library used for sequencing runs.
R1 reads	Contains information about the cell label and UMI.
R2 reads	Contains information about the bioproduct.
RSEC	Recursive substitution error correction.
SAM	Tab-delimited text file with sequence alignment data.
Singlet	A putative cell where more than 75% of sample tag reads are from a single tag.
Singleton	Clustering: Cell not assigned to any of the clusters. UMI correction/adjustment: Molecule that is represented by only one read.
UMI	Unique molecular identifier. A string of eight randomers immediately downstream of the cell label sequence (CLS) 3 of the R1 read that is used to uniquely label a molecule.

References

1. Buus TB, Herrera A, Ivanova E, Mimitou E, Cheng A, Herati RS, Papagiannakopoulos T, Smibert P, Odum N, Koralov SB (2021) Improving oligo-conjugated antibody signal in multimodal single-cell analysis. eLife 10: e61973. https://doi.org/10.7554/eLife.61973
2. Fan HC, Fu GK, Fodor SPA (2015) Combinatorial labeling of single cells for gene expression cytometry. Science 347:1258367. https://doi.org/10.1126/science.1258367
3. Fu GK, Hu J, Wang PH, Fodor SP (2011) Counting individual DNA molecules by the stochastic attachment of diverse labels. Proc Natl Acad Sci U S A 108(22):9026–9031. https://doi.org/10.1073/pnas.1017621108
4. Georg P, Astaburuaga-García R, Bonaguro L, Brumhard S, Michalick L, Lippert LJ, Kostevc T, Gäbel C, Schneider M, Streitz M, Demichev V, Gemünd I, Barone M, Tober-Lau P, Helbig ET, Hillus D, Petrov L, Stein J, Dey HP, Paclik D, et al; PA-COVID-19 Study Group (2022) Complement activation induces excessive T cell cytotoxicity in severe COVID-19. Cell 185(3):493–512.e25. https://doi.org/10.1016/j.cell.2021.12.040
5. Haque A, Engel J, Teichmann SA, Lönnberg T (2017) A practical guide to single-cell RNA-sequencing for biomedical research and clinical applications. Genome Med 9(1):75. https://doi.org/10.1186/s13073-017-0467-4
6. Huang J, Zhu Z, Ji D, Sun R, Yang Y, Liu L, Shao Y, Chen Y, Li L, Sun B (2022) Single-cell transcriptome profiling reveals neutrophil heterogeneity and functional multiplicity in the early stage of severe burn patients. Front Immunol 12:792122. https://doi.org/10.3389/fimmu.2021.792122
7. Perkel JM (2021) Single-cell analysis enters the multiomics age. Nature 595. https://www.nature.com/articles/d41586-021-01994-w
8. Junius S, Mavrogiannis AV, Lemaitre P, Gerbaux M, Staels F, Malviya V, Burton O, Gergelits V, Singh K, Tito Tadeo RY, Raes J, Humblet-Baron S, Liston A, Schlenner SM (2021) Unstable regulatory T cells, enriched for naïve and Nrp1neg cells, are purged after fate challenge. Sci Immunol 6(61). https://www.science.org/doi/10.1126/sciimmunol.abe4723?url_ver=Z39.88-2003&rfr_id=ori:rid:crossref.org&rfr_dat=cr_pub%20%200pubmed
9. Ma KY, Schonnesen AA, He C, Xia AY, Sun E, Chen E, Sebastian KR, Guo YW, Balderas R, Kulkarni-Date M, Jiang N (2021) High-throughput and high-dimensional single-cell analysis of antigen-specific CD8+ T cells. Nat Immunol 22(12):1590–1598. https://doi.org/10.1038/s41590-021-01073-2
10. Russo V, Brasu N, Pace L (2022) Combined measurement of RNA and protein expression on a single-cell level. Methods Mol Biol (Clifton, N.J.) 2386:263–288. https://doi.org/10.1007/978-1-0716-1771-7_16
11. Saigusa R, Ley K (2020) CITE-Seq hits vascular medicine. Clin Chem 66(6):751–753. https://doi.org/10.1093/clinchem/hvaa016
12. Schulte-Schrepping J, Reusch N, Paclik D et al (2020) Severe COVID-19 is marked by a dysregulated myeloid cell compartment. Cell 182(6):1419–1440.e23. https://doi.org/10.1016/j.cell.2020.08.001
13. Siwiak M, Zielenkiewicz P (2010) A comprehensive, quantitative, and genome-wide model of translation. PLoS Comput Biol 6(7): e1000865. https://doi.org/10.1371/journal.pcbi.1000865
14. Triana S, Vonficht D, Jopp-Saile L, Raffel S, Lutz R, Leonce D, Antes M, Hernández-Malmierca P, Ordoñez-Rueda D, Ramasz B, Boch T, Jann JC, Nowak D, Hofmann WK, Müller-Tidow C, Hübschmann D, Alexandrov T, Benes V, Trumpp A, Paulsen M et al (2021) Single-cell proteo-genomic reference maps of the hematopoietic system enable the purification and massive profiling of precisely defined cell states. Nat Immunol 22(12):1577–1589. https://doi.org/10.1038/s41590-021-01059-0
15. Trzupek D, Dunstan M, Cutler AJ, Lee M, Godfrey L, Jarvis L, Rainbow DB, Aschenbrenner D, Jones JL, Uhlig HH, Wicker LS, Todd JA, Ferreira RC (2020) Discovery of CD80 and CD86 as recent activation markers on regulatory T cells by protein-RNA single-cell analysis. Genome Med 12(1):55. https://doi.org/10.1186/s13073-020-00756-z

Chapter 3

Profiling Transcriptional Heterogeneity with Seq-Well S^3: A Low-Cost, Portable, High-Fidelity Platform for Massively Parallel Single-Cell RNA-Seq

Riley S. Drake, Martin Arreola Villanueva, Mike Vilme, Daniela D. Russo, Andrew Navia, J. Christopher Love, and Alex K. Shalek

Abstract

Seq-Well is a high-throughput, picowell-based single-cell RNA-seq technology that can be used to simultaneously profile the transcriptomes of thousands of cells (Gierahn et al. Nat Methods 14(4):395–398, 2017). Relative to its reverse-emulsion-droplet-based counterparts, Seq-Well addresses key cost, portability, and scalability limitations. Recently, we introduced an improved molecular biology for Seq-Well to enhance the information content that can be captured from individual cells using the platform. This update, which we call Seq-Well S^3 (S^3: Second-Strand Synthesis), incorporates a second-strand-synthesis step after reverse transcription to boost the detection of cellular transcripts normally missed when running the original Seq-Well protocol (Hughes et al. Immunity 53(4):878–894.e7, 2020). This chapter provides details and tips on how to perform Seq-Well S^3, along with general pointers on how to subsequently analyze the resultant single-cell RNA-seq data.

Key words Single-cell RNA sequencing, scRNA-seq, Seq-Well, Single-cell genomics, Systems biology, Transcriptomics, RNA-Seq, Picowells

1 Introduction

Single-cell RNA-seq (scRNA-seq) is a powerful approach for examining the cellular composition of healthy and diseased tissues [3–8]. Several high-throughput scRNA-seq methodologies have been developed enabling the characterization of thousands of cells from complex systems in a single experiment. Most rely upon early barcoding of cellular mRNA—for example, during reverse transcription using barcoded mRNA capture beads—to allow ensemble processing while retaining single-cell resolution.

Riley S. Drake, Martin Arreola Villanueva, J. Christopher Love and Alex K. Shalek contributed equally with all other contributors.

Raffaele A. Calogero and Vladimir Benes (eds.), *Single Cell Transcriptomics: Methods and Protocols*,
Methods in Molecular Biology, vol. 2584, https://doi.org/10.1007/978-1-0716-2756-3_3,

Seq-Well, like its predecessor Drop-seq, is one such method [1, 4]. Seq-Well was designed to overcome key cost, portability, and scalability limitations associated with reverse-emulsion droplets-based cell capture and barcoding methods like Drop-seq through combining the throughput and cost-effectiveness of Drop-seq with the simplicity and sampling efficiency of picowells. In Seq-Well, cells are efficiently loaded by gravity into picowells where each cell's transcripts are uniquely barcoded using the aforementioned bead reagent. Engineered surface chemistry allows sealing of the picowell array with a selectively permeable membrane, facilitating rapid buffer exchange and parallel cell lysis, while preventing mRNA escape. Seq-Well works with low inputs (e.g., sorted populations of targeted cells [9] like antigen-specific T or B cells, fine-needle aspirates [10], tissue pinches) and fragile cells (e.g., granulocytes [11], hepatocytes), and it can be implemented in challenging environments with limited peripherals—e.g., the clinic, BSL-3/4 facilities [12]. This alleviates the need to ship samples and associated artifacts/concerns, and enables work across diseases and geographies.

In the original Seq-Well and Drop-seq protocols, as in SMART-seq, a template-switching mechanism was used to append a second PCR priming site during reverse transcription, facilitating subsequent whole transcriptome amplification by PCR [1, 4, 13]. Through a series of experiments, we found that this process had inefficiencies, limiting our ability to characterize cells with high fidelity. To address these issues, we recently developed a new variant of Seq-Well, Seq-Well S^3 (S^3: Second-Strand Synthesis), which incorporates a second-strand-synthesis step after reverse transcription to add a second PCR priming site. This modification allows for the recovery of complementary DNA (cDNA) that is reverse transcribed but for which the template switch reaction failed, enhancing the capture of key transcripts, such as transcription factors, affinity receptors, and signaling molecules by up to tenfold [2].

In this chapter, we provide a step-by-step breakdown of the Seq-Well S^3 (Second-Strand Synthesis) protocol, along with tips and tricks that leverage our collective knowledge to help prevent common pitfalls.

2 Materials

All buffers and solutions are to be prepared with sterile ultrapure water (Gibco Cat. No. 15230170) and stored at room temperature, unless otherwise indicated.

2.1 Array Processing Prior to Reverse Transcription

1. Bead loading buffer (BLB): 10% BSA, 100 mM sodium carbonate, pH 10. Add 2.5 mL BSA (100 mg/mL) to a 50 mL screwtop conical tube. Add water to ~15 mL followed by 1.25 mL 2 M sodium carbonate. Add additional water to achieve a final volume of 25 mL. Titrate with glacial acetic acid to reach pH 10 (*See* **Note 1**).
2. Prelysis buffer: 5 M guanidine thiocyanate (Sigma Cat. No. AM9422), 1 mM EDTA (*See* **Note 2, 3**)
3. Complete lysis buffer: 5 M guanidine thiocyanate, 1 mM EDTA, 0.5% sarkosyl, 1% β-mercaptoethanol. Combine 5 mL prelysis buffer with 25 μL 10% sarkosyl and 50 μL β-mercaptoethanol (*See* **Note 4**).
4. Hybridization buffer: 2 M NaCl, 4% PEG 8000 in PBS. Combine 10 mL 5 M NaCl with 13 mL of PBS, and 2 mL 50% (w/v) PEG 8000 (Fisher Scientific Cat No. BP337-100ML).
5. Wash buffer: 2 M NaCl, 3 mM MgCl2, 20 mM Tris–HCl (pH 8.0), 4% PEG 8000. Combine 20 mL 5 M NaCl, 150 μL 1 M MgCl, 1 mL 1 M Tris–HCl (pH 8.0), and 4 mL 50% (w/v) PEG 8000. Add water to bring volume to 50 mL (*See* **Note 5**).
6. TE-TW: 10 mM Tris–HCl pH 8.0, 1 mM EDTA, 0.01% Tween-20. Combine 49.5 mL water, 0.5 mL 1.0 M Tris pH 8.0, 100 μL 0.5 M EDTA, and 50 μL Tween-20 (Millipore Sigma Cat. No. P9416-50ML).
7. Wafer-Handling Tweezers for 4 in. wafers (Excelta, Cat. No.17-467-333).
8. Fisherbrand Filter/Membrane Forceps (Fisher Scientific, Cat. No. 09-753-50).
9. Polycarbonate (PCTE) Membrane Filters, 0.01 Micron, 62 mm × 22 mm (Sterlitech Corp. Cat. No. PCT00162X22100).
10. Plasma Cleaner (Harrick Plasma Cat. No. PDC-001-HP).
11. Pump for Plasma Cleaner (Harrick Plasma Cat. No. PDC-VPE).
12. VWR Sterile Disposable Serological Pipets, Plugged, Polystyrene 10 mL (VWR Cat. No. 89130-898).
13. VWR Sterile Disposable Serological Pipets, Plugged, Polystyrene 25 mL (VWR Cat. No. 89130-900).
14. mRNA capture beads (*See* **Note 6**). (Chemgenes Cat. No. MACOSCKO-2011-10(V+)) (*See* **Note** 7).
15. Vacuum desiccator (VWR Cat. No. 24988-164).
16. Vacuum tubing, for connecting vacuum desiccator to house vacuum or vacuum pump.

17. Surgical tubing, for connecting disposable aspirators to liquid trap.
18. Sterile 4-well dishes (Thermo Fisher Scientific Cat. No. 267061).
19. VWR Sterile Disposable Aspirating Pipets, Polystyrene 1 mL (VWR Cat. No. 414004-264).
20. Fisherbrand Fine Point High Precision Forceps (Fisher Scientific, Cat. 22-327379).
21. Seq-Well arrays (*See* **Note 8, 9**)
22. RPMI (Gibco Cat. No. 11875093).
23. FBS, Heat Inactivated (Gibco Cat. No. A3840002).
24. RP-10: RPMI with 10% FBS.
25. Neubauer Improved Disposable Hemocytometer (VWR Cat. No. 82030-468).
26. Trypan Blue Solution, 0.4% for Cell Counting (Thermo-Fisher Cat. No. 1520061).
27. X-Y Rotator (Thermo-Fisher Cat. No. 88880025).
28. Platform for X-Y Rotator (Thermo-Fisher Cat. No. 88880106).
29. 1× PBS for washing (Thermo-Fisher Cat. No. 10010049).
30. Glass Slides, precleaned (Thermo-Fisher Cat. No. 12-550-A3).
31. Hybridization Clamp (*See* **Note 10**) (Agilent Cat. No. G2534A).
32. Incubator with a range of 37C-52C.

2.2 Array Storage

1. Array quenching buffer: 100 mM sodium carbonate, 10 mM Tris–HCl (pH 8.0). Combine 2.5 mL 2 M sodium carbonate with 500 μL 1 M Tris–HCl. Add water to bring the total volume to 50 mL. Arrays can be stored in array quenching buffer for up to 1 month at 4C.
2. Aspartic acid solution: 20 μg/mL of L-aspartic acid, 2 M NaCl, and 100 mM sodium carbonate solution (pH 10.0). Arrays can be stored in the aspartic acid solution for up to 6 months at 4 C. If storing arrays longer than 1 month, they should be stored in the aspartic acid solution.
3. Sterile 50 mL Conical Tubes.

2.3 Reverse Transcription

1. VWR cleanroom SpecWipers, for RNA-clean work (VWR Cat. No. 21913-214).
2. RNAse-Zap (Thermo-Fisher Cat. No. AM97800).
3. Safe-Lock Lo-Bind 1.5 mL microcentrifuge tubes, Biopur grade (Eppendorf Cat. No. 022600028).

4. Maxima H-RT with Maxima 5× RT buffer. (Thermo-Fisher Cat. No EP0753).
5. 30% PEG 8000 (mixed from 50% PEG 8k (Fisher Scientific Cat No. BP337-100ML).
6. dNTP mix, 10 mM each (Thermo-Fisher Cat. No. R0193).
7. SuperaseIN RNase Inhibitor (Thermo-Fisher Cat. No. AM2696).
8. Template Switch Oligo (*see* Subheading 2.6).
9. TE-SDS: 10 mM Tris pH 8.0, 1 mM EDTA, 0.05% SDS. Combine 49.5 mL water, 0.5 mL 1 M Tris pH 8.0, 100 μL 0.5 M EDTA, and 250 μL 10% SDS.
10. HulaMixer Sample Mixer (Thermo-Fisher 15920D).

2.4 Exonuclease Treatment and Second Strand Synthesis

1. Exonuclease I (E. coli) with buffer (New England Biolabs Cat. No. M0293).
2. 10 mM Tris–HCl, pH 8.0.
3. Safe-Lock Lo-Bind 1.5 mL microcentrifuge tubes, Biopur grade (Eppendorf Cat. No. 022600028).
4. Maxima 5× RT buffer. (Thermo-Fisher Cat. No EP0753).
5. 30% PEG 8000 (mixed from 50% PEG 8k (Fisher Scientific Cat No. BP337-100ML).
6. dNTP mix, 10 mM each (Thermo-Fisher Cat. No. R0193).
7. SuperaseIN RNase Inhibitor (Thermo-Fisher Cat. No. AM2696).
8. dN-SMRT Oligo (*see* Subheading 2.6).
9. TE-SDS: 10 mM Tris pH 8.0, 1 mM EDTA, 0.05% SDS. Combine 49.5 mL water, 0.5 mL 1 M Tris pH 8.0, 100 μL 0.5 M EDTA, and 250 μL 10% SDS.
10. HulaMixer Sample Mixer (Thermo-Fisher 15920D).

2.5 WTA, Library Preparation and Sequencing

1. 10 mM Tris–HCl, pH 8.0.
2. Thermocycler, 96-well plate compatible.
3. Microseal B adhesive seal (Bio-rad Cat. No. MSB1001).
4. Microseal F foil (Bio-rad Cat. No. MSF1001).
5. Qubit assay tubes (Thermo-Fisher Cat. No. Q32856).
6. Qubit 4.0 fluorometer (Thermo-Fisher Cat. No. Q33238).
7. Qubit DNA High Sensitivity Kit (Thermo-Fisher Cat. No. Q32854).
8. 96-well PCR plates, fully skirted (Thermo-Fisher Cat. No. AB2396).
9. SMART PCR Primer (*see* Subheading 2.6).

10. KAPA HiFi Hotstart Readymix (Roche Cat. No. KK2602).
11. Ampure XP beads (Beckman-Coulter A63881).
12. 80% ethanol (mix fresh day-of: combine ultrapure water with 200-proof ethanol (VWR Cat. No. TX89125-170).
13. Agilent TapeStation 4200 (Agilent Cat. No. G2991BA).
14. Agilent D5000 High Sensitivity DNA Tapes (Agilent Cat. No. 5067-5592).
15. Agilent D5000 High Sensitivity DNA Buffer and Ladder (Agilent Cat. No. 5067-5593).
16. Custom P5-SMART PCR hybrid oligo (*see* Subheading 2.6).
17. Nextera XT kit (Illumina Cat. No. 20024906).
18. NextSeq500 v2.5 Kit (Illumina Cat. No. 20024906).

2.6 Primers

1. Template Switch Oligo: AAGCAGTGGTATCAACGCAGAGTGAATrGrGrG. (HPLC purification, diluted to working concentration of 100 uM).
2. dN-SMART randomer AAGCAGTGGTATCAACGCAGAGTGANNNGGNNNB. (HPLC purification, diluted to working concentration of 1 mM).
3. SMART PCR Primer: AAGCAGTGGTATCAACGCAGAGT. (HPLC purification, diluted to working concentration of 100 uM).
4. Custom P5-SMART PCR hybrid oligo: AATGATACGGCGACCACCGAGATCTACACGCCTGTCCGCGGAAGCAGTGGTATCAACGCAGAGT*A*C. (HPLC purification, diluted to working concentration of 10 uM).
5. Custom Read 1 Primer: GCCTGTCCGCGGAAGCAGTGGTATCAACGCAGAGTAC. (HPLC purification, diluted to working concentration of 100 uM).

3 Methods

3.1 Membrane Functionalization

1. Use membrane forceps or plastic tweezers (supplied with the Agilent Hybridization clamp) to place a precut (22 × 66 mm) polycarbonate membrane onto a glass slide. Be careful not to touch the surface of the membrane, manipulate it by touching the edges only. Using a gloved finger and tweezers, carefully separate the membrane from the paper—be careful to only touch the edges of the paper (Fig. 1).
2. Make sure the shiny side of the polycarbonate membrane is facing up. Discard any membranes that have creases or large visible imperfections. Repeat this process for the number of arrays that you plan to process, plus one extra, in case of error during membrane application (Figs. 2, 3, 4, and 5).

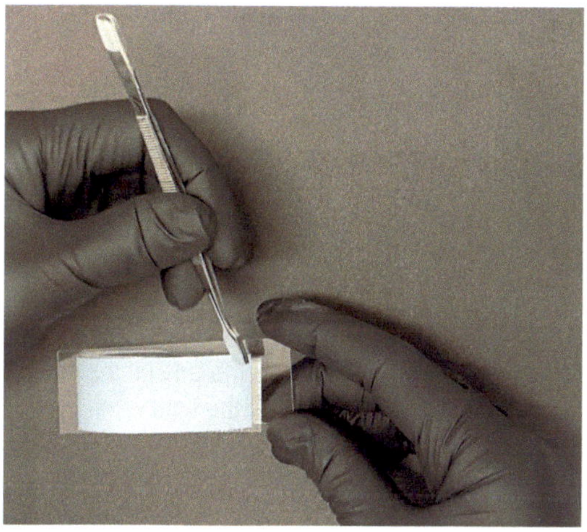

Fig. 1 Removing the separator paper using membrane forceps

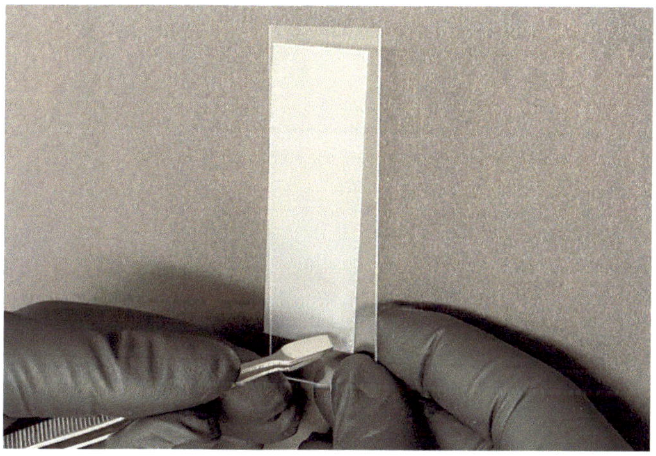

Fig. 2 Using membrane forceps to center the membrane on the slide

Fig. 3 Membrane centered on a slide, shiny side up. This is *correct*

Fig. 4 Membrane centered on slide, dull side up. This is *incorrect.* If you are having difficulty distinguishing between the dull and shiny side, hold the membrane on a side under a light source and gradually change the angle of the slide—the shiny side will catch the light

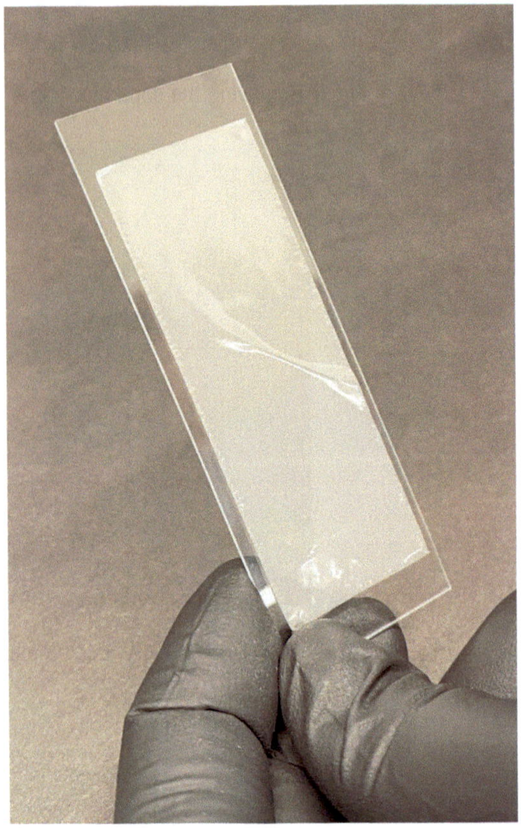

Fig. 5 Membrane with a significant crease; membranes with large imperfections like this should be discarded

3. Place membranes on slides onto a shelf (*See* **Note 11**) in the plasma cleaner (Fig. 6).
4. Close the plasma cleaner door, and make sure that the (three-way main valve) lever is pointed to the left, this is the closed position. Turn on the main power and pump switch to form a vacuum (Fig. 7).

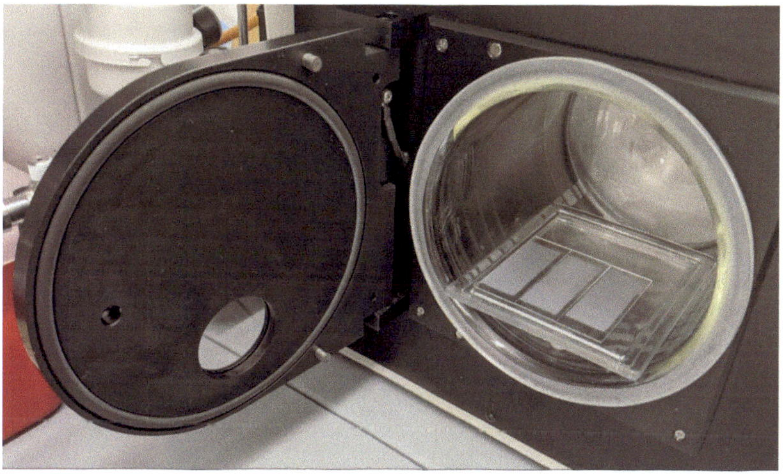

Fig. 6 Place membranes on slides onto a shelf in the plasma cleaner

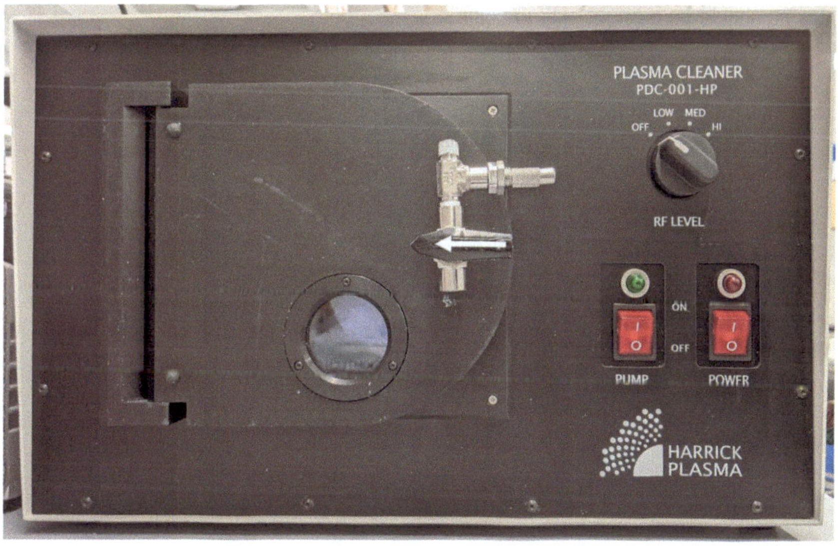

Fig. 7 Turn on the main power and pump switch to form a vacuum

5. Allow a vacuum to form for 2 min. Once the vacuum has formed, simultaneously turn the three-way valve clockwise to 12:00 (pointing up), while turning the power to "High." The plasma should be a bright pink in color (*See* **Note 12**) (Fig. 8).
6. Treat membranes with plasma for 7 min (*See* **Note 13**), start the timer when you see the plasma appear.
7. After treatment, in the following order, turn the RF level valve from HIGH to OFF, and then turn off the power, then turn off the vacuum.

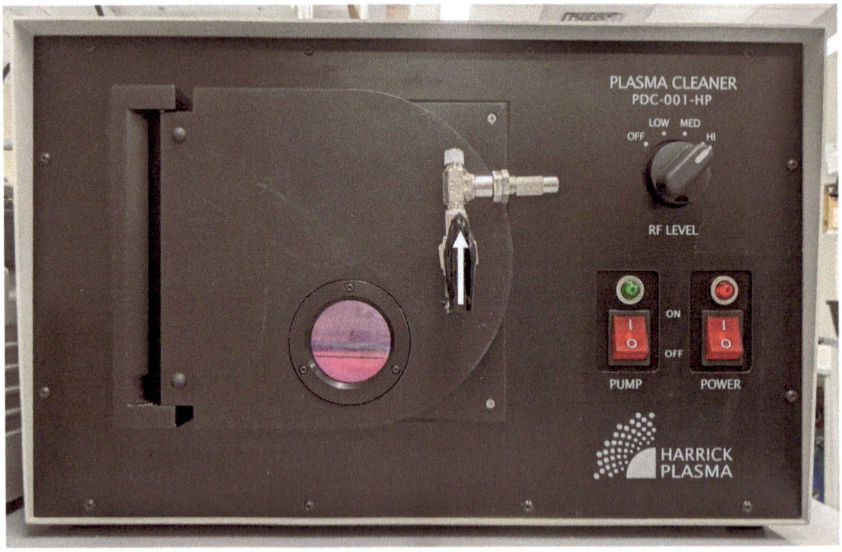

Fig. 8 The plasma should be a bright pink in color

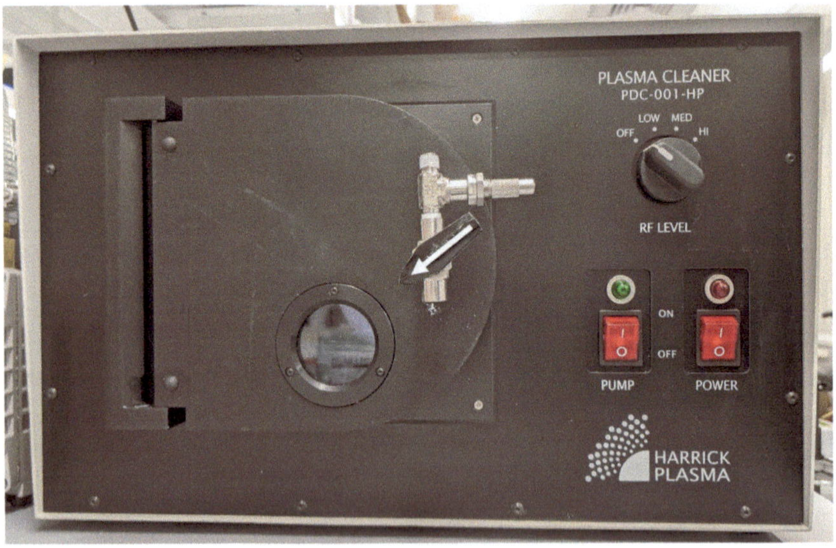

Fig. 9 Allow the chamber to slowly fill with air until the door opens on its own

8. Slowly turn open the air valve until you can just barely hear the air entering the chamber (*See* **Note 14**). Allow the chamber to slowly fill with air until the door opens on its own (*See* **Note 15**) (Fig. 9).
9. Prefill a 4-well dish with 1 mL of PBS in each well. Tilt the dish so that the PBS collects in the bottom right corner of each well.
10. Carefully transfer the membranes-on-slides to individual wells of a 4-well dish (Fig. 10).

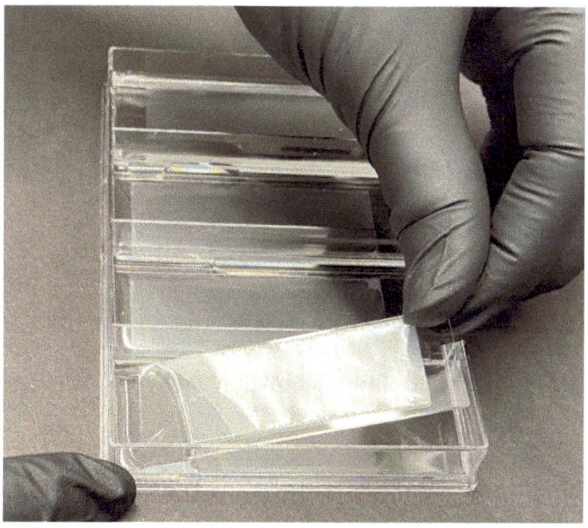

Fig. 10 Carefully transfer the membranes-on-slides to individual wells of a 4-well dish

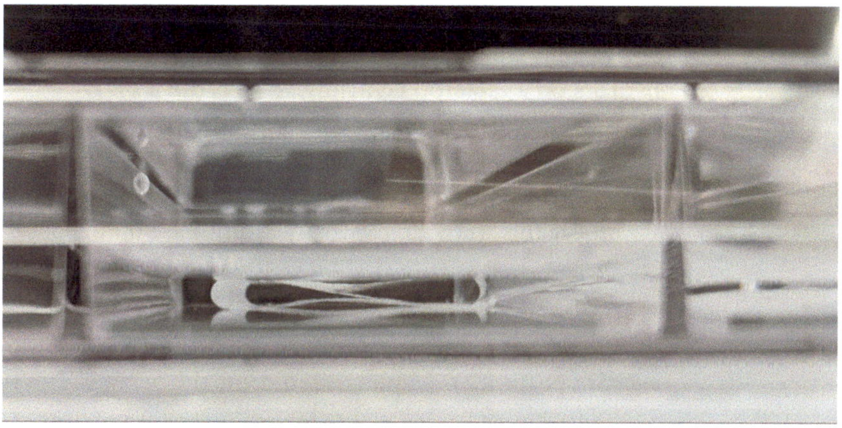

Fig. 11 Side view of 4-well dish showing a fully hydrated membrane (you can tell because it is not floating) that folded over during transport. Because the membrane has fully folded over, it is impossible to know which side was functionalized—membranes like this should be discarded

11. Quickly pipette 5 mL of 1× PBS over the membrane (*See* **Note 16**)—you may need to use tweezers to secure the membrane and prevent the membrane from folding over on itself (*See* **Note 17**).
12. Remove any air bubbles underneath the membrane by gently pressing on the membrane with membrane forceps or the tip of serological pipette used to dispense the PBS (only touch the edges of the membrane, avoid any contact with the center of the membrane) (*See* **Note 18**). Inspect the membranes (*See* **Note 19**). Membranes are now functionalized and ready for use, and solvated membranes should be used within 24 hrs of solvation (Figs. 11 and 12).

Fig. 12 An over-functionalized membrane will usually turn opaque white, and may also shrivel

3.2 Bead Loading

1. Aspirate storage solution and solvate each array with 5 mL of BLB. Inspect arrays (*See* **Note 20**).
2. Set up the house vacuum with an aspirator trap.
3. Aliquot 110,000 beads/array from a bead stock. Do not vortex beads, as this action can fragment them and interfere with bead loading and transcript capture. Resuspend beads by gently rocking and/or flicking the tube. Pellet beads by spinning for 15 s on a tabletop centrifuge (*See* **Note 21**).
4. Resuspend beads (previously in storage buffer) by aspirating the storage buffer and adding 500 μL of BLB. Invert the tube several times to wash the beads, pellet the beads, and then repeat this wash step with an additional 500 μL of BLB.
5. Pellet beads in 1.5 mL microcentrifuge tube, aspirate BLB, and resuspend in 200 μL of BLB per 110,000 beads.
6. Use the vacuum aspirator to aspirate *all* bead loading buffer (*See* **Note 22**). Minimal liquid on the array is crucial for the following steps: work quickly and periodically inspect arrays to ensure that they do not dry out completely.
7. Use a clean P1000 pipette tip, wafer forceps, or membrane forceps to center the array in the 4-well dish, making sure that the two long sides are not touching the sides of the dish (*See* **Note 23**).
8. Before the arrays dry out (*See* **Note 24**), add 200 μL of beads, dropwise, onto each array, in the pattern shown below (*See* **Note 25, 26**), Be very careful not to get bubbles on the surface of the array (*See* **Note 27**).
9. Gently agitate the arrays (*See* **Note 28**) or place on a flat rocker moving very slowly (<50 RPM).
10. Rock the arrays for 5 min, and then leave the arrays stationary for 5 min. Check the arrays frequently during both incubations to ensure that they have not completely dried out (*See* **Note 29**), and that the beads have not fallen off the side of the array (*See* **Note 30**).

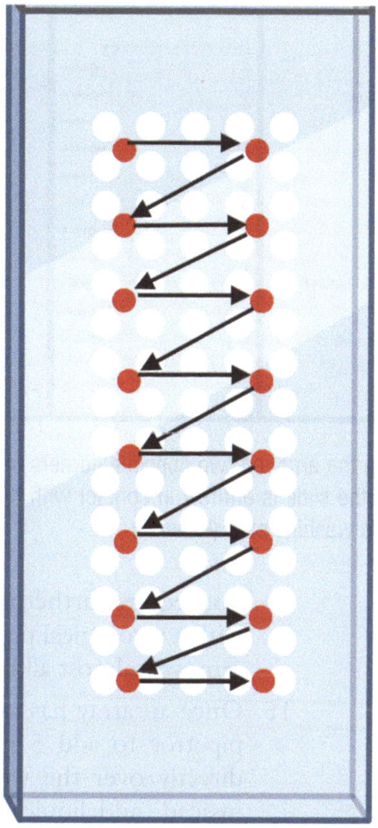

Fig. 13 A very large number of beads on the top of the array (not in wells) is expected and in most fields of view nearly all microwells will be filled with beads

11. Inspect the beads under a microscope. Here, we expect to see a very large number of beads on the top of the array (not in wells) and we also expect that in most fields of view nearly all microwells will be filled with beads (*See* **Note 31**) (Fig. 13).
12. Start washing the array—first, use a 1 mL pipette to deposit 500 μL of BLB on two opposite corners of the array (Fig. 14, Panel a). Then, reposition the array so that the long side of the slide is entirely in contact with the side of the 4-well dish (Fig. 14, Panel b, c). Use a 1 mL pipette to aspirate excess beads from the array surface and the dish and collect them in a 1.5 mL microcentrifuge tube. Rotate the arrays and repeat this washing step with 1 mL pipette (Fig. 14, Panel d).
13. Rotate the arrays again and repeat the washes twice more, using a serological pipette to add the BLB and the aspirator connected to the house vacuum. Do not touch the aspirator on top of the actual wells, stay on the borders of the array.
14. Check the arrays under a microscope to determine whether further washing is necessary. If you can scan across the entire array and find ~10–20 beads still on the surface, this array does

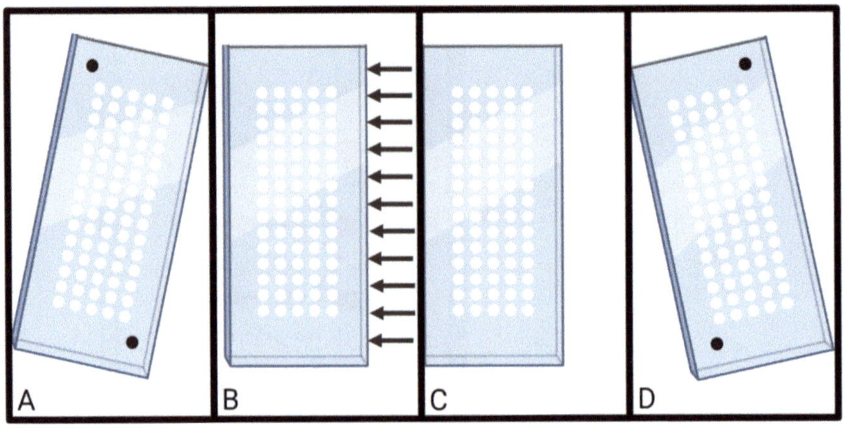

Fig. 14 Start washing the array on two opposite corners of the array (Panel A). Then, reposition the array so that the long side of the slide is entirely in contact with the side of the 4-well dish (Panel B, C). Rotate the arrays and repeat this washing step (Panel D)

not require further washing. Otherwise, repeat the wash steps with a serological pipette and vacuum aspirator until you have removed almost all beads from the surface of the array.

15. Once an array has been sufficiently washed, use a serological pipettor to add 5 mL of BLB to the well. Do not pipette directly over the wells—as this action can displace beads—instead, add liquid from the corner of the 4-well dish (*See* **Note 32**).
16. Spin down the beads collected in a 1.5 mL microcentrifuge tube during the first two washes using a tabletop centrifuge (or for 1 min at 800G). Use a 1 mL pipette to wash the beads twice in TE-TW. Resuspend the beads in 200 μL of TE-TW and store at 4 C (*See* **Note 33, 34**).

3.3 Cell Loading

1. Keep cells on ice until you are ready to load. Before proceeding with cell loading, cells should be counted (*See* **Note 35**). Generally, cells should be loaded in a certified biosafety cabinet in accordance with your approved biosafety protocols.
2. Remove arrays from 4 C. Mark the top and bottom of the dish so that you know which way the lid goes (Fig. 15).
3. Aspirate the bead loading buffer (or array quenching buffer, if left overnight).
4. Add 5 mL of RPMI (*See* **Note 36**) + 10% FBS (*See* **Note 37**) to each array.
5. The RPMI will slowly turn purple, this is because the bead loading buffer (or array quenching buffer) are both very basic, allow this to proceed for ~1 min.
6. Aspirate RPMI + 10% FBS.
7. Add 5 mL Fresh RPMI + 10% FBS.

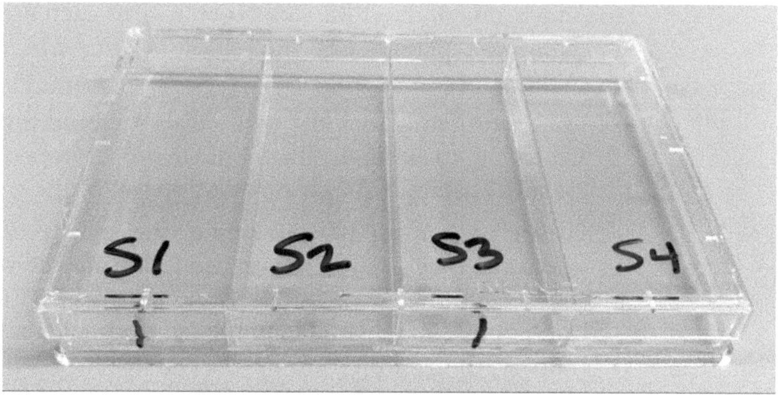

Fig. 15 Mark the top and bottom of the dish so that you know which way the lid goes

8. Incubate arrays room temperature with RPMI + 10% FBS for 5–10 min.
9. While this incubation is happening, prepare a 1.5 mL microcentrifuge tube containing cells at a dilution of 75 k/mL in RPMI + 10% FBS (*See* **Note 38**).
10. Use the vacuum aspirator to aspirate all the liquid in one well. Aspirate both off the top of the array and also from the well. Only perform this action for one array at a time, leaving the rest fully hydrated.
11. Use a P1000 pipette tip to center the array, ensure no side of the array is touching the 4-well dish.
12. Keeping in mind that there are beads in the wells, tip the 4-well dish as little as possible (~1–2 degrees off the table) to make the small amount of liquid on the array pool on the bottom border of the array. Gently aspirate this small remaining volume of liquid.
13. Use a P200 pipette to add 200 μL of cells to the array, dropwise. Again, be sure not to hold the pipette too far above the wells, as the falling drops can displace both beads and also result in uneven cell loading. Watch the pipette tip as you are dispensing liquid, be careful to stop dispensing liquid before you hit a bubble. Refer to the pattern illustrated in **Step 8** of Subheading 3.2.
14. Once all the arrays are loaded, take the arrays out of the biosafety cabinet for 5 min (*See* **Note 39**). This is a critical step, do not leave the arrays unattended (*See* **Note 40**). Instead, watch the arrays to make sure they are not touching the sides of the dish (*See* **Note 41**) or drying out (*See* **Note 42**).
15. After 10 min, bring the arrays back into the hood.

16. Add 5 mL of 1× PBS (*See* **Note 43**) to each array. Be very careful to pipette not on the top of the array, but into the side of the dish.
17. Very, very gently rock the entire dish back and forth; here we are trying to remove all of FBS from the top of the array, without letting the beads or cells pop out of the wells.
18. Use the vacuum aspirator to aspirate the PBS carefully from the sides of the well. Also aspirate the PBS from the very bottom (border) of the array—not from any wells—this procedure can be accomplished by touching the aspirator to the very bottom (border) of the array, quickly, once or twice.
19. Repeat PBS wash 4×.
20. Add 5 mL of RPMI on top of the array (if you accidentally add FBS here the membrane will not seal).

3.4 Membrane Sealing

1. Membranes should be in PBS, fully hydrated, and plasma-treated on the same day. If membranes were hydrated previously and are stored at 4 °C, allow them to come to room temperature.
2. Gather the following additional items: Hybridization Clamps, Clean Slides, VWR SpecWipers, 1000 μL Pipette tips. Ensure that the hybridization clamp is clean, and open it, as below, before you begin the membrane sealing process (Fig. 16).
3. Take off the top cover of a 4-well dish, keep it in the same orientation as it was sitting on the dish.

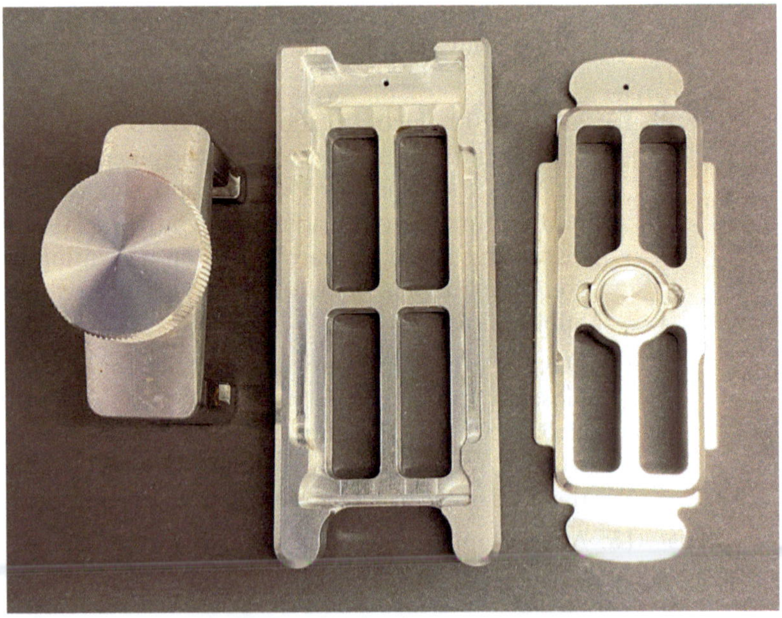

Fig. 16 Ensure that the hybridization clamp is clean

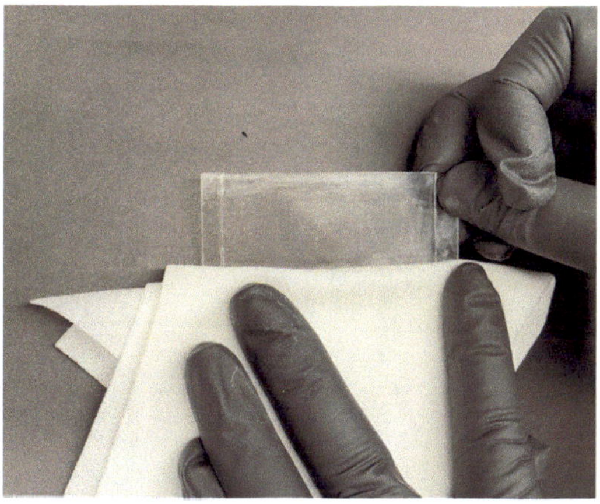

Fig. 17 Blotting excess liquid off the array, loaded with beads and cells, onto a SpecWiper. Note that the array is angled to aid in the removal of liquid, but tilted no more steeply than about 20–30 degrees

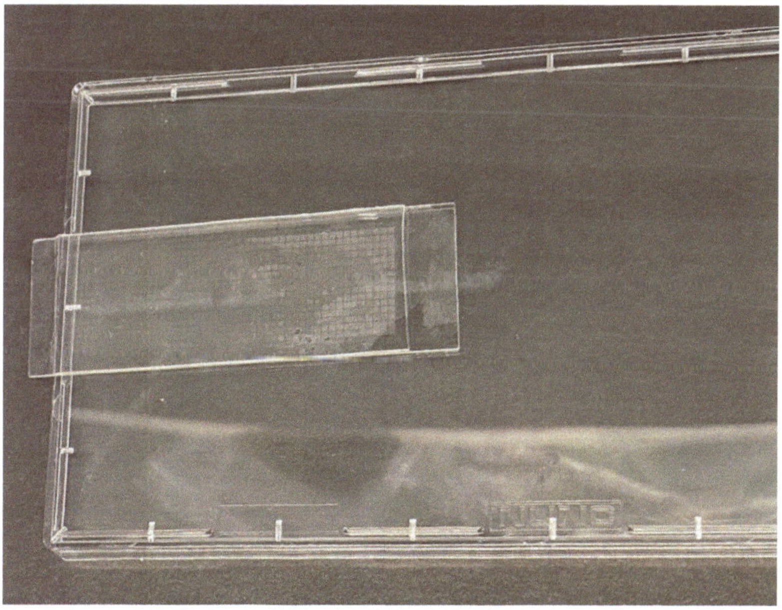

Fig. 18 Blotted array positioned at the edge of 4-well dish

4. Keeping the array as flat as possible, remove the array out from the 4-well dish. Blot excess liquid off the array using a SpecWiper. Place the array on the top of the 4-well dish—leave a little bit of the array overhanging so it does not get stuck on the top (wet glass on wet plastic) (Figs. 17 and 18).
5. Work very quickly—take the slide with the membrane on top of it out of the PBS dish, tap the sides of the slide with the membrane on a SpecWiper to remove excess liquid (Fig. 19).

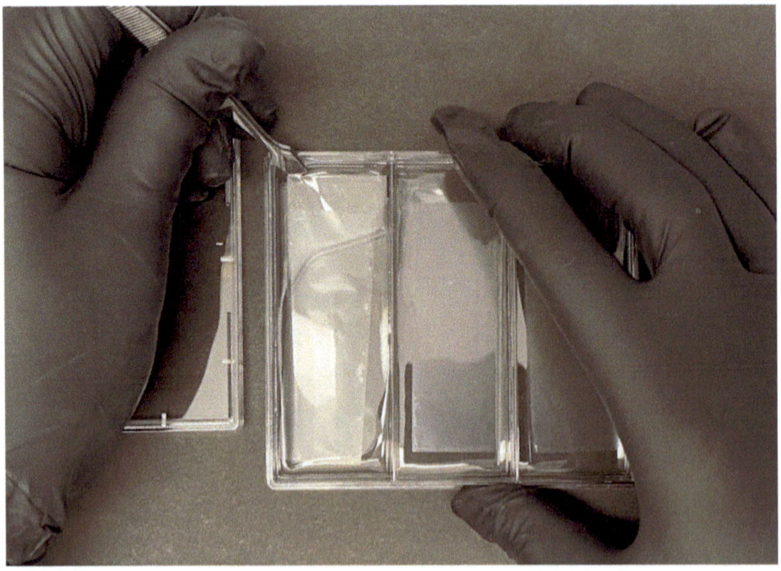

Fig. 19 Using membrane forceps (Wafer forceps can also be used for this step) to remove the solvated functionalized membranes from the 4-well dish

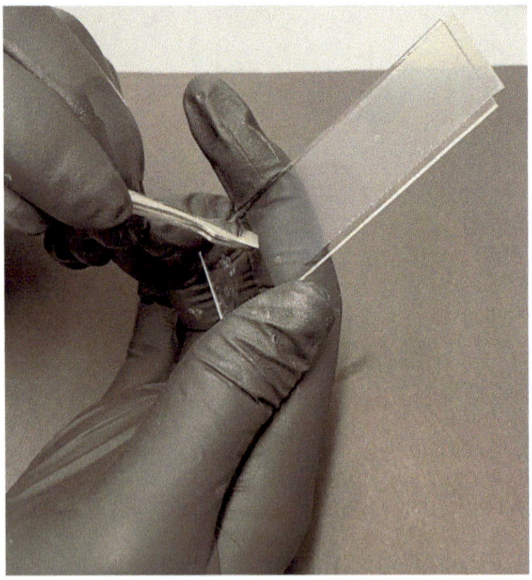

Fig. 20 Turn the glass slide with the membrane over so that the membrane is facing down. Center the slide with the membrane over the array

6. Use membrane forceps to manipulate the membrane so that it is centered on the slide with a 2 mm overhang on the end (Fig. 20).
7. Turn the glass slide with the membrane over so that the membrane is facing down. Center the slide with the membrane over the array (Fig. 21).

Fig. 21 Use another slide to pin the overhang of the membrane to the left side of the array

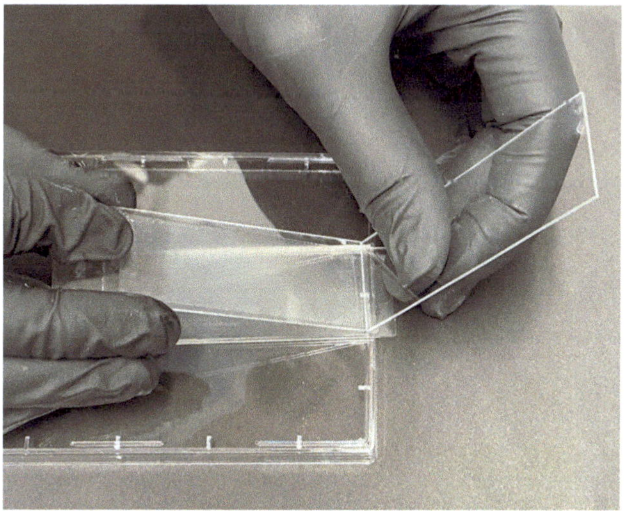

Fig. 22 Gently drag the slide with the membrane on it across the surface of the array, do not press down on the slide

8. Use another slide to pin the overhang of the membrane to the left side of the array (Fig. 22).
9. Gently drag the slide with the membrane on it across the surface of the array, do not press down on the slide. If you see imperfection in the membrane, take note. Do not try to flatten it manually (Figs. 23 and 24).

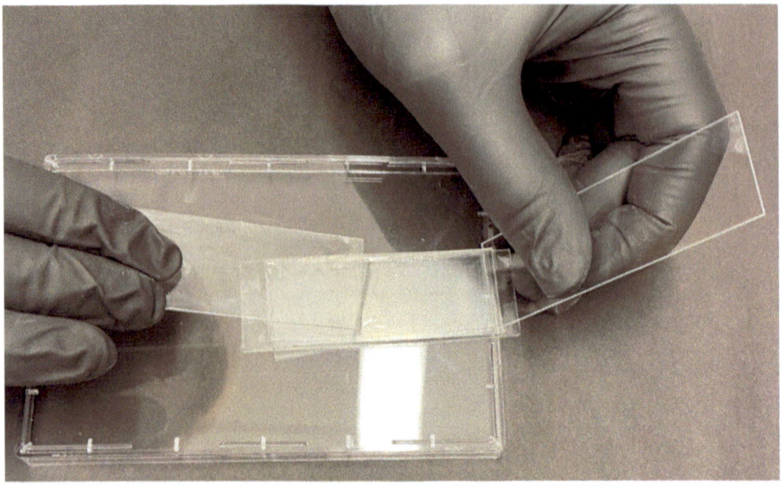

Fig. 23 Gently drag the slide with the membrane on it across the surface of the array, do not press down on the slide

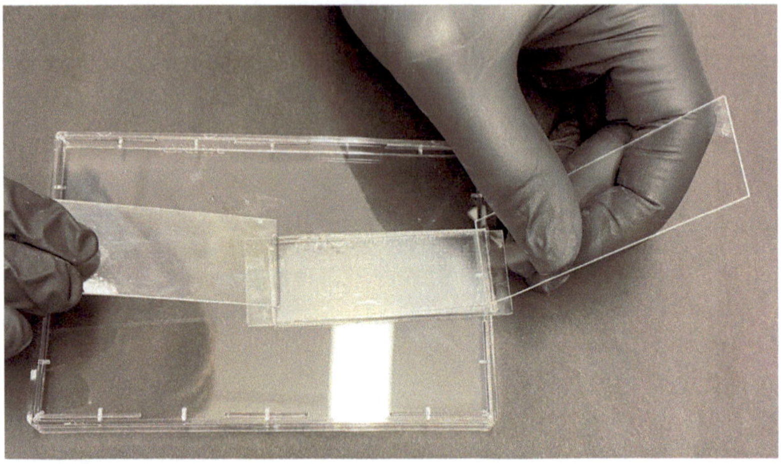

Fig. 24 If you see imperfection in the membrane, take note

10. Keeping the array relatively flat, place the array and membrane in a hybridization clamp (Fig. 25).
11. Gently add another clean slide on top of the membrane (Fig. 26).
12. Assemble the clamp, tighten the screw to the point of resistance (Figs. 27, 28, and 29).
13. Label the lab tape on the array screws.
14. Place clamp in 37 C incubator for 40 min.

3.5 Lysis and Hybridization

1. Prepare lysis buffer plates—in a fresh 4-well dish, add 5 mL of Complete-Lysis buffer to every well of the 4-well dish that will be occupied by an array.

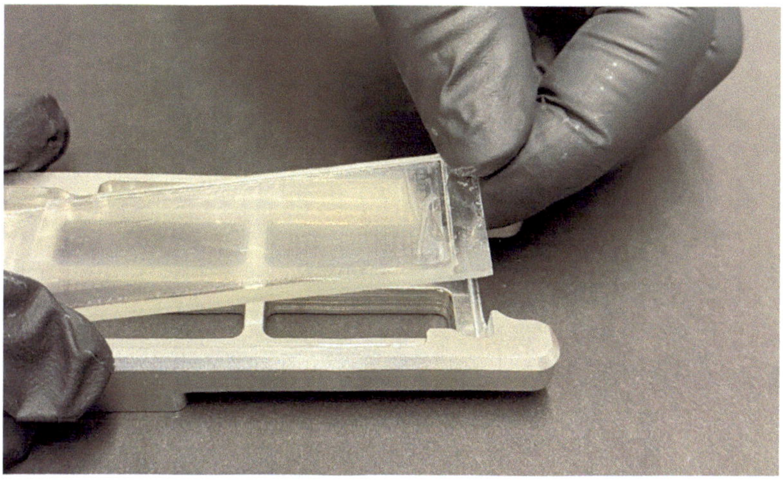

Fig. 25 Keeping the array relatively flat, place the array and membrane in a hybridization clamp

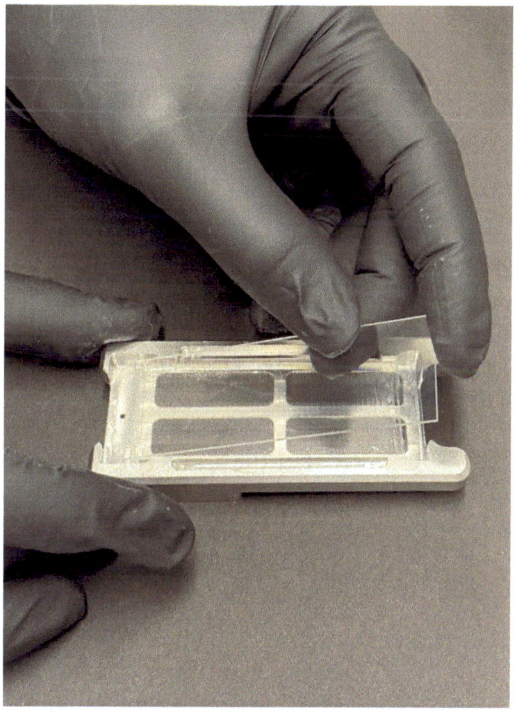

Fig. 26 Gently add another clean slide on top of the membrane

2. Remove the clamp(s) from the incubator. Then, carefully remove the array from the clamp (*See* **Note 44**). Lifting the arrays sandwich by the long sides of the slide, carefully transfer the arrays into the lysis buffer plates (*See* **Note 45**). Do not attempt to remove the top slide (Fig. 30).

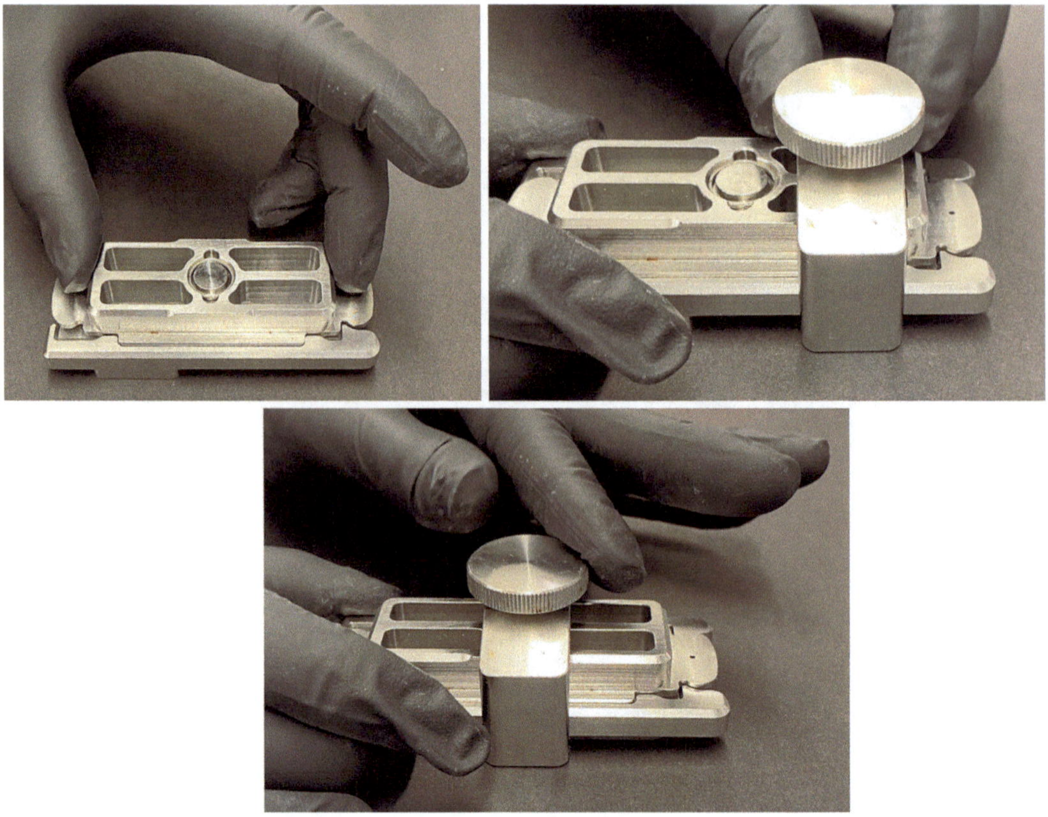

Figs. 27, 28, 29 From left: Placing the top on the clamp, sliding the locker bar on the clamp, and tightening the hybridization clamp screw. The screw on the clamp should be tightened to "finger tightness"—the amount of tightness that can be easily achieved using one finger to tighten the screw (If using two fingers, tighten the screw just to the point of resistance. Be conservative here: overtightening can crack the slides and may also prevent membrane sealing)

3. Let the arrays rock on an X-Y rotator (50 RPM) for 5 min. The top slide should slide off on its own. If any top slide remains stuck, do not pry off the top slide, as this can reverse membrane sealing. Instead, continue rotation for 10 more minutes. If the slide has still not moved, use a P200 pipette tip to gently nudge the top slide (Fig. 31).
4. Once all the top slides have detached, incubate in lysis buffer while rocking for 20 min.
5. Check to make sure the aspirator you are using is connected to a liquid trap that *does not contain bleach. SAFETY:* This step is very important as the interaction between lysis buffer and bleach will create cyanide gas. Aspirate the lysis buffer into a specific, appropriately labeled, waste container.
6. Add 5 mL of hybridization buffer to each well.

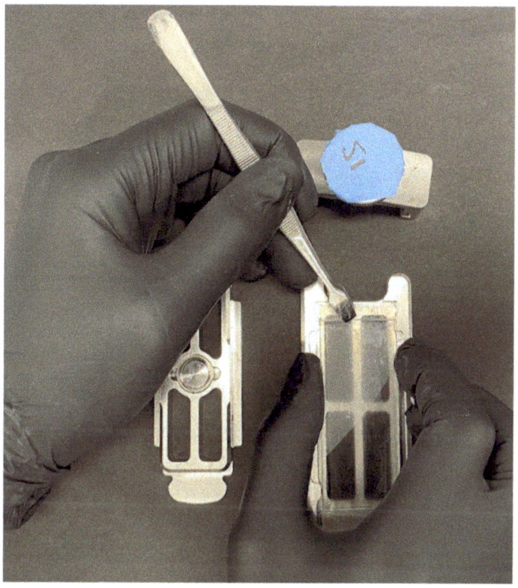

Fig. 30 Using wafer forceps or membrane forceps, remove the sealed array from the hybridization clamp. During this transfer step, it is very important to avoid putting different (or opposing) forces on the top and bottom slide, as this may reverse membrane sealing. A common pitfall here is tightly clamping the end of the top and bottom slides, forcing the slides apart. To avoid this outcome, gently clamp the forceps on the end of the slides, and slowly lift the array until it is high enough to grab the long ends of the slide sandwich with your thumb and forefinger

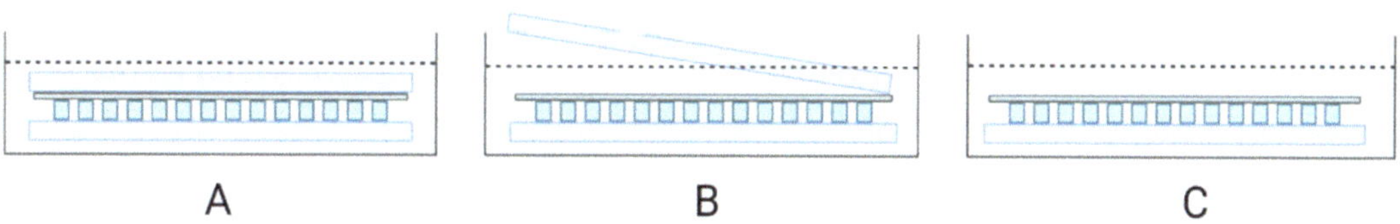

Fig. 31 Illustration showing a sealed array with membrane and top slide (**a**), top slide being removed after sliding off during rocking (**b**), array with sealed membrane in lysis buffer (**c**)

7. Aspirate the hybridization buffer into the same dedicated waste receptacle (*See* **Note 46**).
8. Add 5 mL of hybridization buffer to each well.
9. Rotate arrays for 40 min at 50 RPM. While arrays are rotating, prepare RT mastermix, as described in Subheading 3.6.

3.6 Bead Recovery

1. Create an RNA-clean workspace: Spray down your working area and tools (pipettes, plate sealers etc.) 2× with 70% ethanol, 1× with RNAse Zap (allow the Zap 30 s to react, do not wipe it away until it visibly foams) and 2× with 70% ethanol.

2. Gather one 50 mL conical per array, clean slides, a new 4-well dish, at least 40 mL of wash buffer per array and wafer forceps.
3. Label the conicals with the names of your arrays. Fill each conical with 35–40 mL of Wash Buffer (*See* **Note 47**).
4. Remove the hybridization buffer from one well of the 4-well dish (*See* **Note 48**). Optionally, replace with 1–2 mL of wash buffer (*See* **Note 49**).
5. Take one array out at a time. Use wafer forceps to collect the array from its position out in the 4-well dish.
6. While the array is still in the dish, use pointed forceps to gently remove the membrane and discard (*See* **Note 50**).
7. Place the array at a 45-degree angle above a clean prelabeled falcon tube. About one-quarter of the array should be in the falcon tube.
8. Use a p1000 pipette to carefully yet vigorously squirt 1000 uL of wash buffer directly onto the loaded surface of the array.
9. With a clean glass slide, scrape the array from its top to its middle 3–5 times. In between scrapes dunk the array in a clean wash buffer and, while it is at the bottom of the tube, use clean wafer forceps to partially lift and lower the array (displacing about 1 cm) 3–5 times, before pulling it back up to its original position (Fig. 32).
10. Flip the array in the tube (the side that was inside of the tube becomes the side on top) while maintaining that 45-degree angle.
11. Repeat **step 8**.
12. Wash again with 1 mL wash buffer and a P1000 pipet, this time aiming at the part of the array inside of the Falcon tube. This step washes beads left on the edge (Fig. 33).
13. Return the array to its position in the 4-well dish.
14. Repeat for each array **steps 5–13** for each array.
15. Inspect the arrays under a microscope. For any arrays with more than 10% of beads remaining, repeat **steps 5–13** on the already scraped arrays, and inspect again (*See* **Note 51**).
16. Once all arrays are complete, spin the 50 mL conical for 5 min at 1000G.While this spin is occurring, label microcentrifuge tubes with the array names.
17. Aspirate all but 500 uL from the conicals, be very careful not to disturb the bead pellet.
18. Using a 1 mL pipette, resuspend the bead in the 500 uL of wash buffer and transfer to the prelabeled microcentrifuge tubes.

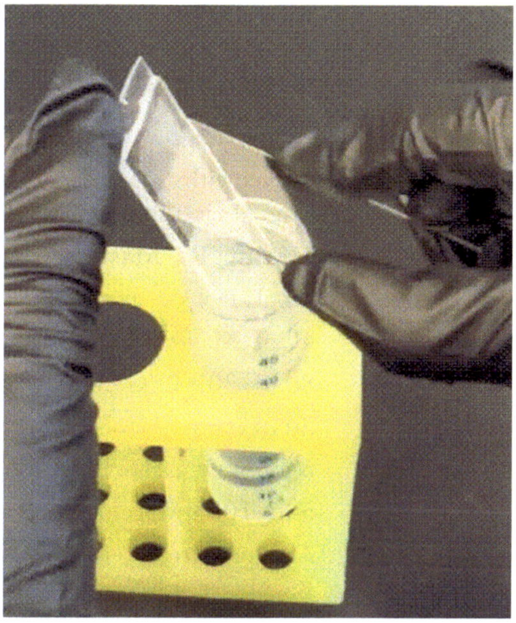

Fig. 32 With a clean glass slide, scrape the array from its top to its middle 3-5 times

Fig. 33 Wash again with 1mL wash buffer and a P1000 pipet, this time aiming at the part of the array inside of the Falcon tube

19. Rinse the 50 mL conicals with 500 uL of fresh wash buffer—add this wash to the same 1.5 mL microcentrifuge tube. Repeat for each array. Proceed immediately to reverse transcription.

3.7 Reverse Transcription

1. During hybridization, prepare the following Maxima RT Mastermix with 1.1× overage (volumes below are for 1 array). Add reagents in the order listed, store the mastermix on ice.

   Maxima RT Mastermix

   40 μL RNAse-free H20

   40 μL Maxima 5× RT Buffer

   80 μL 30% PEG 8000

   20 μL 10 mM dNTPs

   5 μL SuperaseIN RNAse Inhibitor

   5 μL 100 uM Template Switch Oligo
2. Also prepare 1× Maxima RT Buffer, by diluting 5× Maxima RT buffer in RNAse-free H20. Prepare 250 μL 1× Maxima RT Buffer per array, again using 1.1× overage.
3. Centrifuge the 1.5 mL microcentrifuge tubes with collected beads at 800G for 1 min.
4. Aspirate supernatant and resuspend beads in 250 μL of 1× Maxima RT Buffer. Flick the bottom of the tube to resuspend the beads (*See* **Note 52**).
5. Centrifuge the 1.5 mL centrifuges tubes containing the beads collected from the arrays in the previous step at 800G for 1 min.
6. Centrifuge the tubes with collected beads at 800G for 1 min.
7. Aspirate supernatant (*See* **Note 53**) and resuspend beads in 200 μL of RT Mastermix. Flick to mix.
8. Incubate the tubes at room temperature for 30 min with end-over-end rotation on HulaMixer or equivalent device.
9. Incubate the tubes at 52 C for 90 min with end-over-end rotation (*See* **Note 54**).
10. Wash beads 1× with 1 mL of TE-TW, 1× with 1 mL of TE-SDS and 2× with 1 mL of TE-TW (*See* **Note 55**).
11. This step is a stopping point: After the final TE-TW wash, beads can be resuspended in TE-TW and stored at 4 C for up to 2 weeks.

3.8 Exonuclease I Treatment

1. For each sample, prepare the following Exonuclease I Mix with a 1.2× overage:

   Exonuclease I Mix

   20 μL 10× Exo I Buffer

170 μL RNAse Free H20

10 μL Exo I Enzyme

2. Centrifuge beads (stored in TE-TW) at 800G for 1 min.
3. Aspirate TE-TW storage solution and resuspend in 500 μL of 10 mM Tris–HCl pH 8.0.
4. Centrifuge beads again at 800G for 1 min.
5. Aspirate Tris–HCl and resuspend beads in 200 μL Exonuclease I Mix.
6. Incubate beads in tubes at 37 C for 50 min with end-over-end rotation.
7. Wash beads 1× with 500 μL of TE-SDS, 2× with 500 μL TE-TW.
8. Beads can be stored at 4 C in TE-TW for up to 2 weeks.

3.9 Second Strand Synthesis

1. For each sample, prepare the following Second Strand Synthesis Mastermix with 20% additional total volume (1.2× overage):

Second Strand Synthesis Mastermix

40 μL Maxima 5× RT Buffer

80 μL 30% PEG 8k

20 μL 10 mM dNTPS

2 μL 1 mM dN-SMRT oligo

5 μL Klenow Enzyme (add the Klenow enzyme immediately before adding the beads)

53 μL H20

2. Centrifuge samples at 800 G for 1 min at room temperature.
3. While samples are spinning, prepare 550 μL of 0.1 N NaOH from a >1 N stock solution for each sample.
4. Aspirate supernatant and resuspend beads in 500 μL 0.1 N NaOH.
5. Rotate the tube for 5 min (end-over-end) at room temperature
6. Centrifuge for 1 min at 800 g. Aspirate supernatant.
7. Wash samples 1× with 500 μL of TE-TW (*See* **Note 56**) and 1× with 500 μL of TE.
8. Resuspend beads in 200 μL Second Strand Synthesis Mastermix and incubate at 37 C for 1 hour with end-over-end rotation.
9. Wash beads 2× with 500 μL TE-TW and 1× with 500 μL TE.
10. Proceed directly with WTA.

3.10 Whole Transcriptome Amplification (WTA)

1. Wash beads once with 500 μL ultrapure water, pellet beads. After spin down, the bead pellet is easily disturbed—remove the supernatant as much as you can without disturbing the bead pellet—usually this condition means leaving about 20 μL of ultrapure water—and resuspend in 220 μL of water.
2. Prepare the following PCR Mastermix with 1.2× overage. Volumes are per reaction. It is standard protocol to perform 24 reactions per array

   WTA Mastermix

   25 μL 2× Kapa HiFi Hotstart Readymix

   24.6 μL H20

   0.4 μL 100 uM SMART PCR Primer
3. Pellet beads, remove supernatant, and resuspend in 50 μL of PCR Mastermix for every 2000 beads or 1200 μL for 24 reactions.
4. To a 96-well plate, aliquot 50 μL of beads and PCR Mastermix to each well, resuspending periodically to prevent beads from settling.
5. Add 10 uL of water to the bottom of the tube you emptied in the previous step, and use it to wash the bottom of the tube.
6. Distribute the water you used to wash the bottom of the tube dropwise across the 24 reactions for that array, focusing on wells that might have received fewer wells in **step 6** due to beads settling.
7. Use the following cycling conditions to perform whole-transcriptome amplification (*See* **Note 57**) (Table 1).

Table 1
Cycling conditions to perform whole-transcriptome amplification

	Temperature	Time
	95 C	3 min
4 cycles	98 C 65 C 72 C	20 s 45 s 3 min
9–12 cycles	98 C	
	67 C 72 C	20 s 3 min
Final extension	72 C	5 min
	4 C	Infinite hold

3.11 Purification of WTA Products

1. Retrieve Ampure XP beads from 4 C and allow equilibrate to RT for 20 min (*See* **Note 58**).
2. Pool 8 wells of PCR reactions in a 1.5 mL microcentrifuge tube (*See* **Note 59**).
3. Thoroughly mix Ampure XP beads with the product at a 0.6× volumetric ratio (Ampure XP: PCR products) (*See* **Note 60**).
4. Incubate for 3 min at RT before placing on a magnet. Incubate for another 3 min.
5. During incubations, prepare 5 mL of 80% ethanol from a stock solution RNA-clean 200 proof ethanol.
6. Wash beads 3× with freshly made 80% ethanol.
7. After the third wash, allow the beads to dry for 5 min on the magnet.
8. After 5 min, while samples remain on the magnet, use a P200 pipette to remove any liquid that has pooled at the bottom of the microcentrifuge tubes.
9. Incubate on the magnet for another 5 min.
10. Once the beads appear matte in color, but before they develop significant cracks, remove the beads from the magnet.
11. Elute the dried beads in 100 μL RNAse-free water.
12. Incubate for 1 min, then place samples on the magnet.
13. Incubate for 1 min on the magnet, then transfer the 100 μL to new microcentrifuge tubes.
14. Repeat **steps 3–7** with the 100 μL product, combine the beads with the product at 0.8× volumetric ratio.
15. After the third wash, allow the beads to dry for 10 min on the magnet.
16. After 5 min, use a P200 pipette to remove any liquid that has pooled at the bottom of the microcentrifuge tubes.
17. Once the beads appear matte in color, remove the beads from the magnet.
18. Elute the beads in 16 μL RNAse-free water.
19. Incubate for 1 min, then place samples on the magnet.
20. Incubate for 1 min on the magnet, then transfer 15 μL to new Lobind microcentrifuge tubes.
21. It is helpful to analyze the distribution of fragment lengths in the purified WTA product by running a High Sensitivity D5000 Screen Tape by electrophoresis such as by agarose gel or on an Agilent 4200 Tapestation. The distribution of products should extend between 400 bp and 1500 bp with a peak size greater than 900 bp. Distribution of product will vary by sample type. Additionally, Seq-Well S^3 cDNA libraries will

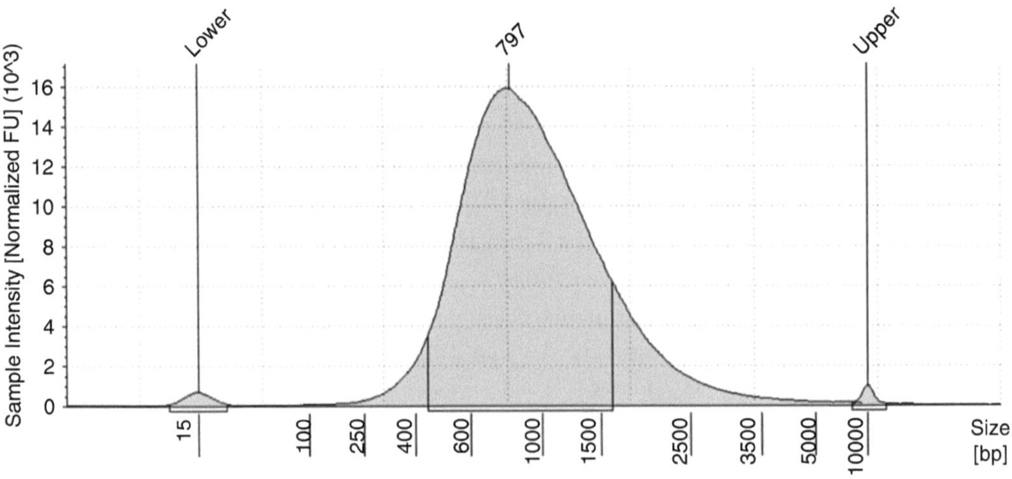

Fig. 34 An ideal WTA product distribution is a smooth bell curve with a peak between 700 and 2000 bp

appear shorter than full-length cDNA libraries as the dN-SMRT primer allows for recovery of transcripts that are not fully extended by the RT Polymerase and thus do not possess a TSO (Fig. 34).

22. Proceed to library preparation or store the WTA product at −20 C.

3.12 Nextera Library Preparation

1. Ensure that your thermocycler has been set to run tagmentation (**step 5**) and Nextera PCR (**step 11**) protocols.
2. For each sample, combine 800 pg of purified cDNA with water in a total volume of 5 μL (*See* **Note 61**).
3. To a new 96-well plate, add 10 μL of Nextera TD Buffer followed by 5 μL of ATM Buffer (*See* **Note 62**). Finally, add 5 μL of diluted PCR product.
4. Mix by pipetting up and down about 5 times. Seal the plate and spin the plate down for 1 min at 800G.
5. Incubate at 55 C for 5 min.
6. Once the thermocycler has finished its program, immediately remove the plate and place it in a bucket of wet ice, with 1" of ice on top of it.
7. Wait for 2 min, or until water condenses onto the plate seal over the wells.
8. Spin down for 30 s at 800G.
9. Immediately unseal and add 5 μL of Neutralization Buffer (NT). Mix by pipetting 10–15 times, try to pipette smoothly to minimize the production of bubbles. Spin samples briefly after mixing.
10. Incubate at room temperature for 5 min.

Table 2
Cycling conditions

	Temperature	Time
	72 C	5 min
	95 C	30 s
12 cycles	95 C 55 C 72 C	10 s 30 s 30 s
Final extension	72 C	5 min
	4 C	Infinite hold

11. Add to each well, in the order listed:

    8 μL RNAse-free H20

    1 μL 10 uM New-P5-SMART PCR hybrid oligo

    1 μL 10 uM Nextera N700× oligo

    15 μL Nextera PCR Mix
12. Seal the plate and centrifuge for 1 min at 800G.
13. Carry out PCR using the following cycling conditions (Table 2):
14. Store tagmentation products at −20 C or proceed to purification steps.
15. To each sample add Ampure XP Beads at 0.6× volumetric ratio.
16. Let samples incubate for 3 min off the magnet.
17. Transfer to a microcentrifuge tube magnet and incubate for another 3 min.
18. Perform three washes with freshly made 80% ethanol.
19. After the third wash, allow the beads to dry for 10 min on the magnet. Cover the open microcentrifuge tubes with the top of a pipette tip box to prevent dust from falling into the tubes while they are incubating.
20. After 5 min, use a P20 pipette to remove any liquid that has pooled at the bottom of the microcentrifuge tubes.
21. Once the beads appear matte in color, but ideally before they develop significant cracks, move the tubes off the magnet to elute.
22. Elute the dried beads in 101 μL RNAse-free water, pipette up and down to mix well, until the solution is uniform in color. Incubate for 1 min, then return the plastic rack to the magnet.

23. Incubate for 1 min on the magnet, then transfer the 100 μL to new microcentrifuge tubes.
24. Repeat **steps 3–7** with the 100 μL product, combine Ampure XP beads with the product at 0.8× volumetric ratio.
25. After the third wash, allow the beads to dry for 5–10 min on the magnet. Watch the beads closely, as overdrying beads can impair DNA recovery. Cover the open microcentrifuge tubes with the top of a pipette tip box to prevent dust from falling into the tubes while they are incubating.
26. After 5 min, use a P20 pipette to remove any liquid that has pooled at the bottom of the microcentrifuge tubes.
27. Once the beads appear matte in color, but ideally before they develop significant cracks, remove from the magnet and add 16 μL of RNAse-free water. Pipette up and down to mix well, until the solution is uniform in color.
28. Incubate for 1 min, then return to the magnet.
29. Incubate for 1 min on the magnet, then transfer 15 μL to new microcentrifuge tubes

3.13 Quality Control and Library Prep

1. Analyze the distribution of fragment lengths in the purified product by running a High Sensitivity D5000 Screen Tape on an Agilent 4200 Tapestation. The distribution should be fairly smooth with an average size of 600–750 bp. We have successfully sequenced tagmentation products with average fragment lengths of 350 bp to 700 bp (Fig. 35).

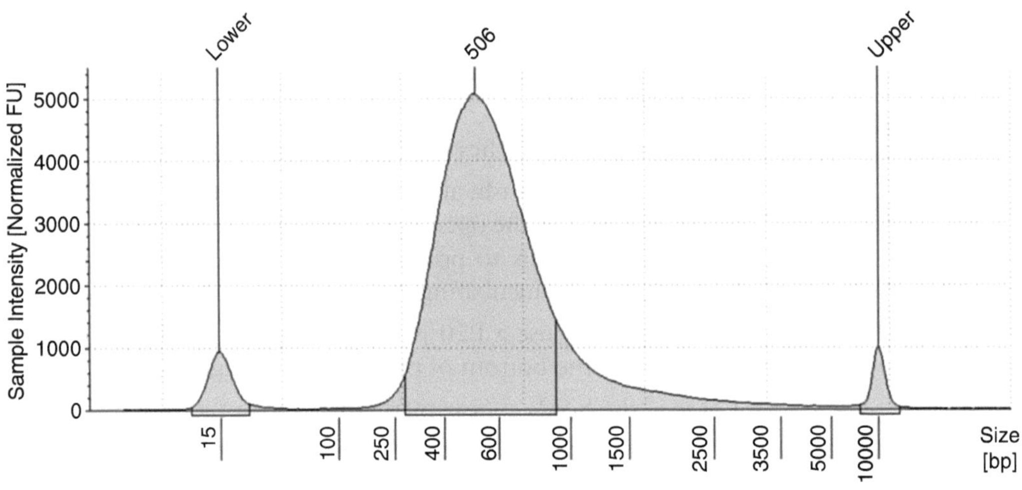

Fig. 35 An ideal NTA product distribution is a smooth bell curve with a peak between 350 and 700 bp

2. Determine the concentration of sequencing libraries using the high-sensitivity dsDNA kit for Qubit Fluorometer.
3. Library can be stored at −20 C until sequencing (*See* **Note 63**). For best results, prepare the 5 μL 4 nM dilution on the day of sequencing (*See* **Note 64**).

3.14 Sequencing

1. Make a 5 μL library pool, diluted to 4 nM.
2. Prepare a fresh aliquot of 0.2 N NaOH by combining 180 μL of ultrapure water and 20 μL of 2 N NaOH in a 1.5 mL Lo-Bind microcentrifuge tube.
3. Program a timer for 5 min. Add 5 μL of the freshly diluted 0.2 N NaOH to 5 μL of the 4 nM library.
4. Pipette to mix, spin briefly, then incubate at RT for 5 min.
5. After 5 min (*See* **Note 65**), add 5 μL of Tris–HCl pH 7.5. Pipette up and down to mix.
6. Add 985 μL freshly defrosted HT1 Buffer to the same tube. This is a 1 mL 20 pM library (solution 1).
7. In a new tube, add 165 μL of Solution 1 to 1.5 mL HT1 buffer for a final library concentration of 2.2 pM (Solution 2).
8. Add 6 μL of Custom Read 1 Primer to 1994 μL of HT1 to make 2 mL of 0.3 μM Custom Read 1 Primer.
9. Load 1.5 mL of Solution 2 to reservoir #10 on the Illumina reagent cartridge, labeled "Load Library Here."
10. Load 2 mL the Custom Read 1 Primer solution prepared in **step 8** to the NextSeq reagent cartridge reservoir # (look up).
11. Follow Illumina's guide for loading a NextSeq500/550 High Output sequencing kit and starting the sequencing run. Seq-Well requires paired end sequencing with the following read structure: Read 1: 20 bp, Read 2: 50 bp, Index 1:8 bp.

4 Preprocessing of Sequencing Data

4.1 Alignment and Preprocessing

This section outlines the general steps from sequencer to digital gene expression matrix (DGE) cell-gene-count matrix and provides a high-level overview of possible methods for downstream analysis of data generated with Seq-Well S^3(Fig. 36).

Upon the completion of a sequencing run, one should first review the run's performance metrics to check for errors and to determine if the run performed within the manufacturer's expectations [15]. Low Cluster Density or low Clusters Passing Filter, for instance, could imply low sample purity, an instrument error, or inaccurate measurements of sample concentrations. Please refer to the manufacturer's guide to troubleshoot such issues [15]. It is still advised to proceed with demultiplexing as the output files may help to identify the source of most issues.

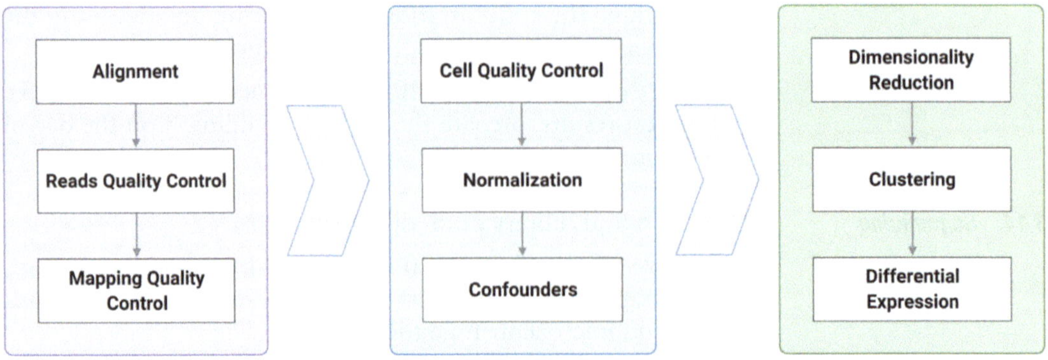

Fig. 36 Flowchart of scRNA-seq analysis with Seq-Well S^3 [14]

Demuliplexing refers to the process of assigning sequencing reads to their sample of origin. For each sequencing run, sequencing reads are first converted from the standard machine output files (in bcl type format) to FASTQ format. Demultiplexing sequencing reads from Illumina machines is performed using the free software bcl2fastq [16]. Briefly, bcl2fastq demultiplexes samples according to sample-specific indices introduced in the post-tagmentation PCR. A helpful representation of the library structure is provided by the Teichmann Group [17]. A user-generated spreadsheet is provided as input to bcl2fastq listing the index-sample pairings. The summary file generated by bcl2fastq lists the number of reads sequenced and the Most Common Unknown Index Sequences which can help troubleshoot issues relating to low quality libraries [18]. Demultiplexed FASTQs are then aligned to the corresponding reference genome using an implementation of the Drop-seq Pipeline maintained by the Broad Institute [19]. The key output file is a DGE cell-gene-count matrix that groups a 12 bp cell barcode to 8 bp unique molecular identifier (UMI) contained in each sequencing fragment.

4.2 Cell Quality Control

Determining the degree of sequencing saturation with a saturation curve, shown below, is a useful initial measure of quality. Sequencing to a depth of 10–50 k reads per cell increases the likelihood of capturing lowly expressed genes and thus reduces the chances of false-dropouts. Tuning the cells used as input to Seq-Well S^3, as well as understanding the sequencing depth required for the experiment, is critical when determining whether or not to resequence a sample [20] (Fig. 37).

At the single-cell level, several tools exist to analyze single-cell RNA-seq data such as Seurat and Scanpy. These platforms can be used to filter data based on quality control metrics, visualize common features, transform and normalize data, and provide an integration point for novel methods of analyses developed by third parties. The modes of analysis will vary based on experimental

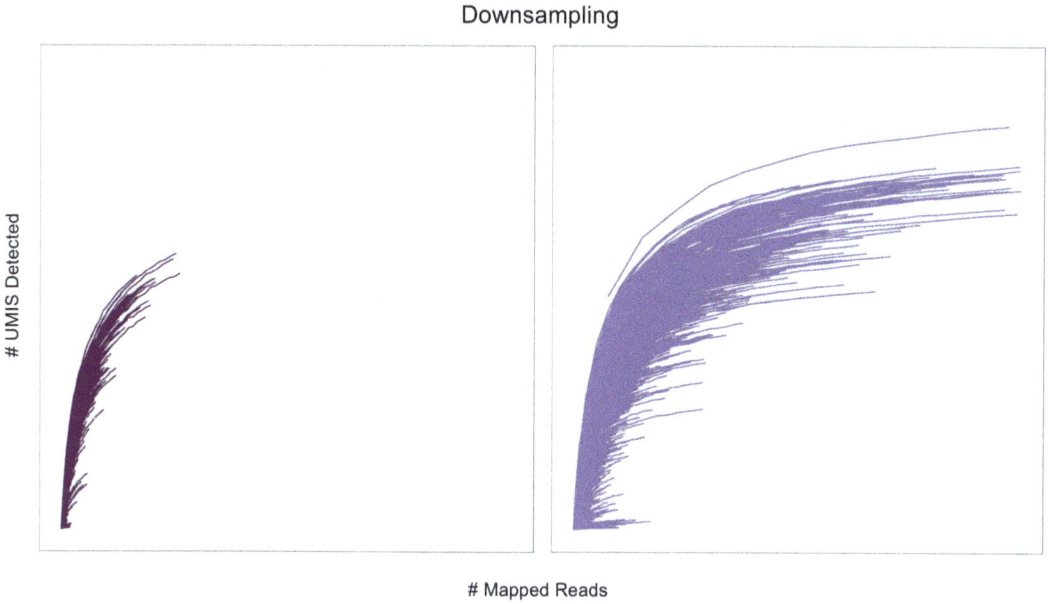

Fig. 37 Estimating Sequence Saturation: In the image to the right, we see that the number of UMIs detected per cell begins to plateau. This is representative of a sample that is adequately sequencing. Further sequencing will not recover a substantial amount of additional UMIs from cells of such a sample. By comparison, the image to the left is representative of a sample that is under-sequenced

design, and it is important to review current literature where this is reviewed comprehensively [21].

A general first step is quality control. Beginning with the DGE matrices generated in **step 4**A, low-quality cells are identified and filtered out from downstream analysis. Second strand synthesis in Seq-Well S^3 increases the expected average number of genes recovered per cell. Visualizing the distribution of genes per cell may be helpful in determining a threshold. The goal is to retain high-quality cells. An initial threshold can be set above the first quartile and adjusted based on experiment needs. It is helpful to remember that different cells with varying transcriptional activity, as well as samples with different levels of integrity, will affect the genes captured per cell. Low quality cells are initially discarded on the basis of gene detection (e.g., <500 genes per cell) and transcript detection (e.g., <700 UMIs per cell). Different thresholds can be tested to see what is lost at a given cut-off, and inform threshold selection (Fig. 38).

In a similar fashion, genes that are not present in more than a handful of cells, due to low expression or difficulties in capturing a gene, are filtered out of the datasets as they do not necessarily serve as reliable or statistically significant carriers of information for differential gene expression analysis. The remaining genes are carried through analysis for further examination. If a rare cell population is expected, the metric can be more permissive and adjusted as

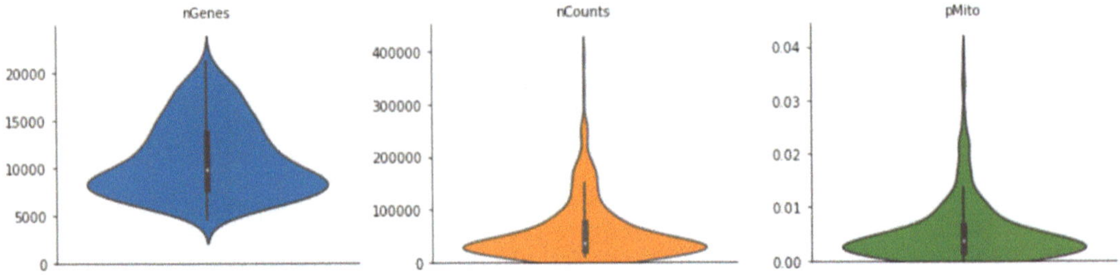

Fig. 38 Violin plots used to visualize QC metrics used to filter low-quality cells. These plots show data from one array

more samples are processed and integrated into the dataset. If certain genes continue to be expressed in a handful of cells and these cells do not belong to a distinct cell type, as discussed below, removing these genes from the dataset is advised. Oftentimes, cells with greater than 10% of reads mapping to mitochondrial genes are also discarded as a high mitochondrial percentage of total transcripts is an indicator of cell death and/or stress. Note that while high mitochondrial mapping percentages can sometimes be due to sample degradation (e.g., during handling of the cell), they can sometimes represent key biology (e.g., very active respiration). It is worth noting that Seq-Well S^3 has been shown to preserve cellular integrity and biological state. Therefore, with new datasets or for samples where cell death or high-mitochondrial activity is expected (in neurons, for example), these cells may be retained in the dataset through clustering, or examined separately.

4.3 Doublets and Ambient RNA Detection and Removal

Occasionally, more than one cell is captured in a single picowell of the Seq-Well array. As the transcripts from these cells will be captured on the same bead, they will be assigned to the same cell barcode, despite originating from separate cells. While rare, this effect can mask true transcriptional heterogeneity in a sample and skew any methods to normalize the data. Thankfully, streamlined protocols exist that are designed to identify doublets within a dataset with a variety of strategies, assumptions, and sample specificities [23]. No method is perfect and it is recommended to review the most up-to-date literature.

Ambient mRNA contamination is another possible source of background noise that may be encountered and removed using computational methods [24]. These methods typically seek to define the nonspecific binding of free-floating RNA to beads loaded on an array. Unlike doublets, ambient RNA does not typically represent a significant portion of the transcripts captured on any particular bead. It is most typically observed when processing primary tissue samples with large amounts of debris and/or low cell-viability. As the background noise is thought to be relatively uniform across the sample, it may not, in some instances, significantly affect downstream analyses that seek to contrast cells. It is always a good idea to check and remove if necessary. Note that ambient RNA is array-specific, and thus methods for removing ambient RNA should be run separately for each Seq-Well S^3 array.

5 Computational Analysis

Once low-quality cells have been filtered from a sample, various computational and statistical tools can be applied to characterize and interpret transcriptional heterogeneity and the underlying cellular processes that inform it. The following sections provide a general approach for interpreting the high-dimensional data generated from Seq-Well S^3. Various methods exist that attempt to resolve what is meaningful variation, find structure within it, and then link gene expression covariation to putative biological drivers (e.g., cell type/differences in activation); these methods are covered extensively elsewhere [21].

5.1 Variable Genes

A critical step in the analysis is the process of choosing variable genes. It is an iterative process. Beginning with higher degrees of variation, such as treatment groups, tissue types, and cell types, the goal is to identify expression profiles that distinguish subsets of the data.

The list of genes will change based on the question that is being asked. As an example, a gene that is expressed in T cells but is not expressed highly in epithelial cells may be highly variable (ranging from 0 in some cells to high expression in others) and of much interest at the first step of analysis when the goal may be to subset T cells from epithelial cells. When identifying heterogeneity between T cells, however, this same gene may not be as informative if it is similarly expressed across these cells.

5.2 Dimensionality Reduction

To identify orthogonal axes of variation in the dataset, a Principal Component Analysis may be performed on the highly variable genes. Ranked in order from small to large, the principal components represent orthogonal linear combinations of variable genes that reflect decreasing degrees of variation.

Critically, not all principal components represent a high degree of variance (or biologically relevant variation) and therefore should not be included in dimensionality reduction. Selecting the number of principal components to include is an iterative process. Visualizations of the percentage variance explained by each principal component may be used to identify the set of principal components that represent the majority of variance present in a sample; this may also be calculated statistically [22] (Fig. 39).

Other methods such as Independent Component Analysis and consensus non-negative matrix factorization (cNMF) exist are reviewed elsewhere [25, 28] (Figs. 40 and 41).

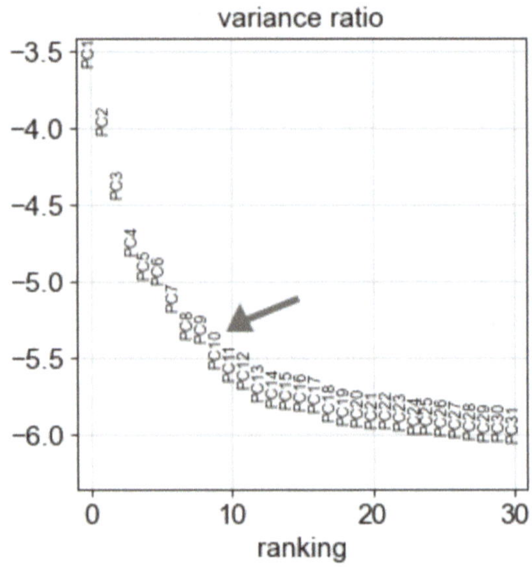

Fig. 39 Plot of ranked principal components based on percent of variance.[10] Arrow points to the elbow

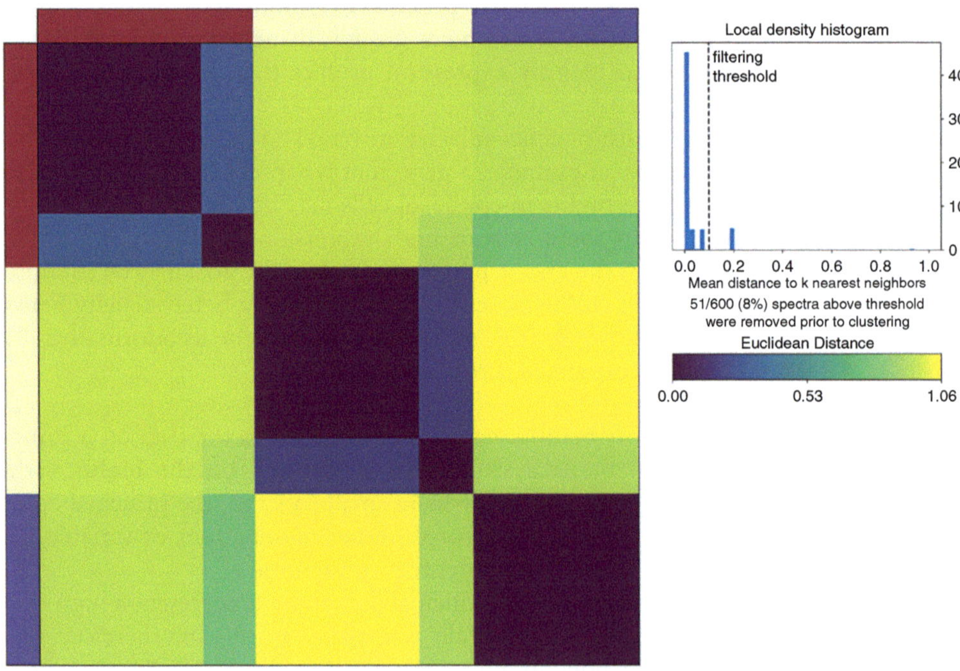

Fig. 40 Plot demonstrating the use of cNMF to identify gene programs

A common approach used to visualize dimensionally reduced data is UMAP (Uniform Approximation and Projection) [27]. Simply, UMAP works by constructing a high-dimensional representation of the dataset and then a low-dimensional analogue that attempts to minimize differences between the two models.

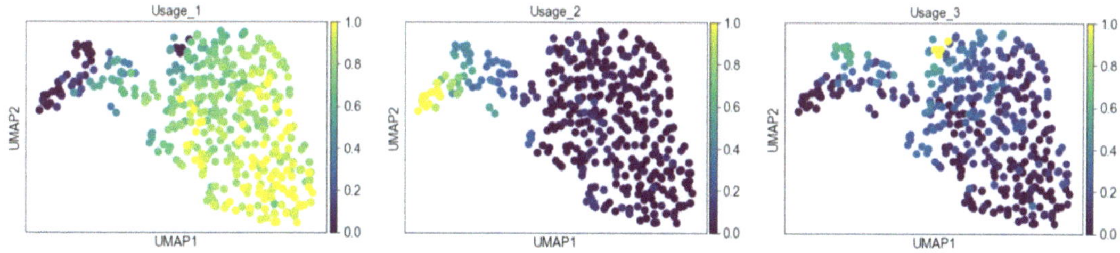

Fig. 41 Gene program usage plotted on UMAP of gene expression profiles

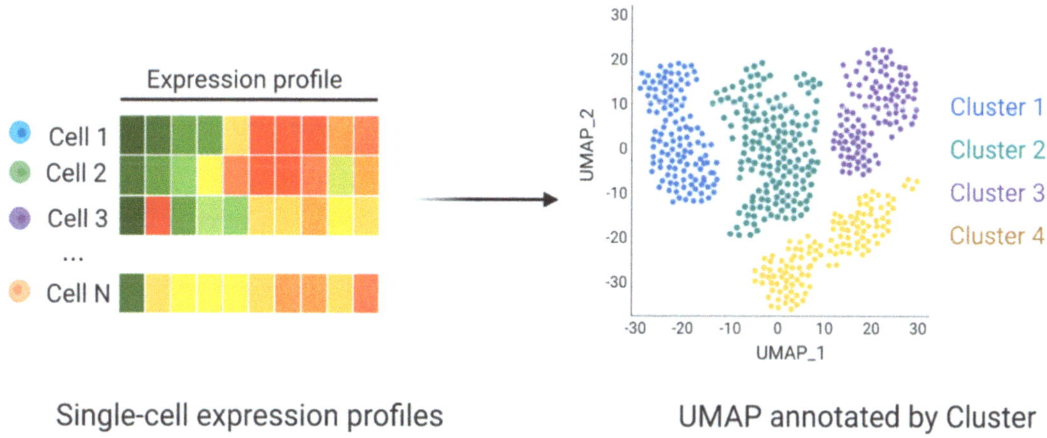

Fig. 42 Visualizing a dataset after dimensionality reduction and clustering

Ultimately, all dimensionality reduction methods serve to improve interpretability of the high-dimensional datasets created with methods like Seq-Well S^3. Identifying the appropriate balance between interpretability and information loss will be largely dataset-dependent (Fig. 42).

5.3 Clustering

Grouping cells based on the similarity of their expression profiles allows one to identify cell types and states, as well as infer differences between groups. This is typically done by clustering or graph-based learning [26]. Unsupervised clustering is a common approach that includes methods such as hierarchical clustering and k-means clustering. These methods compute a distance metric between cells based on a low-dimensional representation of the dataset, such as PCA or UMAP. Cells are then iteratively grouped into clusters based on these distances.

Similarly, graph-based learning methods aim to identify similar groups of cells [21]. Specifically, graph-based learning treats each cell as a node in a graph, with edges connecting each node to its k nearest neighbors. Based on the connectivities, the graph and thus cells are partitioned into clusters. The most widely used graph-based method is the Leiden algorithm which groups cells with high connectivity, compared to the level of connectivity observed throughout the dataset.

One important parameter that is shared across most clustering and graph-based learning approaches is a resolution parameter [26]. Selecting this parameter is important as it establishes the basis for downstream analyses which compares the distribution of gene expression in one cluster or group versus another. An initial strategy could involve tuning this parameter until the dataset is partitioned in a manner that closely resembles the groups observed by a dimensionality reduction method of choice, UMAP for example; we typically select our resolution parameter by maximizing the average silhouette score across all clusters [27, 29]. The resolution can be increased or decreased depending on the level of heterogeneity that is expected from the sample.

6 Notes

1. You will want ~25 mL of bead loading buffer (BLB) for each array. It is important that you do not add the sodium carbonate directly to the BSA, as this will denature the BSA. This solution should be prepared fresh just before loading beads. The stock of BSA should be filtered prior to use with a 0.22 uM filter, and should be kept at 4 C. If the BSA solution begins to turn amber or yellow, it should be discarded.
2. The guanidine thiocyanate will take ~2 hours to go into solution, prepare this solution in advance.
3. Prelysis buffer is photosensitive. Wrap the buffer's container with aluminum foil to protect it. Wrapped prelysis buffer can be stored in a glass bottle at room temperature for approximately 6 months.
4. Prepare complete lysis buffer fresh, just prior to use.
5. Prepare 50 mL of wash buffer per array.
6. Unused beads should be stored at 4 C in TE-TW buffer in 1.5 mL microcentrifuge tubes. When they arrive from the manufacturer, beads should be processed according to the manufacturer's package insert, namely, washed 2× with 30 mL 200 Proof Ethanol for Molecular Biology, 2× with 30 mL TE-TW, passed through a 100 uM filter, and resuspended in 50 mL TE-TW.
7. Currently, Chemgenes is the only supplier manufacturing these beads.
8. Functionalized arrays can be made using the protocol in Hughes et al. 2020 [2] or purchased from the Nanowell Cytometry division of the Biomicro Core at the Koch Institute for Integrative Cancer Research at MIT, Cambridge, Massachusetts, contact: biomicro@mit.edu.

9. There are two ways to transport arrays. The preferred method is to place arrays in 50 mL conical tubes, filled with array quenching buffer or aspartic acid solution. Two arrays can be transported in each conical; arrays should be arranged back-to-back with glass slides touching. Alternatively, the arrays can be dried and transported in a histology slide box. To dry the arrays, remove them from the storage buffer and (being careful not to touch the surface of the array) use a paper towel to wick off excess liquid from the slide. Cover the arrays with the top of a pipette tip box to prevent dust from falling onto them as they dry. Let the arrays dry overnight at room temperature. Dried arrays can be stored at room temperature for 2 weeks. Rehydrate the arrays by placing them in a 4-well dish with either array quenching buffer or aspartic acid solution. Place the arrays under vacuum until (by microscope inspection) there are no air bubbles remaining in the wells of the rehydrated arrays. If no vacuum chamber is available, the arrays can be left to soak overnight and should be ready to use the following day.
10. You will need one hybridization clamp for each array that you plan to process in parallel. Prepare the hybridization clamps by cleaning them with 70% ethanol. You may also find it helpful to apply a small piece of lab tape to the screws—this mark will enable you to keep track of sample identities during membrane sealing.
11. If the plasma cleaner has multiple shelves, place the membranes on the bottom shelf. This placement will reduce the risk that the treated membranes will fly off the slides when the vacuum is released.
12. If your plasma is not bright pink, adjust the bleed valve until the desired color is observed. If you suspect a vacuum has not formed, power off the plasma cleaner—but not the pump—and pull gently on the door to see if a seal has been formed.
13. This amount of time required for membrane functionalization will vary based on the size of the chamber, the power of the instrument, and the settings of the needle valve. When beginning to use a new plasma cleaner, you will need to calibrate to find the correct time for membrane treatment, typically between 3 and 8 min. Anytime you reassemble a plasma cleaner, you will connect NPT tapered threads (3-way valve). For best results, wrap the thread in PFTE (Teflon tape) to ensure a good seal.
14. The sound of the air being released at this low level is very faint, it may be beneficial to put your ear near to the valve as you slowly turn the valve (with appropriate personal protective equipment, like eyeshields), so that you can detect the exact moment when air starts to be released.

15. If you open the door to find that the membranes are slightly folded over, but still on their original slides, slowly flip the membranes back using tweezers. If the membranes have blown off the slides entirely, it is impossible to know which side of the membrane has been exposed to the plasma, and these membranes should be discarded. The plasma-treatment procedure should be repeated with a new set of membranes.
16. Ensure the *entire* membrane is hydrated, usually 5 mL is sufficient. Continue adding PBS over the top of the membrane until it is fully submerged. Avoid touching the top of the membrane with the pipette tip.
17. If any of the arrays have folded over completely and you are unsure which side was UP during the plasma treatment, discard the membrane. This folding can also happen when membranes are being transferred, and it may be helpful to prepare extra membranes.
18. If bubbles remain after the initial solvation step, rock aggressively along the x-axis (i.e., slide the dish across the table) to remove them.
19. Looking down on the 4-well dish, inspect the solvated membranes and discard any which are bright white and opaque. Next, position yourself so that you can view the membranes from the side of the 4-well dish. Membranes which are floating completely on top of the PBS after solvating for 2 min should be discarded.
20. After solvation and before bead loading, use a microscope to inspect wells. If this inspection reveals air bubbles in the wells of the arrays, place arrays under vacuum with rotation (50 RPM) for 10 min to remove air bubbles in wells. The house vacuum in most laboratories should be sufficient to remove bubbles from the wells of the array.
21. If a tabletop centrifuge is unavailable, spin beads for 1 min at 800–1000G.
22. In order to remove all the BLB, you may find it helpful to pipette directly off the top of the array. Be sure to remove excess BLB under the array, as this will allow the array to shift and slide, making bead loading more difficult.
23. This placement is very important, if the long sides of the array are touching the wells of the dish, the loaded beads will slide off the array before they are able to fall into wells.
24. A completely dry array can be identified by its surface finish. Arrays with a small amount of liquid on top of them will remain shiny, while those that have dried out completely will appear matte.

25. During this step, do not lift the pipette tip more than 0.5 cm above the surface of the array, as the force of the beads falling from a greater height will result in a pattern of uneven bead loading.
26. Place a black background behind the 4-well dish to better visualize the bead coverage of the array while loading.
27. If you do get bubbles on the surface of the array, use a fresh pipette tip to carefully pop them. Alternatively, use a pipettor to pipette air out of the bubbles.
28. Manually agitate the arrays by pushing the entire 4-well dish slowly across the benchtop in the X and Y direction. Do not pick up the dish during this step, as tilting the 4-well dish relative to the ground will result in the beads sliding off the array.
29. If the array starts to dry out and less than 25% of the surface of the array looks dry (matte finish), carefully add more BLB to the top of the array dropwise. If the array starts to dry out and more than 25% of the array is dry, add 5 mL of BLB to the well and place the array under vacuum for 5 min. You will likely have to add more beads to this array.
30. If the beads have fallen off—usually this is caused by the long side of the array touching the side of the 4-well dish—use a pipette tip to re-center the array, pipette up beads from the side/bottom of the well, and redeposit the beads on the top of the array. Be careful not to introduce bubbles to the surface of the array during this step. If you do see bubbles in your pipette tip, you may find it useful to transfer the beads you have aspirated into a clean microcentrifuge tube, and spin them down on a tabletop centrifuge to remove bubbles.
31. The edges of the array will not generally have full bead occupancy, this observation is not reason for concern.
32. Loaded arrays can be stored for up to 12 hours at 4 °C. If you are preparing the arrays more than 12 hours in advance, place the loaded array in array quenching buffer and store at 4 °C.
33. For every step of the protocol where beads meet TE-TW, the beads take longer to pellet when they are spun. When pipetting to remove supernatant during these steps, watch closely to ensure that you are not also removing beads. We recommend that you collect the supernatant two steps—spin down for 1 min, remove the first 200 uL of supernatant, spin for another minute, then aspirate the remainder of the supernatant.
34. Ensure that tubes containing beads are stored upright. If tubes are knocked over and beads are incompletely submerged, they may dry out, rendering them unusable.

35. While automated cell counters are time-efficient, we find that the most replicable results are achieved with manual cell counting. We typically mix 20 μL of rigorously resuspended cell suspension with 20 μL of 0.4% Trypan Blue, and load 10 μL into each chamber of a Neubauer Improved disposable hemocytometer. If cells are plentiful, we count samples in triplicate. If counting on a flow cytometer, be sure to include both a live stain (like Calcein) and a dead stain (such as an amine dye).
36. Do not use RPMI with any additives.
37. It is very important that this solution contains FBS, because FBS allows the cells to fall into the wells instead of just sticking to the surface of the array. The FBS minimizes nonspecific interactions at the surface of the array, thereby preventing cells from sticking to the surface of the array.
38. Cells can also be loaded in DMEM + 5% FBS or PBS + 0.05% BSA.
39. The fan in biosafety cabinets tends to quickly dry arrays out. If safety protocols allow, place the arrays on the benchtop for 5 min or turn off the fan in the biosafety cabinet (only if feasible with your biosafety protocol requirements).
40. Once the cells are loaded, do not put the arrays under vacuum.
41. If the array is touching the side and the liquid spills off, collect cells, re-center, and pipette cells directly onto the top of the array again.
42. If the array begins to dry, use P200 pipette to add a little bit (~50 μL) of RPMI + 10% FBS.
43. It is *critical* that nothing is added to this PBS.
44. Sometimes the array will stick to the top piece of the clamp. In this case, carefully separate the array sandwich from the top piece. Hold the array sandwich from the sides to prevent forces being applied unequally to the two sides. *Do not pry the array off the metal*, as this action can reverse membrane sealing.
45. During this transfer, be careful to maintain the orientation that the slide sandwich had when it entered the clamp—with the slide attached to the array down.
46. The hybridization buffer may contain trace amounts of guanidine thiocyanate and should not be mixed with bleach. Aspirate the hybridization buffer into the same container you used to collect lysis buffer in **Step 5**.
47. The amount of wash buffer you used in this step should be sufficient to fill the 50 mL conical to the top of the array.
48. The hybridization buffer may contain trace amounts of guanidine thiocyanate and should be collected in the lysis waste container.

49. If you are only processing a few arrays and working quickly, you can elect to not replace the aspirated hybridization buffer with a wash buffer. This exchange will prevent the beads that float from being lost, and may make the arrays easier to image under the microscope. Once the arrays have dried out completely, it will be very difficult to remove beads from them.
50. You may find it helpful to make a note of how much resistance you felt when removing the membrane—was it stuck onto the array, or did it slide right off? Membrane sealing is one of the primary points of failure for this protocol.
51. This step is optional. Once air enters the microwells, it can be very difficult to identify beads in the microwell array and distinguish them from empty wells. We recommend this step if previous experimental runs have had low yields of cDNA, or when samples are irreplaceable.
52. Pipetting up and down at this step will cause beads to get stuck in the pipette tip, leading to a loss of product.
53. At this point, the bead pellet will be easily disturbed, and you may not be able to completely aspirate the supernatant. Remove the supernatant as completely as possible. For best results, do not leave more than 10 μL of 1× RT buffer covering the beads.
54. Alternatively, you can let the RT continue overnight at 52 C and wash the beads the following morning.
55. We start with a single TE TW wash because salts in the RT buffer can cause SDS to precipitate, making it difficult to remove.
56. For every step of the protocol where beads meet TE-TW, we recommend collecting the supernatant in two steps—spin down for 1 min, remove the first 200 uL of supernatant, spin for another minute, then aspirate the remainder of the supernatant.
57. The optimal number of PCR cycles depends on the cell type used. In general, 13 cycles are optimal for cell lines or larger cells (e.g., macrophages) while 16 cycles are optimal for primary cells.
58. To quickly gauge if the Ampure SPRI beads are ready to use, quickly invert the bottle. If you observe no beads stuck to the bottom, the beads are ready to use. If you see a significant number of beads stuck to the bottom, incubate another 10 min at RT.
59. It is especially important to use Lo-Bind 1.5 mL centrifuge tubes for the following steps, using non-Lo-Bind tubes will result in the loss of a significant amount of WTA product.
60. For example, if you have 200 μL of PCR product add 120 μL of SPRI beads for a 0.6× volumetric ratio.

61. Dilute your PCR product in a separate tube/plate so that you can transfer 5 μL of this dilution to the tagmentation reaction.
62. Change pipette tips as you add the ATM buffer to prevent the ATM stock from being contaminated with TD buffer.
63. Final libraries are compatible for all Illumina sequencing machines that allow for a custom Read 1 primer. It is standard protocol to allot 200 million reads per sample (per Seq-Well Array). The following protocol is prepared for sequencing using a High-Output 75 Cycle v2.5 Kit for the NextSeq 500/550 platform. This kit produces 500 million reads and thus provides sufficient sequencing depth for 2 arrays per sequencing run.
64. Use the average length from the tape station and the average concentration from the Qubit fluorometer (take measurements in triplicate) to calculate the required dilution.
65. The timing of this step is important, an overly denatured library may significantly reduce the quality of your sequence data. If you accidentally allow the denaturing to proceed for longer than 10 min, make a fresh 4 nM library and repeat the denaturation.

Acknowledgments

This work was supported in part by the NIH (P30CA14051, 5P01AI039671, 1DP2GM1194192, U19AI0899922, R01HL0957911, U54CA2173772, P01AI 0396715, U24AI1186722, RM1HG0061931, R33CA2028201, R01AI1385461, R01HL1265541, R01DA0462771, U2CCA23319501, 5U19AI089992, R01A R060802, R01AI30025, P30AR075043, R01AI022553, R01AR040312, F30AI143160, and R01AR074302), the Bridge Project (a partnership between the Koch Institute for Integrative Cancer Research at MIT and the Dana-Farber/Harvard Cancer Center), and the Food Allergy Science Initiative at the Broad Institute. A.K.S. was supported by the Searle Scholars Program, the Beckman Young Investigator Program, the Pew-Stewart Scholars Program for Cancer Research, a Sloan Fellowship in Chemistry, and the Bill and Melinda Gates Foundation (OPP1139972, OPP1202327, OPP1137006, and OPP1202327).

Declaration A.K.S. and J.C.L. have received compensation for consulting and SAB membership from Honeycomb Biotechnologies, which makes a commercial variant of Seq-Well. A.K.S. and J.C.L. are co-inventors on a provisional patent application filed by MIT relating to the improved methodology described in this chapter.

References

1. Gierahn TM, Wadsworth MH 2nd, Hughes TK, Bryson BD, Butler A, Satija R, Fortune S, Love JC, Shalek AK (2017) Seq-Well: portable, low-cost RNA sequencing of single cells at high throughput. Nat Methods 14(4):395–398. https://doi.org/10.1038/nmeth.4179
2. Hughes TK, Wadsworth MH 2nd, Gierahn TM, Do T, Weiss D, Andrade PR, Ma F, de Andrade Silva BJ, Shao S, Tsoi LC, Ordovas-Montanes J, Gudjonsson JE, Modlin RL, Love JC, Shalek AK (2020) Second-Strand synthesis-based massively parallel scRNA-Seq reveals cellular states and molecular features of human inflammatory skin pathologies. Immunity 53(4):878–894.e7. https://doi.org/10.1016/j.immuni.2020.09.015
3. Klein AM, Mazutis L, Akartuna I, Tallapragada N, Veres A, Li V, Peshkin L, Weitz DA, Kirschner MW (2015) Droplet barcoding for single-cell transcriptomics applied to embryonic stem cells. Cell 161(5): 1187–1201. https://doi.org/10.1016/j.cell.2015.04.044
4. Macosko EZ, Basu A, Satija R, Nemesh J, Shekhar K, Goldman M, Tirosh I, Bialas AR, Kamitaki N, Martersteck EM, Trombetta JJ, Weitz DA, Sanes JR, Shalek AK, Regev A, McCarroll SA (2015) Highly parallel genome-wide expression profiling of individual cells using Nanoliter droplets. Cell 161(5): 1202–1214. https://doi.org/10.1016/j.cell.2015.05.002
5. Montoro DT, Haber AL, Biton M, Vinarsky V, Lin B, Birket SE, Yuan F, Chen S, Leung HM, Villoria J, Rogel N, Burgin G, Tsankov AM, Waghray A, Slyper M, Waldman J, Nguyen L, Dionne D, Rozenblatt-Rosen O, Tata PR, Mou H, Shivaraju M, Bihler H, Mense M, Tearney GJ, Rowe SM, Engelhardt JF, Regev A, Rajagopal J (2018) A revised airway epithelial hierarchy includes CFTR-expressing ionocytes. Nature 560(7718):319–324. https://doi.org/10.1038/s41586-018-0393-7
6. Ordovas-Montanes J, Dwyer DF, Nyquist SK, Buchheit KM, Vukovic M, Deb C, Wadsworth MH 2nd, Hughes TK, Kazer SW, Yoshimoto E, Cahill KN, Bhattacharyya N, Katz HR, Berger B, Laidlaw TM, Boyce JA, Barrett NA, Shalek AK (2018) Allergic inflammatory memory in human respiratory epithelial progenitor cells. Nature 560(7720): 649–654. https://doi.org/10.1038/s41586-018-0449-8
7. Smillie CS, Biton M, Ordovas-Montanes J, Sullivan KM, Burgin G, Graham DB, Herbst RH, Rogel N, Slyper M, Waldman J, Sud M, Andrews E, Velonias G, Haber AL, Jagadeesh K, Vickovic S, Yao J, Stevens C, Dionne D, Nguyen LT, Villani AC, Hofree M, Creasey EA, Huang H, Rozenblatt-Rosen O, Garber JJ, Khalili H, Desch AN, Daly MJ, Ananthakrishnan AN, Shalek AK, Xavier RJ, Regev A (2019) Intra- and inter-cellular rewiring of the human colon during ulcerative colitis. Cell 178(3):714–730.e22. https://doi.org/10.1016/j.cell.2019.06.029
8. Vento-Tormo R, Efremova M, Botting RA, Turco MY, Vento-Tormo M, Meyer KB, Park JE, Stephenson E, Polański K, Goncalves A, Gardner L, Holmqvist S, Henriksson J, Zou A, Sharkey AM, Millar B, Innes B, Wood L, Wilbrey-Clark A, Payne RP, Ivarsson MA, Lisgo S, Filby A, Rowitch DH, Bulmer JN, Wright GJ, Stubbington MJT, Haniffa M, Moffett A, Teichmann SA (2018) Single-cell reconstruction of the early maternal-fetal interface in humans. Nature 563(7731):347–353. https://doi.org/10.1038/s41586-018-0698-6
9. Monian B, Tu AA, Ruiter B, Morgan DM, Petrossian PM, Smith NP, Gierahn TM, Ginder JH, Shreffler WG, Love JC (2022) Peanut oral immunotherapy differentially suppresses clonally distinct subsets of T helper cells. J Clin Invest 132(2):e150634. https://doi.org/10.1172/JCI150634
10. Genshaft AS, Subudhi S, Keo A, Sanchez Vasquez JD, Conceição-Neto N, Mahamed D, Boeijen LL, Alatrakchi N, Oetheimer C, Vilme M, Drake R, Fleming I, Tran N, Tzouanas C, Joseph-Chazan J, Villanueva MA, van de Werken HJG, van Oord GW, Groothuismink ZMA, Beudeker BJ, Osmani Z, Nkongolo S, Mehrotra A, Feld J, Chung RT, de Knegt RJ, Janssen HLA, Aerssens J, Bollekens J, Hacohen N, Lauer GM, Boonstra A, Shalek AK, Gehring A (2021) Clinical implementation of single-cell RNA sequencing using liver fine needle aspirate tissue sampling and centralized processing captures compartment specific immuno-diversity. BioRxiv. https://doi.org/10.1101/2021.11.30.470634
11. Morgan DM, Ruiter B, Smith NP, Tu AA, Monian B, Stone BE, Virk-Hundal N, Yuan Q, Shreffler WG, Love JC (2021) Clon-

ally expanded, GPR15-expressing pathogenic effector TH2 cells are associated with eosinophilic esophagitis. Sci Immunol 6(62): eabi5586. https://doi.org/10.1126/sciimmunol.abi5586

12. Kotliar D, Lin AE, Logue J, Hughes TK, Khoury NM, Raju SS, Wadsworth MH 2nd, Chen H, Kurtz JR, Dighero-Kemp B, Bjornson ZB, Mukherjee N, Sellers BA, Tran N, Bauer MR, Adams GC, Adams R, Rinn JL, Melé M, Schaffner SF, Nolan GP, Barnes KG, Hensley LE, McIlwain DR, Shalek AK, Sabeti PC, Bennett RS (2020) Single-cell profiling of Ebola virus disease in vivo reveals viral and host dynamics. Cell 183(5):1383–1401.e19. https://doi.org/10.1016/j.cell.2020
13. Trombetta JJ, Gennert D, Lu D, Satija R, Shalek AK, Regev A (2014) Preparation of single-cell RNA-Seq libraries for next generation sequencing. Curr Protoc Mol Biol 107: 4.22.1–17. https://doi.org/10.1002/0471142727.mb0422s107
14. Figure "Created with BioRender.com
15. Illumina (2016) Optimizing cluster density on illumina sequencing systems (https://www.illumina.com/content/dam/illumina-marketing/documents/products/other/interview_cmason.pdf)
16. Illumina (2019) bcl2fastq2 Conversion Software v2.20 Software Guide (https://support.illumina.com/content/dam/illumina-support/documents/documentation/software_documentation/bcl2fastq/bcl2fastq2-v2-20-software-guide-15051736-03.pdf)
17. Sarah Teichmann's group at EMBL-EBI (2022) Seq-Well S3 library structure (https://teichlab.github.io/scg_lib_structs/methods_html/SeqWell_S3.html)
18. Illumina (2016) Troubleshooting demultiplexing issues (https://support.illumina.com/bulletins/2016/08/troubleshooting-demultiplexing-issues-using-basespace-sequence-hub-and-bclfastq-v.html)
19. The Broad Institute, Inc. and The General Hospital Corporation (2019) Drop-seq Pipeline (https://cumulus-doc.readthedocs.io/en/0.12.0/drop_seq.html)
20. Zhang MJ, Ntranos V, Tse D (2020) Determining sequencing depth in a single-cell RNA-seq experiment. Nat Commun. https://doi.org/10.1038/s41467-020-14482-y
21. Luecken MD, Theis FJ (2019) Current best practices in single-cell RNA-SEQ analysis: a tutorial. Mol Syst Biol. https://doi.org/10.15252/msb.20188746
22. Chung NC, Storey JD (2015) Statistical significance of variables driving systematic variation in high-dimensional data. Bioinformatics 31(4):545–554. https://doi.org/10.1093/bioinformatics/btu674
23. Xi NM, Li JJ (2021) Benchmarking computational doublet-detection methods for single-cell RNA sequencing data. Cell Syst. https://doi.org/10.1016/j.cels.2020.11.008
24. Young MD, Behjati S (2020) SoupX removes ambient RNA contamination from droplet-based single-cell RNA sequencing data. GigaScience. https://doi.org/10.1093/gigascience/giaa151
25. Sun S, Zhu J, Ma Y, Zhou X (2019) Accuracy, robustness and scalability of dimensionality reduction methods for single-cell RNA-seq analysis. Genome Biol. https://doi.org/10.1186/s13059-019-1898-6
26. Hie B, Peters J, Nyquist SK et al (2020) Computational methods for single-cell RNA sequencing. Ann Rev Biomed Data Sci 3: 339–364. https://doi.org/10.1146/annurev-biodatasci-012220-100601
27. Becht E, McInnes L, Healy J et al (2018) Dimensionality reduction for visualizing single-cell data using UMAP. Nat Biotechnol 37:38–44. https://doi.org/10.1038/nbt.4314
28. Kotliar D, Veres A, Nagy MA et al (2019) Identifying gene expression programs of cell-type identity and cellular activity with single-cell RNA-seq. elife. https://doi.org/10.7554/elife.43803
29. Ziegler CGK, Miao VN, Owings AH, Navia AW, Tang Y, Bromley JD, Lotfy P, Sloan M, Laird H, Williams HB, George M, Drake RS, Christian T, Parker A, Sindel CB, Burger MW, Pride Y, Hasan M, Abraham GE 3rd, Senitko M, Robinson TO, Shalek AK, Glover SC, Horwitz BH, Ordovas-Montanes J (2021) Impaired local intrinsic immunity to SARS-CoV-2 infection in severe COVID-19. Cell 184(18):4713–4733.e22. https://doi.org/10.1016/j.cell.2021.07.023

Chapter 4

A MATQ-seq-Based Protocol for Single-Cell RNA-seq in Bacteria

Christina Homberger, Antoine-Emmanuel Saliba, and Jörg Vogel

Abstract

Microbes exhibit an extraordinary capacity to adapt their physiology to different environments using phenotypic heterogeneity. However, the majority of gene regulation studies are conducted in bulk reflecting only averaged gene expression levels across millions of cells. Bacterial single-cell RNA-seq (scRNA-seq) can overcome this by enabling whole transcriptome and unbiased analysis of microbes at the single-cell level. Here, we describe a detailed workflow of single-cell RNA-seq based on the multiple annealing and dC-tailing-based quantitative single-cell RNA-seq (MATQ-seq) protocol. Following adjustments to the original eukaryotic protocol, the workflow was applied to two major human pathogens *Salmonella enterica* serovar *Typhimurium* (henceforth *Salmonella*) and *Pseudomonas aeruginosa* (henceforth *Pseudomonas*). The development of bacterial scRNA-seq protocols offers promising avenues to explore the molecular programs underlying phenotypic heterogeneity on the transcriptome level in different settings such as infection, persistence, ecology, and biofilms.

Key words Bacterial single-cell RNA-seq, Single bacteria sorting, MATQ-seq, *Salmonella enterica*, *Pseudomonas aeruginosa*, Gene expression heterogeneity

1 Introduction

Phenotypic heterogeneity is ubiquitously observed within bacterial communities [1]. For prominent examples, cell-to-cell variability has been described in biofilms [2, 3], antibiotic tolerance [4], and stress responses [5, 6]. However, the study of bacterial heterogeneity has been hindered by the lack of unbiased and transcriptome-wide methods that can report bacterial physiology at the single-cell level. Tracking bacterial heterogeneity has mostly relied on using fluorescent-based reporter systems, which are restricted to visualizing the expression of only few genes in genetically tractable organisms [7].

The development of single-cell RNA-seq (scRNA-seq) has revolutionized transcriptomic studies by allowing the study of cell-to-cell gene expression variability [8]. While these technologies

Raffaele A. Calogero and Vladimir Benes (eds.), *Single Cell Transcriptomics: Methods and Protocols*,
Methods in Molecular Biology, vol. 2584, https://doi.org/10.1007/978-1-0716-2756-3_4,

have been applied to various eukaryotic species and revealed a wealth of information on cell heterogeneity, their application to prokaryotes is in its infancy [9].

Compared to eukaryotic single-cell RNA-seq, establishing a similar workflow in bacteria faces several challenges due to the intrinsic characteristics of bacterial transcriptomes. Bacterial transcripts are labile (half-life of only few minutes), present in low copy numbers (on average less than one per cell for a given mRNA), and generally lack functional poly-A tails [9]. Thus, a direct transfer of poly-A-dependent scRNA-seq protocols from eukaryotic to bacterial cells is far from straightforward. Recently, however, several laboratories have published protocols for bacterial scRNA-seq [10–14], one of which is based on Multiple Annealing and dC-tailing-based Quantitative single-cell RNA-seq (MATQ-seq).

MATQ-seq, which was originally developed for eukaryotic single-cell RNA-seq [15], lends itself to application to bacteria as it does not require poly-A tailed mRNA. MATQ-seq has been initially reported to be highly sensitive with respect to RNA capture and robust with respect to detecting changes on transcript levels. Another unique feature of the protocol is the use of so-called MALBAC primers, which reduces the biases of PCRs by quasi-linear amplification [16]. This enables successful adaptation for bacterial single-cell RNA-seq [14]. Nonetheless, major modifications are required for the cell lysis step, and these will depend on the bacterium of interest to achieve maximal lysis efficiency and therefore minimal dropout rate.

Here, we provide a detailed protocol to study gene expression heterogeneity at the single-cell level. In detail, this bacterial scRNA-seq workflow based on an improved MATQ-seq protocol starts with single-cell isolation of bacteria using Fluorescence Activated Cell Sorting (FACS), followed by cell lysis, cDNA synthesis, library preparation, sequencing, and data analysis (Fig. 1).

The protocol has proven to be able to report the physiological state of bacteria, as inferred from the detection of transcripts of 170 genes on average across all classes of RNA in *Salmonella* and *Pseudomonas* [14]. Importantly, generalizing the method and adaptation to other microorganisms are possible by modifying lysis as well as sorting strategy. The MATQ-seq workflow itself as well as library preparation is generic and can be performed with eukaryotic as well prokaryotic RNA samples.

2 Materials

Prepare all solutions with ddH_2O and use certified DNase/RNase-free plasticware only. Use molecular biology grade reagents and store all reagents and kits according to manufacturer's instructions. For all pipetting steps, suitable DNase/RNase-free filter tips are

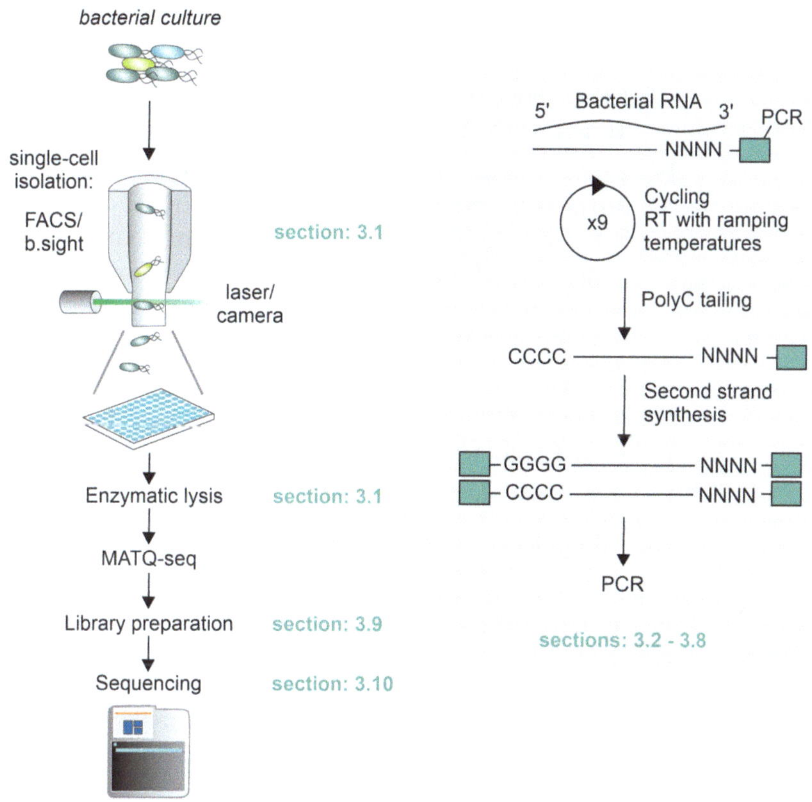

Fig. 1 Schematic representation of bacterial scRNA-seq workflow and the associated protocol steps. First, single bacterial cells are FACS-sorted into a plate prefilled with lysis buffer. The lysis step is followed by MATQ-seq that integrates reverse transcription, poly(C)-tailing, second strand synthesis, and amplification. The obtained cDNA is used as input for library preparation. Libraries are pooled and subsequently sequenced. (Modified from [14] with permission from Nat. Microbiol.)

highly recommended to prevent contaminations. All equipment, especially the workbench and pipettes should be wiped with RNase removal spray prior working followed by high cleaning standards (*see* **Note 1**). Consumables and reagents, which are used several times through the protocol, are listed only once.

2.1 Single Cell Isolation and Lysis

1. 0.5, 1.5, and 2.0 ml plastic tubes (preferably low binding).
2. Non-skirted 48 or 96-well plates (Brand).
3. Vortex mixer.
4. 10 ml glass tubes.
5. 50 ml of Erlenmeyer flasks.
6. FACS Aria III Flow Cytometry Cell Sorter (BD).
7. Benchtop centrifuge for tubes and 96-well plates.
8. Sonorex Digitec DT 52 Sonicator (Bendelin).

9. Accudrop Beads (BD): Add 1 droplet of beads to 1 ml of PBS 1 ×.
10. LB (Lysogeny Broth) medium: 5 g yeast extract, 5 g NaCl, 10 g Tryptone. Fill up with H_2O up to 1 l and autoclave.
11. LBA (Luria Bertani agar) plates: 5 g yeast extract, 5 g NaCl, 10 g Tryptone, 18 g agar. Fill up with H_2O up to 1 l and autoclave. Let cool down in a 60 °C water-bath. For agar plates containing 20 μg/ml Chloramphenicol (Cm) add 1 ml of the antibiotic (20 mg/ml in 100% Ethanol) and mix well before pouring into petri dishes (25 ml per dish). Let plates solidify and store them at 4 °C (*see* **Note 2**).
12. ddH_2O (aliquots of 2 ml tubes).
13. Lysis buffer 10 × (Takara).
14. Recombinant RNase Inhibitor (Takara).
15. PBS pH 7.4 10 × (Gibco).
16. PBS pH 7.4 1 × (Gibco).
17. 50 mM EDTA (Thermo Fisher).
18. RNAlater (Invitrogen).
19. DTT 100 mM (Invitrogen).
20. Microseal "B" and "F" PCR Plate Sealing Film (Biorad) (*see* **Note 3**).
21. 50 U/μl Ready-Lyse Lysozyme Solution (Lucigen): Add 1 μl of Lysozyme (30'000 U/μl) to 599 μl lysozyme storage buffer containing 50 mM Tris–HCl (pH 7.5), 0.1 M NaCl, 0.1 mM EDTA, 1 mM DTT, 0.1% Triton X-100 and 50% glycerol. Prepare aliquots of 5 μl and store them at −20 °C (*see* **Note 4, 5**).

2.2 Reverse Transcription

1. GAT27 dT primer 100 μM (Table 1, *see* **Note 6**).
2. GAT27 5N3G primer 100 μM (Table 1).
3. GAT27 5N3T primer 100 μM (Table 1).
4. dNTP (NEB).
5. SuperScript III Reverse Transcriptase (Invitrogen).
6. Superscript III buffer (Invitrogen).
7. Thermocycler compatible with non-skirted 96-well plates.
8. Cooling block for 2 ml tubes.
9. Cooling block for 96-well plates.

2.3 Primer and RNA Digestion

1. T4 DNA Polymerase (NEB).
2. RNase H (NEB).
3. RNase If (NEB).

Table 1
List of primers used for MATQ-seq protocol

Primer name	Sequence
GAT27 dT	5′-GTG AGT GAT GGT TGA GGA TGT GTG GAG NNN NN TTTTTTTTTTTTTTTTTTTTTT
GAT27 5N3G	5′-GTG AGT GAT GGT TGA GGA TGT GTG GAG NNN NNG GG
GAT27 5N3T	5′-GTG AGT GAT GGT TGA GGA TGT GTG GAG NNN NNT TT
GAT21 6N3G	5′-GAT GGT TGA GGA TGT GTG GAG NNN NNN GGG
GAT27 PCR	5′-GTG AGT GAT GGT TGA GGA TGT GTG GAG

2.4 Tailing

1. Terminal Transferase (NEB).
2. TdT buffer 10 × (NEB).
3. dCTP (Invitrogen).

2.5 Second Strand Synthesis

1. Deep Vent (exo-) DNA Polymerase (NEB).
2. Thermopol Buffer (NEB).
3. GAT21 6N3G primer 100 μM (Table 1).

2.6 PCR Amplification

1. GAT27 PCR primer 100 μM (Table 1).

2.7 Cleanup with Magnetic Beads

1. AMPure XP beads (Beckman Coulter) (*see* **Note 7**).
2. 80% fresh prepared undenatured ethanol (360 μl per sample).
3. Sterile 25 ml pipetting reservoirs.
4. Magnetic plate for 96-well plates.

2.8 Quality Control of cDNA

1. Qubit assay tubes (Invitrogen).
2. Qubit dsDNA High Sensitivity (HS) Kit (Invitrogen).
3. Qubit fluorometer (Invitrogen).
4. Bioanalyzer (Agilent).
5. Bioanalyzer Priming Station (Agilent).
6. Vortex mixer compatible with Bioanalyzer chips (IKA).
7. Bioanalyzer High Sensitivity (HS) DNA Kit (Agilent) (*see* **Note 8**).

2.9 Library Preparation and Pooling

1. Nextera XT DNA Library Preparation Kit (Illumina).
2. Nextera XT Index Kit v2 (Illumina).

2.10 Sequencing

1. PhiX Control Spike-in (Illumina) (*see* **Note 9**).
2. NextSeq or NovaSeq sequencing platform (Illumina).
3. Sequencing kit (compatible with sequencing platform) with at least 100 cycles (Illumina).

3 Methods

All procedures for *Salmonella* and *Pseudomonas* are identical unless otherwise stated.

3.1 Single Cell Isolation and Lysis

1. Day 1—Streak *Salmonella* and *Pseudomonas* wild type on LBA and *Salmonella* GFP-tagged strain on LBA plates containing Chloramphenicol. Grow them overnight at 37 °C.
2. Day 2—Start overnight cultures: Pick a single colony from agar plate and inoculate 10 ml glass tubes filled with 2 ml LB media (supplemented with 20 μg/ml Chloramphenicol for GFP strain).
3. Day 3—Prepare 50 ml Erlenmeyer flasks with 20 ml LB. Add 200 μl of overnight culture (OD = 1) – starting OD = 0.01
4. Incubate at 37 °C with agitation.
5. Shock conditions: Grow until OD 0.3, then add NaCl to a final concentration of 0.3 M and incubate for 10 min at 37 °C with agitation (for NaCl shock), respectively, add a layer of mineral oil and incubate for 30 min at 37 °C without agitation (*see* **Note 10**, according to [17]).
6. Late stationary phase: Grow until OD 2.0, followed by further 6 hours growth at 37 °C with agitation (according to [17]).
7. Prepare lysis mix meanwhile cultures grow.
8. Lysis mix for 20 reactions: 39 μl ddH_2O, 5.2 μl 10 × lysis buffer, 5.2 μl PBS 10 ×, 2 μl Lysozyme (50 U/μl) and 0.6 μl RNase Inhibitor. Vortex briefly and spin down (*see* **Note 11**).
9. For *Pseudomonas*: Add EDTA to a final concentration of 0.5 mM and DTT to a final concentration of 100 mM. Adjust volume of ddH_2O in previous step accordingly.
10. Use either 48 or 96-well plate and add 2.6 μl of lysis buffer per well (*see* **Note 12**). Seal plates properly and store at 4 °C until sorting.
11. As soon as cultures are ready, take 1 ml of each culture and spin down immediately for 4 min at 17,000 × g, washing procedure is done at room temperature.
12. Take off supernatant carefully without touching pellet.
13. Resuspend pellet in 1 ml 1 × PBS.

14. Spin down immediately for 4 min at 17,000 × g, room temperature.
15. Repeat washing procedure (**steps 13** and **14**) once again.
16. Prepare 1:1 1 × PBS, RNAlater mix (1 ml per sample).
17. Resuspend pellet in PBS, RNAlater mix.
18. Store samples on ice until sorting.
19. Start FACS (*see* **Note 13**) and cooling system (*see* **Note 14**). Perform fluidics start-up.
20. Insert 70 nozzle and perform Accudrop delay after stream is stable.
21. Adjust automated cell deposition unit (ACDU) adaptor and home device settings to 96-well plate format.
22. Dilute samples in 1 × PBS, at least 1:10 (*see* **Note 15**).
23. Load GPF-tagged strain (*see* **Note 16**).
24. Increase the flow rate to max for about 30 s to ensure stable detection of events.
25. Lower flow rate to achieve about 1000–1500 events/s.
26. Check events on Forward/ Sideward Scatter (SSC/ FSC) plot and check counts for FITC-positive cells as well (Fig. 2).
27. FITC-positive cells are used as marker to identify bacterial cells.
28. Perform selection of fluorescein isothiocyanate (FITC-) positive cells and apply selection on SSC/ FSC plot to identify events resulting from GFP-tagged bacteria.
29. The sorting gate on SSC/ FSC plot is defined as population 1 and contains FITC-positive cells only.
30. Save gating strategy as global worksheet.
31. Unload GFP-tagged strain and load WT strains (*Salmonella* or *Pseudomonas*).
32. Increase the flow rate to max for about 30 s to ensure stable detection of events.
33. Lower flow rate to achieve about 1000–1500 events/s.
34. Create sort scheme: Use single-cell precision. Include at least one negative (NC) and one positive control (PC) on each sorted plate. Exclude those wells from sorting (*see* **Note 17**).
35. Spin down plate containing lysis buffer and put on 96-well plate adapter on FACS.
36. Sort cells with about 1000–1500 events/s.
37. Seal plate properly using Microseal "F", spin down quickly, keep on ice, and store at −80 °C for up to 6 months (*see* **Note 18**).

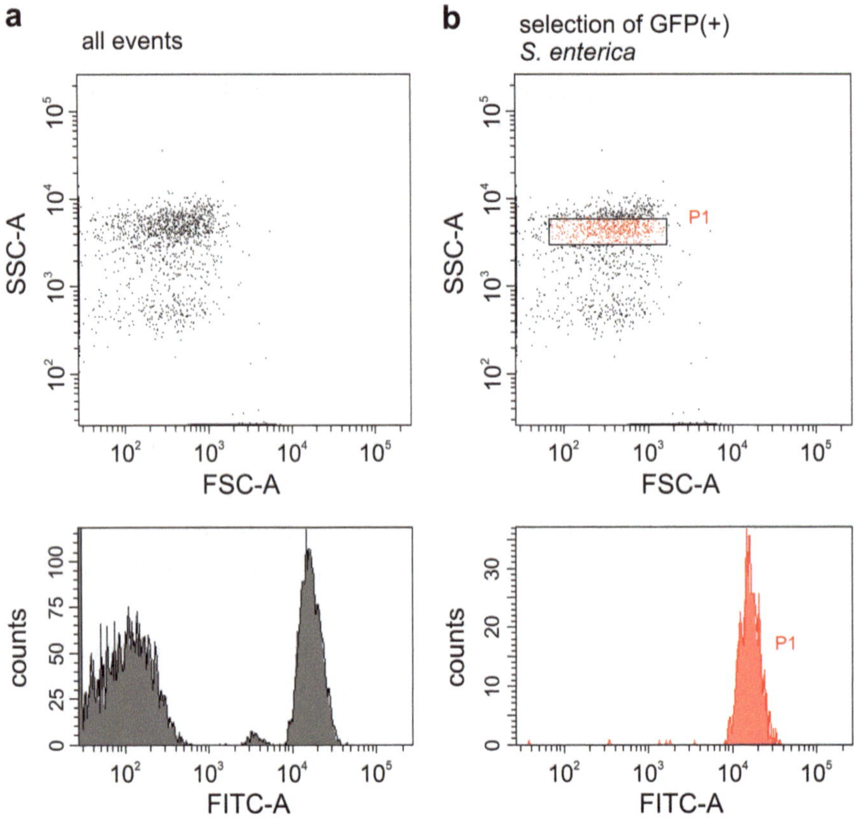

Fig. 2 Optimization of FACS sorting strategy of *Salmonella* using GFP (+) strain. (**a**) All events captured by sorting 10,000 events are shown on a Forward Scatter (FSC-A) and Sideward Scatter (SSC-A) plot and detected GFP signal in FITC-A channel is shown as a count plot. (**b**) GFP-positive cells from (**a**) are used as reference marker to identify bacterial cells on the Forward and Sideward scatter plot. Based on these parameters, the gating of population 1 (P1) is performed which defines a selection of solely GFP-positive cells as shown in FITC-A plot

38. For *Pseudomonas*: Sonicate plate for 10 s after spin down, keep on ice, and store afterwards at −80 °C.

To reduce possible contaminations, a physical separation of reverse transcription (RT) and post-polymerase chain reaction (PCR) workstations is highly recommended (*see* **Note 19**). The given reaction volume is for one well only if not stated otherwise (*see* **Note 20**). Prepare all reagent mixes on ice and keep samples on ice throughout all procedures. Besides, prevent long incubation times outside PCR cyclers (*see* **Note 21**). Before proceeding with next PCR step: seal multi-well plate properly, vortex for 3 s and spin briefly down.

3.2 Reverse Transcription

1. Prepare the primer mix by adding 7.08 µl ddH_2O, 0.12 µl GAT27 dT primer, each 0.4 µl of GAT27 5N3G and GAT27 5N3T into 0.5 ml tube (Table 1, *see* **Note 22**). Vortex and spin down.
2. Mix 0.4 µl of primer mix with 0.05 µl DTT and 0.12 µl dNTP (*see* **Note 23**). Vortex and spin down pre-RT mix.
3. Positive control: Add 0.5 µl of DNase treated total RNA (100 pg/µl, total input: 50 pg).
4. Negative control: Add 0.5 µl ddH_2O.
5. Add 0.57 µl of pre-RT mix to each well containing sorted cell in lysis buffer. Vortex and spin down.
6. Incubate for 3 min at 72 °C, transfer on ice for at least 1 min.
7. Prepare RT mix: 1.15 µl ddH_2O, 0.8 µl Superscript III buffer, 0.2 µl DTT, 0.1 µl RNase Inhibitor, and 0.15 µl Superscript III.
8. Add 2.4 µl of RT mix to each well. Vortex and spin down.
9. Run PCR program I (Table 2). Proceed immediately with Primer Digestion.

3.3 Primer and RNA Digestion

1. Incubate 1 min at 50 °C (*see* **Note 24**).
2. Perform this step at room temperature: Add 0.2 µl T4 DNA Polymerase (*see* **Note 25**). Vortex and spin down.
3. Run PCR program II (Table 2).
4. Put plate on ice or cooling block. Add 0.1 µl of RNase H and RNase If. Vortex and spin down.
5. Run PCR program III (Table 2).

3.4 Tailing

1. Add 3.13 µl ddH_2O, 0.4 µl dCTP, 0.4 µl TdT-buffer, and 0.1 µl ddH_2O Terminal Transferase. Vortex and spin down.
2. Run PCR program III (Table 2).

3.5 Second-Strand Synthesis

1. Add 12.925 µl ddH_2O, 1.5 µl Thermopol buffer, 1.25 µl dNTP, and 0.125 µl GAT21 6NG3 primer. Vortex and spin down.
2. Incubate 1 min at 95 °C, hold at 48 °C.
3. Perform this step at room temperature: Add 0.4 µl Deep Vent (exo-) DNA Polymerase. Vortex and spin down.
4. Run PCR program IV (Table 2).

3.6 PCR Amplification

1. Add 114 µl ddH_2O, 13 µl Thermopol buffer, 3 µl dNTP, 3 µl Deep Vent (exo-) DNA polymerase, and 0.8 µl GAT27 PCR primer. Mix 10 times up and down (*see* **Note 26**).

Table 2
PCR programs for MATQ-seq

I		
8 °C	11x	12 sec
15 °C		45 sec
20 °C		45 sec
30 °C		30 sec
42 °C		2 min
50 °C		3 min
50 °C		15 min
4 °C		∞

II	
37 °C	40 min
75 °C	20 min
37 °C	40 min
80 °C	20 min
4 °C	∞

III	
37 °C	15 min
72 °C	15 min
4 °C	∞

IV		
48 °C	11x	20 sec
72 °C		1 min
72 °C		2 min
4 °C		∞

V		
95 °C		30 sec
95 °C	25x	15 sec
62 °C		20 sec
72 °C		2 min
72 °C		5 min
4 °C		∞

(In program I, the 11x cycle covers 8 °C through 50 °C (3 min); in program IV, the 11x cycle covers 48 °C and 72 °C (1 min); in program V, the 25x cycle covers 95 °C (15 sec) through 72 °C (2 min).)

2. Distribute each reaction into 4 wells (40 μl each) (*see* **Note 27**). Vortex and spin down.
3. Run PCR program V (Table 2).
4. Safe stopping point: Plate can be stored at −20 °C for up to two days. Otherwise continue directly with cleanup procedure.

3.7 Cleanup with Magnetic Beads

All steps are performed at room temperature.

1. Bring AMPure XP beads at room temperature at least 30 min in advance.
2. Take out sample plate and keep at room temperature for 15 min. Spin down briefly before continuing.
3. Vortex beads vigorously and make sure beads are well suspended.

4. Add 40 μl of beads to every well and mix up and down several times (*see* **Note 28**).
5. Incubate for 10 min (*see* **Note 29**).
6. Place the plate onto a 96-well magnet and incubate for 2 min or until the supernatant is completely clear.
7. Remove the supernatant carefully without touching the pellet.
8. Add 180 μl of fresh 80% ethanol to each well and incubate for 30 s on the magnetic stand.
9. Remove and discard ethanol without removing the PCR plate from the magnet.
10. Repeat washing procedure (**steps 8, 9**) once again.
11. Remove residual ethanol with 10 or 20 μl pipette. Make sure no ethanol is remaining in wells (*see* **Note 30**). Keep plate on magnet.
12. Air-dry on magnet for about 8 min. Carefully check when the beads are dry. The best indicator is lack of shininess and appearing cracks (*see* **Note 31**).
13. Remove plate from magnet.
14. Add 13.1 μl ddH_2O to each well and mix by pipetting up and down until all beads are eluted (*see* **Note 32**).
15. Incubate for 2 min at room temperature.
16. Place plate back on magnet and incubate for 2 min or until the supernatant is completely clear.
17. Transfer 11 μl to a new 48-/96-well plate.
18. Safe stopping point: Plate can be stored at −20 °C for up to 6 months.

3.8 Quality Control of cDNA

Quality control includes Qubit assay for cDNA concentration measurement and Bioanalyzer for quality check of cDNA.

1. Perform Qubit dsDNA High Sensitivity assay for every sample and check DNA concentration (*see* **Note 33**, according to manufacturer's protocol).
2. Dilute samples to 1 ng/μl and run them on Bioanalyzer DNA HS chip (Fig. 3, *see* **Note 34**, according to manufacturer's protocol).

3.9 Library Preparation and Pooling

Library preparation is performed according to Nextera XT workflow with following modifications:

1. Sample input: 1.25 μl cDNA (normalized to 0.4 ng/μl, total of 0.5 ng).
2. Only 1/4 of reaction/ indices volume is used (*see* **Note 35**).
3. Tagmentation time: 10 min.

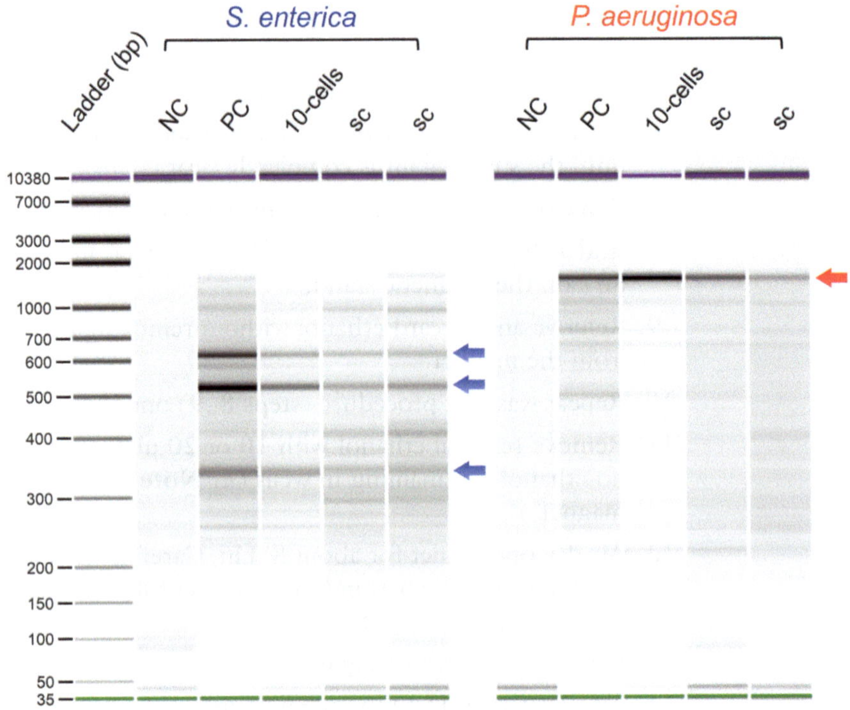

Fig. 3 Representative cDNA profiles of *Salmonella* and *Pseudomonas* after MATQ-seq. Electrophoretic run of the cDNA libraries using Bioanalyzer DNA High Sensitivity measuring a gradient from positive control (PC, 50 pg total RNA of *Salmonella* or *Pseudomonas* used as a spike-in) down to single cell (sc). Arrows indicate species-dependent characteristic bands for *Salmonella* (blue) and *Pseudomonas* (red) which are obtained independent on cell input. The presence of these patterns is used as a quality control before proceeding with library preparation. NC, negative control. 10-cells, 10-pooled bacteria. (Data were obtained from [14])

4. PCR cycles: 13.
5. Quality control of libraries: follow QC of cDNA (*see* Subheading 2.8).
6. Determine average base pair size according to Bioanalyzer profile of cDNA (*see* **Note 36, 37**).
7. Calculate molarity of libraries:

$$\text{concentration in nM} = \frac{\left(\text{concentration in } \frac{\text{ng}}{\mu\text{l}}\right) \times 10^6}{\left(660 \frac{\text{g}}{\text{mol}} \times \text{average library size in bp}\right)}$$

   660 g/mol is the average molecular weight of a nucleotide pair.
8. Dilute libraries to 5 nM in ddH_2O and pool samples equimolar (*see* **Note 38**).
9. Quality control of pool: follow QC of cDNA (*see* Subheading 2.8).

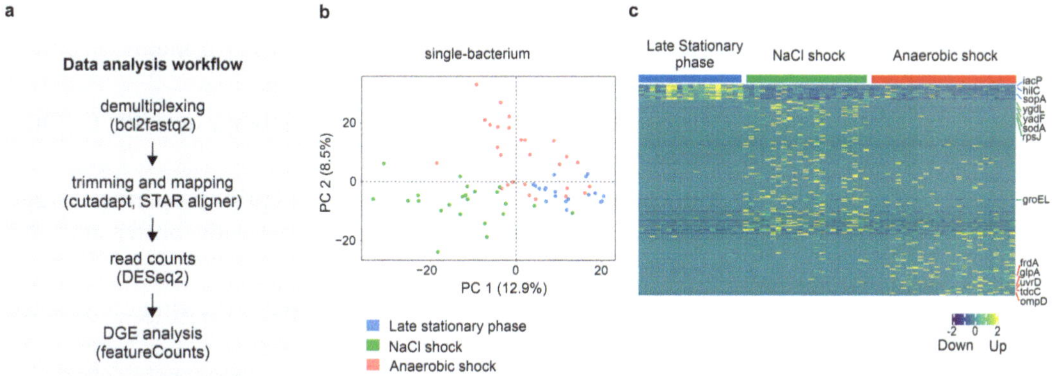

Fig. 4 Data analysis pipeline and visualization of scRNA-seq data obtained from single *Salmonella* cells under different growth conditions. (**a**) Workflow of data analysis pipeline for single-cell data sets. DGE: differential gene expression. (**b**) Principal component analysis (PCA) and projection in the first two dimensions shows growth-condition-specific distinction of single bacteria. Each dot represents a single bacterium color-coded according to the growth condition. (**c**) Heatmap showing RNA expression profile of DE genes between bacteria. Data are arranged according to three growth conditions and each column represents a single bacterium. A selection of growth-specific genes is highlighted. (Modified from [14] with permission from Nat. Microbiol.)

3.10 Sequencing and Demultiplexing

Sequencing procedures according to Illumina protocols using following parameters:

1. Sequencing is performed in paired-end mode—2 × 50 (NovaSeq) or 2 × 75 cycles (NextSeq 500) (*see* **Note 39**, **40**).
2. PhiX Spike-in: 1% (*see* **Note 41**).
3. Sequencing depth: at least 20 M reads per single cell.
4. Demultiplexing of raw sequencing data is performed with bcl2fastq2 v2.20.0.422 (Illumina).
5. Data analysis after demultiplexing includes quality control (FastQC), adaptor trimming (cutadapt), and mapping (STAR aligner). Differential gene expression analysis is performed using DESeq2 package and read counts for each gene are determined using featureCounts algorithm (Fig. 4, [14]). Thereby about 93% of mapped reads account for ribosomal as well as transfer RNA.

4 Note

1. High cleaning standards are required for every step starting with lysis buffer preparation: certified DNAse/RNase filter tips, RNase removal procedures before and after sample preparation, and frequent decontamination of work space using 70% ethanol and UV-light. Dedicated pipettes and distinct PCR cyclers for RT and PCR steps are recommended. Change gloves as soon as they get in touch with potential contaminated surfaces.

2. For antibiotic plates add a magnetic stir bar before autoclaving. After adding chloramphenicol, put bottle on magnetic stirrer for 2 min to ensure proper mixing.
3. Microseal “F” films are only used for storage at −80 °C. Otherwise always use microseal “B” films.
4. Autoclave Tris–HCl, NaCl and glycerol prior preparing storage buffer. Perform sterile filtration of the ready buffer and add lysozyme afterwards.
5. Aliquots are required to prevent any freeze thawing cycles since lysozyme efficiency is directly affected.
6. Use HPLC-purified primers to ensure high purity.
7. Make sure AMPure XP beads are at room temperature and mixed vigorously before usage.
8. Prepare gel according to manufacturer’s protocol. Storage at 4 °C and protected from light.
9. Dilute PhiX to 1 or 2 nM according to manufacturer’s protocol. The dilution factor depends on the used sequencing platform.
10. Add mineral oil very slowly and carefully along the edge of the tube. Prevent any mixing of culture with oil and ensure there is a proper phase separation visible.
11. Plates containing lysis buffer should be prepared freshly at the day of experiment.
12. Omit left and right row of 48-well plate because of potential evaporation due to small frame.
13. In order to streamline the sorting process using FACS, the instrument should be started and set-up before the bacterial cultures are ready. This simplifies the sorting procedure and enables fast processing of cells.
14. The FACS should be equipped with a cooling system for the input as well as output source.
15. For the sorting process, the samples have to be diluted again in 1 × PBS. A high concentration of RNAlater leads to artifacts and shifted scatter plots. We recommend a dilution factor of at least 1:10. The dilutions should be prepared right before sorting.
16. As a valid quality control for the sorting procedure, a fluorescent-tagged sample should always be included in any FACS experiment.
17. For the evaluation and establishment of new lysis protocols, a gradient of sorted cells (e.g., from 100 cells down to single cell) is useful.

18. Storage up to 6 months has no impact on the cDNA quantity as well as quality.
19. For the pipetting steps, a dedicated PCR hood is recommended to protect the samples from possible contamination sources. The cleanup of the cDNA, QC steps as well as the library preparation should be conducted at a second dedicated workstation which does not require a PCR hood.
20. Unless otherwise stated, reagents for each PCR step can be prepared as a mix. Add always highest volume first.
21. Pipetting steps between PCR steps have to be performed on cooling block and samples should be processed fast. Required reagents except of enzymes can be thawed on ice for 30–60 min before each PCR step. We observed a direct correlation between cDNA integrity and processing time. PCR cycler should be preheated to fasten the PCR steps. Lid temperature has to be above 100 °C.
22. Primer mix is sufficient for 20 reactions.
23. Preparation of master mix is highly recommended since several volumes are below 0.2 μl.
24. Procedure can be shortened by taking plate out directly after last step of PCR II before the cycler cools down to 4 °C.
25. T4 DNA Polymerase as well as RNase H/ If is very viscous. Pipette with special care and check for successful pipetting by eye.
26. For adding the PCR mix, a multichannel pipette is recommended to accelerate the procedure.
27. Instead of running all four replicates, the reaction optionally can be split directly after Second-Strand Synthesis and only a quarter of the total reaction is further processed. Remaining reaction can be stored at −20 °C as a backup. In that case, the reaction volume for the PCR amplification has to be adjusted accordingly.
28. Optional: Cleanup procedure can be performed using multichannel pipettes.
29. Optional: Incubation can be performed on a shaker. Handle plate with care and seal it properly before putting on shaker. This can improve the binding of cDNA to beads which leads in higher yield.
30. It is really crucial that remaining ethanol completely evaporated before the cDNA is eluted with ddH_2O. Otherwise, the elution is extremely inefficient and further processing of cDNA is not possible.

31. Depending on how careful residual ethanol was removed, the air-drying might take up to 15 min. Check frequently if cracks appear and if so, proceed straight with elution. Otherwise, the elution might be hindered and result in low yield.
32. Make sure all beads are well resuspended which results in a homogenous brown solution. If drops stick to the side, spin down plate quickly before proceeding.
33. Typical cDNA concentrations for single cells are in the range of 0.5 to 3 ng/μl.
34. The profile of the cDNA samples is highly species dependent and should always be determined by a positive control (spike-in of total RNA) or 100-cell pool.
35. In order to reduce the costs per library, the Nextera XT protocol is performed with only ¼ of reagents. An overall impact on the library quality or performance was not observed.
36. For the determination of the average library size, the Bioanalyzer region table should be set from 150 to 1500 bp. This allows a precise determination of the pool molarity and prevents over-clustering.
37. A precise pooling requires the exact average library size as well as the concentration of the libraries. There are several options available for the determination of both parameters. Since the concentrations are in the range of 5–20 ng/μl, high sensitivity assays are required. For the concentration measurement, Qubit dsDNA High Sensitivity assay gives very accurate results for cDNA, libraries as well as pool. The average library size can be checked on various instruments based on capillary electrophoresis like Bioanalyzer, Tape-Station, or Fragment-Analyzer. For less precise pooling or larger experiments, the average library sizes can also be checked for a subset of samples and estimated accordingly.
38. Equimolar pooling can be simply conducted by pooling same volume of each normalized library.
39. The sequencing can be performed on non-patterned and patterned Illumina flow-cell systems, and essential differences were not observed in data comparison.
40. In case the flow cell clustering is not ideal, the tagmentation time can be extended which results in higher fragmentation rate and therefore smaller fragments. However, this can also lead to the exclusion of smaller fragments which primarily origin from sRNAs. If the average library size is not higher than 900 bp, the flow cell clustering should not be an issue.
41. For the QC of sequencing run, a spike-in of 1% PhiX is highly recommended.

References

1. Ackermann M (2015) A functional perspective on phenotypic heterogeneity in microorganisms. Nat Rev Microbiol 13(8):497–508. https://doi.org/10.1038/nrmicro3491
2. Beebout CJ, Eberly AR, Werby SH, Reasoner SA, Brannon JR, De S, Fitzgerald MJ, Huggins MM, Clayton DB, Cegelski L, Hadjifrangiskou M (2019) Respiratory heterogeneity shapes biofilm formation and host colonization in Uropathogenic Escherichia coli. MBio 10(2). https://doi.org/10.1128/mBio.02400-18
3. Yannarell SM, Velickovic D, Anderton CR, Shank EA (2021) Direct visualization of chemical cues and cellular phenotypes throughout Bacillus subtilis biofilms. mSystems 6(6): e0103821. https://doi.org/10.1128/mSystems.01038-21
4. Lyu Z, Yang A, Villanueva P, Singh A, Ling J (2021) Heterogeneous flagellar expression in single *Salmonella* cells promotes diversity in antibiotic tolerance. mBio 12:e0237421. https://doi.org/10.1128/mBio.02374-21
5. Jones EC, Uphoff S (2021) Single-molecule imaging of LexA degradation in Escherichia coli elucidates regulatory mechanisms and heterogeneity of the SOS response. Nat Microbiol 6(8):981–990. https://doi.org/10.1038/s41564-021-00930-y
6. Merida-Floriano A, Rowe WPM, Casadesus J (2021) Genome-wide identification and expression analysis of SOS response genes in Salmonella enterica Serovar Typhimurium. Cells 10(4):943. https://doi.org/10.3390/cells10040943
7. Roche B, Bumann D (2021) Single-cell reporters for pathogen responses to antimicrobial host attacks. Curr Opin Microbiol 59:16–23. https://doi.org/10.1016/j.mib.2020.07.013
8. Tanay A, Regev A (2017) Scaling single-cell genomics from phenomenology to mechanism. Nature 541(7637):331–338. https://doi.org/10.1038/nature21350
9. Imdahl F, Saliba AE (2020) Advances and challenges in single-cell RNA-seq of microbial communities. Curr Opin Microbiol 57: 102–110. https://doi.org/10.1016/j.mib.2020.10.001
10. Avital G, Avraham R, Fan A, Hashimshony T, Hung DT, Yanai I (2017) scDual-Seq: mapping the gene regulatory program of Salmonella infection by host and pathogen single-cell RNA-sequencing. Genome Biol 18(1): 200. https://doi.org/10.1186/s13059-017-1340-x
11. Betin V, Penaranda C, Bandyopadhyay N, Yang R, Abitua A, Bhattacharyya RP, Fan A, Avraham R, Livny J, Shoresh N, Hung DT (2019) Hybridization-based capture of pathogen mRNA enables paired host-pathogen transcriptional analysis. Sci Rep 9(1):19244. https://doi.org/10.1038/s41598-019-55633-6
12. Kuchina A, Brettner LM, Paleologu L, Roco CM, Rosenberg AB, Carignano A, Kibler R, Hirano M, RW DP, Seelig G (2021) Microbial single-cell RNA sequencing by split-pool barcoding. Science 371(6531):eaba5257. https://doi.org/10.1126/science.aba5257
13. Blattman SB, Jiang W, Oikonomou P, Tavazoie S (2020) Prokaryotic single-cell RNA sequencing by in situ combinatorial indexing. Nat Microbiol 5(10):1192–1201. https://doi.org/10.1038/s41564-020-0729-6
14. Imdahl F, Vafadarnejad E, Homberger C, Saliba AE, Vogel J (2020) Single-cell RNA-sequencing reports growth-condition-specific global transcriptomes of individual bacteria. Nat Microbiol 5(10):1202–1206. https://doi.org/10.1038/s41564-020-0774-1
15. Sheng K, Cao W, Niu Y, Deng Q, Zong C (2017) Effective detection of variation in single-cell transcriptomes using MATQ-seq. Nat Methods 14(3):267–270. https://doi.org/10.1038/nmeth.4145
16. Zong C, Lu S, Chapman AR, Xie XS (2012) Genome-wide detection of single-nucleotide and copy-number variations of a single human cell. Science 338(6114):1622–1626. https://doi.org/10.1126/science.1229164
17. Kroeger C, Colgan A, Srikumar S, Handler K, Sivasankaran SK, Hammarlof DL, Canals R, Grissom JE, Conway T, Hokamp K, Hinton JC (2013) An infection-relevant transcriptomic compendium for Salmonella enterica Serovar Typhimurium. Cell Host Microbe 14(6): 683–695. https://doi.org/10.1016/j.chom.2013.11.010

Chapter 5

Full-Length Single-Cell RNA-Sequencing with FLASH-seq

Vincent Hahaut and Simone Picelli

Abstract

The single-cell RNA-sequencing (scRNA-seq) field has evolved tremendously since the first paper was published back in 2009 (Tang et al. Nat Methods 6:377–382, 2009). While the first methods analyzed just a handful of cells, the throughput and performance rapidly increased over a very short time span. However, it was not until the introduction of emulsion droplets methods, such as the well-known kits commercialized by 10x Genomics, that the robust and reproducible analysis of thousands of cells became feasible (Zheng et al Massively parallel digital transcriptional profiling of single cells. Nat Commun 8:14049, 2017). Despite generating data at a speed and a cost per cell that remains unmatched for full-length protocols like Smart-seq (Hagemann-Jensen et al Single-cell RNA counting at allele and isoform resolution using Smart-seq3. Nat Biotechnol 38:708–714, 2020; Picelli et al Smart-seq2 for sensitive full-length transcriptome profiling in single cells. Nat Methods 10:1096–1098, 2013), scRNA-seq in droplets still comes with the drawback of addressing only the terminal portion of the transcripts, thus lacking the required sensitivity for comprehensively analyzing the entire transcriptome.

Building upon the existing Smart-seq2/3 workflows (Hagemann-Jensen et al Single-cell RNA counting at allele and isoform resolution using Smart-seq3. Nat Biotechnol 38:708–714, 2020; Picelli et al Smart-seq2 for sensitive full-length transcriptome profiling in single cells. Nat Methods 10:1096–1098, 2013), we developed FLASH-seq (FS), a new full-length scRNA-seq method capable of detecting a significantly higher number of genes than previous versions, requiring limited hands-on time and with a great potential for customization (Hahaut et al. Lightning Fast and Highly Sensitive Full-Length Single-cell sequencing using FLASH-Seq. http://biorxiv.org/lookup/doi/10.1101/2021.07.14.452217. https://doi.org/10.1101/2021.07.14.452217, 2021). Here, we present three variants of the FS protocol.

Standard FLASH-seq (FS), which builds upon Smart-seq2 developed in the past, is non-stranded and does not use unique molecular identifiers (UMIs) but still remains the easiest method to measure gene expression in a cell population.

FLASH-seq low-amplification (FS-LA) represents the fastest method, which generates sequencing-ready libraries in 4.5 h, without sacrificing performance.

FLASH-seq with UMIs (FS-UMI) builds upon the same principle as Smart-seq3 and introduces UMIs for molecule counting and isoform reconstruction. The newly designed template-switching oligonucleotide (TSO) contains a 5-bp spacer, which allows the generation of high-quality data while minimizing the amount of strand-invasion artifacts.

Key words FLASH-seq, FLASH-seq low-amplification, FLASH-seq with UMI, RNA-seq, single cell, full-length, automation, high-throughput

Raffaele A. Calogero and Vladimir Benes (eds.), *Single Cell Transcriptomics: Methods and Protocols*, Methods in Molecular Biology, vol. 2584, https://doi.org/10.1007/978-1-0716-2756-3_5,

1 Introduction

Performing scRNA-seq in emulsion droplets by using 10X Genomics technology is nowadays routine practice for many research groups around the world. The system is robust, reproducible, and offers an unparalleled throughput at a very low cost per cell. The main limitation remains that available kits enable the characterization of either the 3′- or the 5′-end of each RNA, at least when using short read sequencing. This is generally sufficient for the investigation of cellular heterogeneity and the identification of population sub-structures. However, sequencing full-length transcripts would also allow the analysis of single nucleotide polymorphisms (SNPs) and enable splice isoform reconstruction, the characterization of transcriptional start sites (TSSs), as well the identification of mono-allelic and imprinted genes [6–8].

Smart-seq is probably the most well known and robust among the full-length RNA-seq methods and has now reached the third generation with the recent publication of Smart-seq3 [3]. Although differences exist between Smart-seq workflows, they all rely on the SMART technology (switching mechanism at the 5′-end of the RNA transcript), a peculiar property of the Moloney murine leukemia virus reverse transcriptase (MMLV-RT), which couples reverse transcription (RT) and template switching (TS) in a single reaction [9]. Template switching represents the ability of the MMLV-RT to introduce a few untemplated nucleotides, often 2-5 cytosines, upon reaching the 5′-end of the RNA template during RT reaction. These extra nucleotides provide an entry point for a TSO carrying three riboguanosines at its 3′-end. The bond cytosines-riboguanosines (deoxyribonucleotide-ribonucleotide) is key for a stable annealing and ensures that the MMLV-RT can "switch template," synthesizing a complementary DNA strand using the TSO as template. Therefore, TS and oligo-dT priming introduce known sequences at both ends of each transcript, allowing PCR amplification with a single pair of primers.

In FLASH-seq, we combined these two steps in a single RT-PCR reaction, which resulted in a faster, more efficient, and easy-to-use protocol, also thanks to a novel combination of enzymes, buffers, and additives [5]. Like Smart-seq2/3, FLASH-seq can be performed entirely with off-the-shelf reagents keeping the costs around $1/cell. We have shown that FLASH-seq can be easily miniaturized and automated by using the I.DOT (Dispendix) nanodispenser and the Fluent 780 workstation (Tecan) [5]. Several other instruments are most likely going to result in high-quality data; therefore, the choice is rather based on what is available in each lab. We always recommend choosing an automated solution whenever possible, in order to improve reproducibility, increase processing speed, minimize reagent waste and, ultimately, reduce personnel and reagents' cost. However, if automated solutions are not available in the lab, then all FLASH-seq steps can be carried out in larger volumes by using manual pipettes and still returning data of satisfactory quality.

Like all Smart-seq methods, FLASH-seq also requires the cDNA to be further prepared for sequencing, a step carried out by a hyperactive Tn5 transposase. We leave it to the reader to choose which Tn5 (commercial or in-house produced) better suits the experimental needs. Commercial alternatives like the Nextera® XT kit (Illumina) are a good option for inexperienced users due to their robustness and consistency but come with a higher price tag. Alternatively, the plexWell kit (seqWell) offers the same ease of use and performance as the Nextera® XT kit but is significantly faster (especially when processing multiple plates at once) and has a lower price tag. Finally, the cheapest alternative remains the in-house Tn5 transposase, with all its advantages and disadvantages. Protein production might be challenging to optimize at first, but once it has been mastered and tagmentation protocols have been optimized [10, 11], it offers unparalleled flexibility and virtually unlimited room for customization.

Both the commercial and in-house Tn5 can be interchangeably used in the protocols below, as they will generate data of comparable quality. However, in the case of FS-UMI, we recommend using a commercial alternative to ensure higher reproducibility due to the lower batch-to-batch variation, making experiments more comparable even if performed months apart from each other.

A common limitation shared among Smart-seq2/3 and FLASH-seq is that all use an oligo dT-based strategy for priming exclusively polyadenylated RNAs, thus neglecting other potentially relevant RNA species such as microRNAs (miRNAs), piwi-interacting RNAs (piRNAs), and non-polyadenylated long non-coding RNAs (lncRNAs). If retaining strand-specificity is key for the experiment at hand, like when analyzing overlapping genes transcribed by opposite strands, then FS-UMI should be preferred, as the generated UMI reads will be stranded. Although technically easier and faster, both FS and FS-LA will not retain strand information, similar to what happens with Smart-seq2.

We believe that FLASH-seq has the potential to become the tool of choice when looking for an efficient, robust, modular, affordable, and automation-friendly full-length scRNA-seq protocol.

2 Materials

- All solutions should be prepared by using RNase- and DNase-free water and analytical grade reagents.
- It is very important to work as cleanly as possible, especially when the samples are still at the RNA stage.
- All working surfaces, racks, and pipettes should be wiped with 0.5% sodium hypochlorite (NaClO) followed by a rinse with nuclease-free water. Do not neglect this last step for pipettes, to avoid corrosion and damage over time.
- Use separate pre- and post-PCR working areas, in order to avoid contaminations. Be especially careful when handling the index primers used in the final enrichment PCR step.
- All reagents except enzymes can be thawed at room temperature. All master mixes can be briefly vortexed after preparation. However, do not vortex the enzyme stock solutions but rather mix them by (gentle) inversion.

Standard FLASH-seq (FS)

2.1 Cell Lysis Mix

1. 10% v/v Triton X-100. Store at +4 °C (*see* **Note 1**).
2. Pre-mixed dNTP solution (25 mM each). Store at −20 °C.
3. Recombinant RNase inhibitor (40 U/μL). Store at −20 °C.
4. Betaine (5 M). Store at +4 °C.
5. Dithiothreitol (DTT, 100 mM; part of the SuperScript™ IV kit). Store at −20 °C.
6. dCTP (100 mM). Store at −20 °C.
7. Template-switching oligonucleotide, FS-TSO (5′ Bio- AAG CAGTGGTATCAACGCAGAGTAC rGrGrG-3′, 100 μM). "Bio" = biotin. Store working aliquots at −20 °C and stock tubes at −80 °C (*see* **Note 2**).
8. SMART $dT_{30}VN$ oligonucleotide (5′Bio-AAGCAGTGGTAT CAACGCAGAGTACT $_{30}$VN-3′, 100 μM). "Bio" = biotin. Store at −20 °C.
9. Nuclease-free water. Store at room temperature.

2.2 RT-PCR Mix

1. Superscript™ IV kit: dithiothreitol (DTT, 100 mM). Store at −20 °C.
2. Superscript™ IV kit: Superscript™ IV reverse transcriptase (200 U/μL). Store at −20 °C (*see* **Note 3**).
3. Betaine (5 M). Store at +4 °C.
4. Magnesium chloride ($MgCl_2$, 1 M). Store at +4 °C.

5. Recombinant RNase inhibitor (40 U/μL). Store at −20 °C.
6. KAPA HiFi HotStart ReadyMix (2×). Store at −20 °C.
7. Nuclease-free water. Store at room temperature.

2.3 Magnetic Beads Preparation

Here, we exclusively use a Sera-Mag SpeedBeads™ solution containing 18% w/v polyethylene glycol (PEG), prepared as described below (*see* **Note 4**). For 50 ml of working bead solution:

1. Withdraw 1 ml of Sera-Mag SpeedBeads™ suspension (carboxyl magnetic beads, hydrophilic, 5% suspension) and transfer it into a 1.5 ml tube.
2. Pellet the beads by placing the tube on a magnetic stand, wait until the solution is clear, and discard the supernatant.
3. Add 1 ml 10 mM Tris–HCl pH 8.0 and 1 mM EDTA (TE buffer) and resuspend the beads off the magnet. Pellet the beads again, wait until the solution is clear, discard the supernatant, and repeat one more time.
4. Pellet the beads once more, wait until the solution is clear, discard the supernatant, and resuspend off the magnet with 0.9 ml TE buffer.
5. In a beaker, mix 2.92 g NaCl, 500 μL Tris–HCl pH 8.0 (1 M), 100 μL EDTA (500 mM), and 9.5 g polyethylene glycol (PEG, MW = 8000).
6. Dissolve completely the reagents with the help of a magnetic stirrer and by heating the mix to 37 °C.
7. Add 50 μL Tween-20 (10% v/v, (*see* **Note 5**)), 250 μL sodium azide (NaN_3, 10% w/v (*see* **Note 6**)), and the cleaned-up beads prepared in **step 4** to the PEG solution.
8. Adjust the volume to 50 ml with nuclease-free water. Store at +4 °C. Do not freeze (*see* **Note 7**).

2.4 Library Preparation with Commercial Illumina Reagents

1. Nextera® XT DNA sample preparation kit. Store at −20 °C.
2. Nextera® XT index kit v2. Store at −20 °C (*see* **Note 8**).

2.5 Library Preparation with In-House Tn5 Transposase

In-house Tn5 transposase (~20 μM stock solution, diluted 1:10 with storage buffer according to [10]), pre-loaded with the following oligonucleotides: Tn5MErev: 5′-[phos]CTGTCTCTTATACACATCT-3′. "Phos" = phosphate; this oligonucleotide should be annealed either with Tn5ME-A: 5′-GTCTCGTGGGCTCGGAGATGTGTATAAGAGACAG-3′ or with Tn5ME-B: 5′-TCGTCGGCAGCGTCAGATGTGTATAAGAGACAG-3′.

Store working aliquots at −20 °C. Store stock solution at −80 °C to preserve transposase activity in the long term (*see* **Note 9**).

1. 5× TAPS-$MgCl_2$ buffer: 50 mM TAPS-NaOH pH 7.3 at 25 °C, 25 mM $MgCl_2$. Adjust the pH as indicated above with sodium hydroxide (NaOH). Filter with 0.22 μm filters and store at +4 °C.
2. Dimethylformamide (98–100% v/v): store at room temperature in a ventilated cabinet and away from light (*see* **Note 10**).
3. 0.2% w/v SDS: store at room temperature (*see* **Note 11**).
4. KAPA HiFi non-HotStart kit: includes KAPA HiFi DNA polymerase (1 U/μL), KAPA HiFi high-fidelity buffer (5×), and dNTP mix (10 mM each). Store at −20 °C.
5. Nuclease-free water. Store at room temperature.
6. Additional index primer sets. Store at −20 °C (*see* **Note 12**).

2.6 Sample QC and Sequencing

1. Qubit™ dsDNA high-sensitivity assay kit. Store at +4 °C protected from light.
2. Qubit™ assay tubes.
3. Quant-iT™ PicoGreen™ dsDNA assay kit. Store at +4 °C protected from light.
4. Nunc™ F96 MicroWell™ polystyrene plate, black.
5. Agilent high-sensitivity DNA assay. Store at +4 °C.
6. Sera-Mag SpeedBeads™ working solution (for details about preparation, *see* Subheading 2.3). Store at +4 °C.
7. 80% v/v ethanol. Store at room temperature.
8. Nuclease-free water. Store at room temperature.
9. NextSeq™ 550 high output v2.5 kit (75 or 150 cycles), NextSeq™ 1000/2000 (100 or 200 cycles), or NovaSeq 6000 (100 or 200 cycles), depending on sample throughput, desired yield, and required read mode. Other instruments and kits can also be used. Follow the manufacturer's instructions regarding storage.

2.7 Instruments and Consumables

Depending on the throughput, the protocol can be carried out in single PCR tubes, 96- or 384-well plates. Below we list the consumables required for an experiment in 384-well plates.

1. Fully skirted 384-well plates (*see* **Note 13**).
2. Aluminum seal and adhesive plastic foil (*see* **Note 14**).
3. Plate centrifuge.
4. Thermocycler with 384-well block and heated lid set to 105 °C for all programs.

5. I.DOT (Dispendix) nanodispenser. Other (contact as well as non-contact) nanodispenser can also be used.
6. Fluent 780 workstation (Tecan). Other liquid handling robots can also be used.
7. 384 post-magnet plate (Alpaqua).
8. FACSAria™ Fusion (BD Biosciences). Other cell dispensers can also be used.
9. Agilent 2100 Bioanalyzer.
10. Nextseq™ 550/1000/2000 or Novaseq™ 6000 Sequencing Systems, depending on sample throughput and desired read output. Other Illumina instruments can also be used.

3 Methods

3.1 Cell Lysis Mix Preparation

1. For each sample, prepare 1 μL of the following cell lysis mix: 0.020 μL Triton X-100, 0.240 μL dNTP Mix, 0.018 μL SMART $dT_{30}VN$ oligonucleotide, 0.030 μL recombinant RNase inhibitor, 0.012 μL DTT, 0.092 μL FS-TSO, 0.090 μL dCTP, 0.200 μL betaine, Add nuclease-free water reach the final indicated volume, see above.
2. Mix well and gently spin down.
3. Dispense the cell lysis mix into a 384-well plate.
4. If used immediately, keep the plate on ice until needed. The cell lysis mix can be prepared and dispensed on plates several months in advance without any performance loss. The plates can be stored at −20 °C until needed. Avoid multiple freezing-thawing cycles.
5. Collect the cells by FACS using a FACSAria™ Fusion, preferably with a larger nozzle (85 μm or, even better, 100 μm) to minimize shearing forces upon sorting (*see* **Note 15**).
6. Once the sorting is completed, seal the plate with aluminum seal and snap-freeze it at −80 °C or by placing it on dry ice, especially if not proceeding immediately with RT-PCR. Lysed cells can be stored in these conditions for several months without appreciable decrease in RNA quality. The plate should never undergo freeze-thaw cycles for any reason.

3.2 RT-PCR Mix Preparation and RT-PCR Reaction

1. For each sample, prepare 4 μL of the following RT-PCR mix: 0.238 μL DTT, 0.800 μL betaine, 0.046 μL $MgCl_2$, 0.096 μL recombinant RNase inhibitor, 0.050 μL Superscript™ II reverse transcriptase, 2.500 μL KAPA HiFi HotStart ReadyMix, Add nuclease-free water reach the final indicated volume, see above.
2. Briefly vortex and gently spin down the mix. Keep it on ice until needed.

3. Take the plate out of the −80 °C freezer and leave it for a couple of minutes at room temperature to thaw.
4. Briefly spin down to collect eventual drops that might have condensed on the lid.
5. Place the plate on a thermocycler block and perform cell lysis and mRNA denaturation at 72 °C for 3 min.
6. Remove from the thermocycler and keep it on ice for a couple of minutes to cool. Briefly spin down to collect the lysate at the bottom.
7. Dispense 4 μL of RT-PCR Mix into each well. The final volume is now 5 μL.
8. Seal the plate with adhesive transparent film, vortex, spin down, place it in a thermocycler, and carry out the RT-PCR reaction: 50 °C for 60 min, 98 °C for 3 min, then cycles of (98 °C for 20 s, 67 °C for 20 s, 72 °C for 6 min), 15 °C hold. The number of cycles should be adjusted according to the cell type used in the experiment. We recommend 18–19 cycles for HEK 293 T cells and 21 cycles for hPBMC. As a rule of thumb, start with 1–2 fewer PCR cycles compared to Smart-seq2. Avoid over-amplification of the cDNA.
9. It is safe to stop here and store the PCR product in a −20 °C freezer until needed.

3.3 Magnetic Beads Cleanup After Preamplification

Remove the Sera-Mag SpeedBeads™ working solution from the +4 °C storage and equilibrate it at room temperature for 15 min.

1. **OPTIONAL**: We generally add 10 μL of nuclease-free water into each well with the I.DOT to increase the reaction volume and make sample handling a bit easier when the cleanup is performed on the Fluent 780 workstation.
2. Add a 0.8× ratio of Sera-Mag SpeedBeads™ working solution (0.8:1 = beads: cDNA) and pipet up and down 20–30 times to mix.
3. Incubate off the magnetic stand for 5 min at room temperature.
4. Place the plate on the magnetic stand and leave it there for 5 min or until the solution appears clear.
5. Carefully remove the supernatant without disturbing the beads. Proceed immediately to the next step.
6. Remove the plate from the magnetic stand, add 15 μL nuclease-free water, and mix well by pipetting or vortexing to resuspend the beads.
7. Incubate 2 min off the magnetic stand.

8. Place the plate back on the magnetic stand and incubate for 2 min or until the solution appears clear.
9. Carefully remove 14 μL of the supernatant trying to minimize the bead carryover and transfer it to a new plate. It is safe to stop here and store the cDNA in a −20 °C freezer until needed (*see* **Note 16**).
10. Check the cDNA quality on the Agilent Bioanalyzer instrument. Follow the instructions as described in the high sensitivity DNA chip user manual. FLASH-seq yields generally fragments with an average size of 1.8–2.2 Kb and a near-to-complete absence of fragments <500 bp.

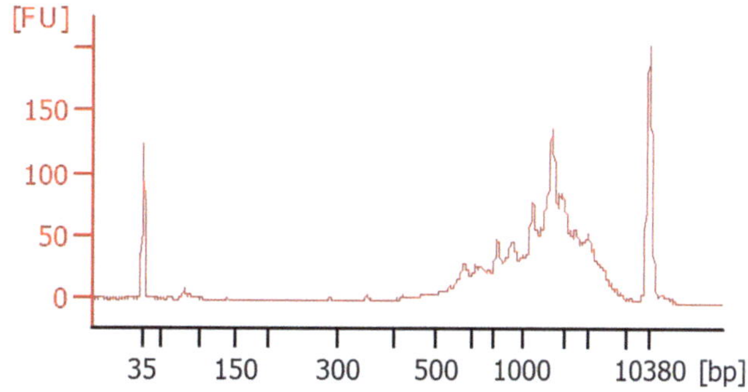

Example of amplified cDNA from a single hPBMC (21 PCR cycles)

3.4 cDNA Quantification and Sample Normalization

The preamplified cDNA is generally too concentrated and needs to be diluted before proceeding with tagmentation.

1. Allow the Quant-iT PicoGreen reagent to reach room temperature before opening the vial. PicoGreen is light sensitive; while thawing, wrap in aluminum foil.
2. Prepare a 1× working solution of TE using 20× TE (supplied) and nuclease-free water.
3. Prepare a 1:400 dilution of PicoGreen solution and always use a plastic vessel. Do not use glass, as PicoGreen may adsorb to glass.
4. Prepare the standard curve using the Lambda DNA standard (supplied with the PicoGreen kit at a concentration of 100 ng/μL) and 1× TE in 8 tubes, as below. The stock tubes can be used multiple times, so keep any leftover in the fridge at +4 °C between experiments.
5. Vortex well and spin down the DNA standards before every use. Not vortexing thoroughly the standards is going to negatively affect the standard curve and your readings! Serial dilutions should be prepared as shown in the table below.

Tube no.	Content	Concentration	Final volume
1	90 μL TE + 10 μL Lambda DNA stock	10 ng/μL	100 μL
2	50 μL from Tube 1 + 50 μL TE	5 ng/μL	100 μL
3	50 μL from Tube 2 + 50 μL TE	2.5 ng/μL	100 μL
4	50 μL from Tube 3 + 50 μL TE	1.25 ng/μL	100 μL
5	50 μL from Tube 4 + 50 μL TE	0.625 ng/μL	100 μL
6	50 μL from Tube 5 + 50 μL TE	0.3125 ng/μL	100 μL
7	50 μL from Tube 6 + 50 μL TE	0.15625 ng/μL	100 μL
8	TE only	blank	–

6. Prepare the PicoGreen solution by pipetting 0.5 μL PicoGreen dye +99.5 μL 1× TE for each sample. Vortex to mix.
7. Pipette 1 μL of each of the 7 standards +1 Blank into a black, flat-bottom Nunc™ F96 MicroWell™ plate. Place the standards on one column.
8. Pipet 1 μL of your samples into the center of each well of the Nunc™ F96 MicroWell™ polystyrene plate.
9. Add 99 μL PicoGreen + TE mix into every well. There is no need to mix.
10. Cover the plate with the provided plastic (transparent) lid to prevent possible contaminations.
11. Allow 2 min for the dye to bind the DNA. Protect from light. For optimal results, read the plate within 1 h.
12. Use a plate reader to measure fluorescence (excitation: 485 nm; emission: 530 nm; read from top; endpoint reading).
13. Prepare a normalization plate by adding 1 μL purified cDNA and nuclease-free water to a final concentration of 150 pg/μL using the PicoGreen readings.

3.5 Library Preparation

This step can be carried out by using either the commercially available Nextera XT kit or the in-house Tn5 transposase. Both protocols are described below.

Indexing primers can be purchased from Illumina or ordered from your local oligo manufacturer.

3.5.1 Library Preparation with the Nextera® XT Kit

1. Remove the Amplicon Tagment Mix (ATM) and the Nextera® PCR mix (NPM) from the −20 °C storage and place them on ice. Remove the Tagment DNA buffer (TD) and the adaptor index plate from the −20 °C storage and keep them at room

temperature. Remove the neutralization buffer (NT) from the +4 °C storage and keep it at room temperature.

2. For each sample, prepare 1.25 μL tagmentation mix: 0.25 μL ATM and 1 μL TD.
3. Briefly vortex and gently spin down the mix. Keep it on ice until needed.
4. Transfer 1 μL of normalized cDNA into a new 384-well plate.
5. Add 1.25 μL of tagmentation mix to each well.
6. Seal the plate with adhesive transparent film, vortex, spin down, place it in a thermocycler, and carry out the tagmentation reaction: 55 °C for 8 min, 4 °C hold. Once the reaction is completed, proceed immediately to the next step.
7. Add 0.5 μL NT buffer to each well.
8. Seal the plate with adhesive transparent film, vortex, spin down, and incubate for 5 min at room temperature. Do not put the plate back on ice.
9. Add 1 μL N7xx + S5xx indices from the adapter index plate (5 μM each).
10. Add 1.5 μL NPM to each well.
11. Seal the plate with adhesive transparent film, vortex, spin down, place it in a thermocycler, and carry out the enrichment PCR reaction: 72 °C for 3 min, 95 °C for 30 s, then N cycles of (95 °C for 10 s, 55 °C for 30 s, 72 °C for 30 s), 72 °C for 5 min, 4 °C hold. The number of cycles "N" should be adjusted according to the amount of cDNA used in the tagmentation and the number of cells that are going to be pooled before cleanup. When starting from 150 pg input cDNA, 14 cycles are sufficient even if pooling only 32–48 cells.
12. It is safe to stop here and store the final library in a −20 °C freezer until needed.

3.5.2 Library Preparation with In-House Tn5 Transposase

Please note that the Tn5 amount indicated below is merely a starting point for the tagmentation of 150 pg cDNA. Optimization might be necessary, depending on the specific activity of each batch of Tn5. If needed, reactions can be miniaturized, paying attention to keep the same ratio of cDNA and Tn5, in order to avoid under- or over-tagmentation.

1. Remove the Tn5 transposase and the KAPA HiFi polymerase from the −20 °C storage and place them on ice. Remove the dNTP mix, high-fidelity KAPA HiFi buffer and the adapter index plate from the −20 °C storage, and keep them at room temperature. Remove the TAPS-$MgCl_2$ buffer from the +4 °C storage and keep it at room temperature. Get the SDS and the DMF solutions.

2. For each sample, prepare 3 μL tagmentation mix: 0.8 μL TAPS-$MgCl_2$ buffer, 0.8 μL DMF, 0.125 μL in-house Tn5 transposase, nuclease-free water to volume.
3. Briefly vortex and gently spin down the tagmentation mix. Keep it on ice until needed.
4. Transfer 1 μL normalized cDNA in a new 384-well plate.
5. Add 3 μL tagmentation mix to each well.
6. Seal the plate with adhesive transparent film, vortex, spin down, place it in a thermocycler, and carry out the tagmentation reaction: 55 °C for 8 min, 4 °C hold. Once the reaction is completed, proceed immediately to the next step.
7. Add 1 μL 0.2% SDS solution to each well.
8. Seal the plate with adhesive transparent film, vortex, spin down, and incubate for 5 min at room temperature. Do not put the plate back on ice.
9. Add 2 μL N7xx + S5xx indices from the adaptor index plate (5 μM each).
10. Add 3 μL enrichment PCR mix: 0.2 μL KAPA HiFi DNA polymerase, 0.3 μL dNTP mix, 2 μL high-fidelity buffer, nuclease-free water to volume.
11. Seal the plate with adhesive transparent film, vortex, spin down, place it in a thermocycler, and carry out the enrichment PCR reaction: 72 °C for 3 min, 98 °C for 30 s, then N cycles of (98 °C for 10 s, 55 °C for 30 s, 72 °C for 30 s), 72 °C for 5 min, 4 °C hold. The number of cycles "N" should be adjusted according to the amount of cDNA used for the tagmentation reaction. When starting from 150 pg input cDNA, 14 cycles are sufficient even if pooling only 32–48 cells.
12. It is safe to stop here and store the final library in a –20 °C freezer until needed.

3.6 Pooling and Bead Cleanup of the Final Library

1. Remove the Sera-Mag SpeedBeads™ working solution from the +4 °C storage and equilibrate it at room temperature for 15 min.
2. Pool the entire volume or just part of the tagmentation reaction from each well into a 2 ml LoBind tube. Vortex to mix and briefly spin down.
3. Add a 0.8× ratio of Sera-Mag SpeedBeads™ working solution, vortex thoroughly, and incubate for 5 min off the magnetic stand.
4. Place the tube on the magnetic stand and leave it there for 5 min or until the solution appears clear.
5. Carefully remove the liquid paying attention not to disturb the beads.

6. Add 1 ml 80% w/v ethanol and incubate for 1 min without removing the tube from the magnet.
7. Remove any trace of ethanol and let the bead pellet dry for 2 min (tube always on the magnetic stand and with the lid open!).
8. Remove the tube from the magnetic stand; add 100 μL nuclease-free water and mix well to resuspend the beads.
9. Incubate for 2 min off the magnetic stand.
10. Place the tube back on the magnetic stand and leave it there for 2 min or until the solution appears clear.
11. Carefully remove the supernatant trying to minimize the bead carryover and place it in a new 1.5 ml LoBind tube. This is the final pool that is going to be used for sequencing.

3.7 Sequencing Library QC and Sequencing

1. Use 1 μL of the purified sequencing pool to assess the concentration on a Qubit instrument.
2. Use 1 μL to assess the average size on a high-sensitivity DNA chip.

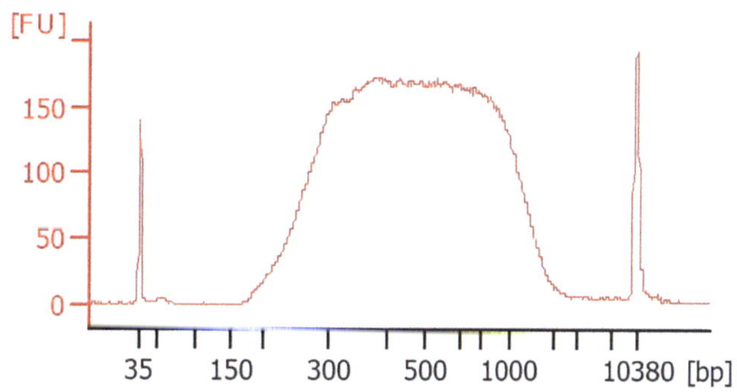

Example of sequencing-ready library (pool of 384 HEK 293 T cells)

3. Single-end 75 or 100 bp mode is generally sufficient but longer read modes or paired-end sequencing can be an option, depending on the question at hand.

4 Data Preprocessing

The data processing will depend on the final goal. The following description provides an example of data preprocessing. We assume that the data were generated using single-end sequencing (75 bp) on an Illumina instrument and are processed on a unix-based machine.

4.1 Prerequisites

Download the following tools. Versions indicated are the ones tested in *Hahaut* et al. [5].

- bcl2fastq (Illumina, v2.20)
- STAR [12] (v2.7.3)
- FeatureCounts [13] (v1.6.5)
- BBMAP [14] (v38.86)
- samtools [15] (v1.9)
- IGV

When working on hundreds of samples, it is recommended to use virtual machines or cloud platforms to process the data, as it will typically require >48 Gb of RAM, >12 cores, and >500 Gb of free space. The footprint of the most time- and resource- consuming step (= read alignment) can be reduced using pseudoalignment tools (i.e., salmon [16] or kallisto [17]), not described in this manuscript.

4.2 Optional: Demultiplexing

Skip this step if the sequencing data are directly provided as separated FASTQ files.

In case samples are sequenced on an illumina sequencer and only the raw sequencing files are provided (.blc2), bcl2fastq can be used for demultiplexing.

```
# 0. Define paths
BASECALL_DIR="/path/to/folder/Data/Intensities/BaseCalls/"
OUTPUT_DIR="/path/to/output_directory/"
SAMPLESHEET="/path/to/demultiplexing_sampleSheet.csv"

# 1. Run bcl2fastq
ulimit -n 10000
bcl2fastq --input-dir $BASECALL_DIR --output-dir $OUTPUT_DIR
--sample-sheet $SAMPLESHEET --create-fastq-for-index-reads --
no-lane-splitting
```

The sample sheet description is provided in the bcl2fastq manual. Alternatively, the Illumina Experiment Manager (IEM) software can be used to generate it.

After demultiplexing, check if any index combinations were missed using:

```
# 2. What are the remaining undemultiplexed indexes?
# Returns the number of reads associated with each index
combination in the Undetermined fastq file.
zcat "$OUTPUT_DIR"/Undetermined_S0_I1_001.fastq.gz | awk -F'
1:N:0:' 'NR%4==1{print $2}' | sort | uniq -c > left_index.txt
sort -k1,1 left_index.txt
```

The top I5 + I7 combinations should be associated with a limited number of unassigned reads. When sequencing on a NextSeq500 instrument, the undetermined index combinations often contain “GGGGGGGG” I5 or I7 indexes. These represent indexes that could not be read by the sequencer (= dark cycles).

Finally, the number of reads per sample can be obtained with:

```
for file in "$OUTPUT_DIR"/*R1*
do
echo $(zcat $file|wc -l)/4|bc
done
```

The read distribution should be relatively uniform. Large variations can indicate sample or library pooling issues.

4.3 Optional: Remove Sequencing Adapter Leftovers

In case sequencing adapter leftovers are observed, bbduk [14] or trimmomatic [18] can be used to trim them.

```
# BBDUK sequencing adapters
# Provided with bbmap
BBDUK_REF="/path/to/BBDUK/adapters.fa"

# Remove left-over adapter (left-side)
$BBDUK -Xmx12g in=sample.fastq.gz out=cleaned.left.fastq t=32
ktrim=l ref="$BBDUK_REF" k=23 mink=7 hdist=1 hdist2=0 tbo
# Remove left-over adapter (right-side)
$BBDUK -Xmx12g in=cleaned.left.fastq out=cleaned.fastq t=32
ktrim=r ref="$BBDUK_REF" k=23 mink=7 hdist=1 hdist2=0 tbo
```

4.4 Data Alignment

4.4.1 Genome Indexing

Read aligners (e.g., STAR or HISAT2 [19]) or pseudo-aligners (e.g., salmon or kallisto) are used to align the data against a reference genome/transcriptome. The following example is demonstrated using STAR.

The first step of the data alignment process is to generate an index for the reference genome. This procedure is generally performed only once.

```
# 0. Define paths
OUTPUTREF="/path/to/output_reference_index/"
FASTA="/path/to/fasta_reference_genome.fa"
GTF="/path/to/genome_annotation.gtf"

# 1. Create the index repository
mkdir $OUTPUTREF
cd $OUTPUTREF

# 2. OPTIONAL: Save the fasta/GTF files to the reference folder
scp $FASTA .
scp $GTF .
```

```
# 3. Genome indexing
# runThreadN: Adapt the number of threads to your device
# sjdbOverhang: Usually set to read length -1
STAR --runThreadN 30 \
--runMode genomeGenerate \
--genomeDir $OUTPUTREF \
--genomeFastaFiles $FASTA \
--sjdbGTFfile $GTF \
--sjdbOverhang 74
```

The reference genome/transcriptome and gene annotation files can be obtained from ENSEMBL, UCSC, or gencode.

4.4.2 Align Reads

```
MYFASTQ="/path/to/fastq_file.fq.gz"

mkdir STAR
STAR --runThreadN 10 --limitBAMsortRAM 20000000000 --genome-
Load LoadAndKeep --genomeDir "$OUTPUTREF" --readFilesIn "$MY-
FASTQ" --readFilesCommand zcat --limitSjdbInsertNsj 2000000 --
outFilterIntronMotifs RemoveNoncanonicalUnannotated --outSAM-
type BAM SortedByCoordinate --outFileNamePrefix STAR/"$ID"_
```

This command will output an alignment file (.bam) and a summary of the alignment results (.log.final.out)

4.4.3 Optional: Filter Reads

Even though they should not influence the rest of the pipeline, unmapped reads (sam flag 4) and secondary alignments (sam flag 256) can be discarded using samtools.

```
samtools view -@ 10 -Sb -F 260 STAR/"$ID"_Aligned.sortedBy-
Coord.out.bam > STAR/"$ID"_Aligned.sortedByCoord.filtered.bam
samtools index STAR/"$ID"_Aligned.sortedByCoord.filtered.bam
```

4.5 Assign Reads to Feature Using Featurecounts

Sequencing libraries are generated by tagmentation, which randomly cut and tag cDNA molecules. The reads generated in FLASH-seq are therefore unstranded (-s 0). Given the mild non-ionic detergent used during cell lysis, nuclear unspliced mRNAs molecules are not expected to be efficiently captured by FLASH-seq. We therefore advise counting reads assigned to each feature using only the exonic features (-t exon).

```
mkdir FEATURECOUNTS

featureCounts -T 1 -t exon -g gene_name --fracOverlap 0.25 -a
"$GTF" -o FEATURECOUNTS/"$ID"_ReadCount.featureCounts.txt
STAR/"$ID"_Aligned.sortedByCoord.filtered.bam
```

4.6 Recommended: Quality Checkups

4.6.1 Gene-Body Coverage (ReSQC)

FLASH-seq is a full-length RNA-sequencing protocol. The gene-body coverage should be uniformly distributed from 5′ to 3′. Deviations can, among others, indicate RNA degradations or reverse transcription issues.

```
# BED12 Reference file.
# This file should contain the same information as the GTF file
used for read alignment / feature assignment.
ReSQCBED="/path/to/annotation.bed"

mkdir ReSQC

geneBody_coverage.py -r "$ReSQCBED" -i STAR/"$ID"_Aligned.
sortedByCoord.filtered.bam -o ReSQC/"$ID"_geneBody.all

Warning: This step is single-threaded (= time-consuming) and
requires a lot of RAM.
```

4.6.2 Read Distribution (ReSQC)

The majority of the reads should be assigned to the gene coding sequences followed by 5′ / 3′ UTRs and intronic sequences. Fewer reads should be assigned to intergenic regions (<10% read tags).

```
read_distribution.py -i STAR/"$ID"_Aligned.sortedByCoord.fil-
tered.bam -r "$ReSQCBED" > ReSQC/"$ID"_readDistribution.txt
```

4.6.3 STAR Mapping Statistics (STAR)

We highly recommend exploring the STAR "Log.final.out" file. It contains a summary of the mapping statistics of each sample.

Important values to monitor include the following:

- Uniquely mapped reads % [70–90%]
- Average mapped length [≃ average read length]
- Mismatch rate per base, % [<1.5%]
- Percentage of reads mapped to multiple loci + too many loci [<20%]
- Percentage of reads unmapped: too many mismatches + too short [<10%]

The thresholds are given as examples and can highly vary from one cell type to another. The best indicator of a good quality sample is a high percentage of uniquely mapped reads.

4.7 Recommended: Processing Multiple Samples

When dealing with multiple samples, it is advisable to run the steps described above in a loop and/or in parallel to speed up the analysis process. Dozens of alternative approaches can be used to perform this operation. The following line describes one of them:

```
# 1. Create a sample sheet containing the prefix of each sample
- one per line. As an example:

cd "$OUTPUT_DIR"
ls *R1*fastq.gz | perl -pe 's|_S.{1,3}_R1_001\.fastq\.gz||' >
mysamplesheet.txt
```

which results in:

```
head -n 3 mysamplesheet.txt

sample_1
sample_2
sample_3
```

This sample sheet can then be used to serially process each sample.

```
# Load the reference genome into memory to greatly improve
speed
STAR --genomeLoad LoadAndExit --genomeDir $OUTPUTREF --run-
ThreadN 12

cat mysamplesheet.txt |
while IFS=$'\t' read -r -a myArray; do

 # 0. Get the sample ID and create a repository
 ID="${myArray[0]}"
 echo "$ID"
 mkdir "$ID"
 cd "$ID"

 # 1. Get the FASTQs associated with this sample
 FASTQ=`ls "$OUTPUT_DIR"/${ID}*R1*fastq.gz`
 cp $FASTQ .
 cat *fastq.gz > sample.fastq.gz

 # 2. Align the data
 mkdir STAR
STAR --runThreadN 10 --limitBAMsortRAM 20000000000 --genome-
Load LoadAndKeep --genomeDir "$OUTPUTREF" --readFilesIn sam-
ple.fastq.gz --readFilesCommand zcat --limitSjdbInsertNsj
2000000 --outFilterIntronMotifs RemoveNoncanonicalUnannotated
--outSAMtype BAM SortedByCoordinate --outFileNamePrefix STAR/
"$ID"_
```

```
 # 3. Filter BAM file
samtools view -@ 10 -Sb -F 260 STAR/"$ID"_Aligned.sortedBy-
Coord.out.bam > STAR/"$ID"_Aligned.sortedByCoord.filtered.bam
 samtools index STAR/"$ID"_Aligned.sortedByCoord.filtered.bam

 # 4. Assign Reads to features
 mkdir FEATURECOUNTS
featureCounts -T 1 -t exon -g gene_name --fracOverlap 0.25 -a
"$GTF" -o FEATURECOUNTS/"$ID"_ReadCount.featureCounts.txt
STAR/"$ID"_Aligned.sortedByCoord.filtered.bam

 # 5. Quality check-ups
 mkdir ReSQC

geneBody_coverage.py -r "$ReSQCBED" -i STAR/"$ID"_Aligned.
sortedByCoord.filtered.bam -o ReSQC/"$ID"_geneBody.all

read_distribution.py -i STAR/"$ID"_Aligned.sortedByCoord.fil-
tered.bam -r "$ReSQCBED" > ReSQC/"$ID"_readDistribution.txt

done

# Unload STAR Reference
STAR --genomeLoad Remove --genomeDir "$OUTPUTREF"
```

5 Data Postprocessing

The postprocessing steps will vary depending on the question at hand. In our experience, most tools working for Smart-seq2 can be used as well with FLASH-seq data. For instance, differential expression can be performed with Seurat [20] (R), scanpy [21] (python), or scatter/scran [22] (R) packages/modules.

As FLASH-Seq does not contain UMI sequences, the sample's library size must be normalized.

6 FLASH-seq Low-Amplification (FS-LA)

We have previously shown that FLASH-seq guarantees a higher gene detection than Smart-seq2/3, thanks to its novel combination of buffers, enzymes, and additives [5]. By combining RT and PCR in a single step, FLASH-seq significantly shortens the reaction and hands-on time required for an experiment. In an attempt to further streamline the original protocol, we developed FLASH-seq low-amplification (FS-LA) which, to our knowledge, is the fastest full-length RNA-seq protocol ever designed.

We reasoned that the entire amount of cDNA generated after 20 or more PCR cycles is never entirely used in the tagmentation (even if the reaction has to be repeated multiple times!), causing unnecessarily longer waiting times and generating a higher number of PCR duplicates.

Therefore, we decreased the number of PCR cycles, aiming to obtain sufficient cDNA for direct tagmentation without intermediate steps. This amount of cDNA remains well below the detection limit of most instruments and cannot be reliably assessed, making intermediate QC unfeasible.

Another factor to take into account is that skipping the cleanup step requires a predilution of the cDNA ahead of tagmentation. Residual salt and primers might otherwise completely inhibit Tn5 transposase activity.

This can be achieved in two ways:

- Either by taking directly an aliquot of the unpurified cDNA and performing the tagmentation reaction in a larger volume, as we chose to do in FS-LA. When using our in-house Tn5 transposase, the costs associated with a larger reaction volume are negligible.
- Alternatively, the cDNA can be prediluted [23] and the tagmentation carried out in a smaller volume, which makes the use of commercial kits (more) financially sustainable. The use of in-house transposase would of course remain an option also in this case.

The removal of QC and normalization steps allowed us to carry out the entire FS-LA protocol, from cell sorting to sequencing-ready libraries, in just 4.5 h, of which <1 h represents hands-on time.

Data analysis of FS-LA samples shows an increasing number of detected genes associated with the lower number of PCR cycles as well as excellent gene coverage [5].

Our experiments showed that once a critical amount of cDNA has been generated, high-quality libraries can be prepared using the RT-PCR product directly. This threshold has to be determined empirically for each cell type in a pilot experiment. Elsewhere we provided general guidelines for avoiding under- or overamplification of the cDNA, which might negatively impact the tagmentation and, ultimately, data quality [5].

As a general guideline, the cell RNA content determines the number of PCR cycles that are required. Large cells, such as HEK 293T, contain about ~16 pg of total RNA, and 10 to 12 PCR cycles are sufficient to generate enough cDNA for direct tagmentation and high-quality libraries. On the other hand, human peripheral blood mononuclear cells (hPBMCs) typically contain 2–10 times less RNA compared to HEK 293T cells and might require a higher number of PCR cycles.

6.1 Cell Lysis Mix

Same reagents as the standard FLASH-seq protocol (Fig. 1).

6.2 RT-PCR Mix

Same reagents as the standard FLASH-seq protocol.

6.3 Magnetic Beads Preparation

Same reagents as the standard FLASH-seq protocol.

6.4 Library Preparation with Commercial Illumina Reagents

Same reagents as the standard FLASH-seq protocol.

6.5 Library Preparation with In-House Tn5 Transposase

Same reagents as the standard FLASH-seq protocol.

6.6 Sample QC and Sequencing

1. Qubit™ dsDNA high-sensitivity assay kit. Store at +4 °C protected from light.
2. Qubit™ assay tubes.
3. Agilent high-sensitivity DNA assay. Store at +4 °C.
4. Sera-Mag SpeedBeads™ working solution (for details about preparation, *see* Subheading 2.3). Store at +4 °C.
5. 80% v/v ethanol. Store at room temperature.
6. Nuclease-free water. Store at room temperature.
7. NextSeq™ 550 high output v2.5 kit (75 or 150 cycles), NextSeq™ 1000/2000 (100 or 200 cycles), or NovaSeq 6000 (100 or 200 cycles), depending on sample throughput, desired yield, and required read mode. Other instruments and kits can also be used. Follow the manufacturer's instructions regarding storage.

6.7 Instruments and Consumables

To appreciate the speed and throughput that FS-LA can offer, we recommend performing the experiment only in 384-well plates.

1. Fully skirted 384-well plates (*see* **Note 13**).
2. Aluminum seal and adhesive plastic foil (*see* **Note 14**).
3. Plate centrifuge.
4. Thermocycler with 384-well block.
5. I.DOT (Dispendix) nanodispenser.
6. Fluent 780 workstation (Tecan).

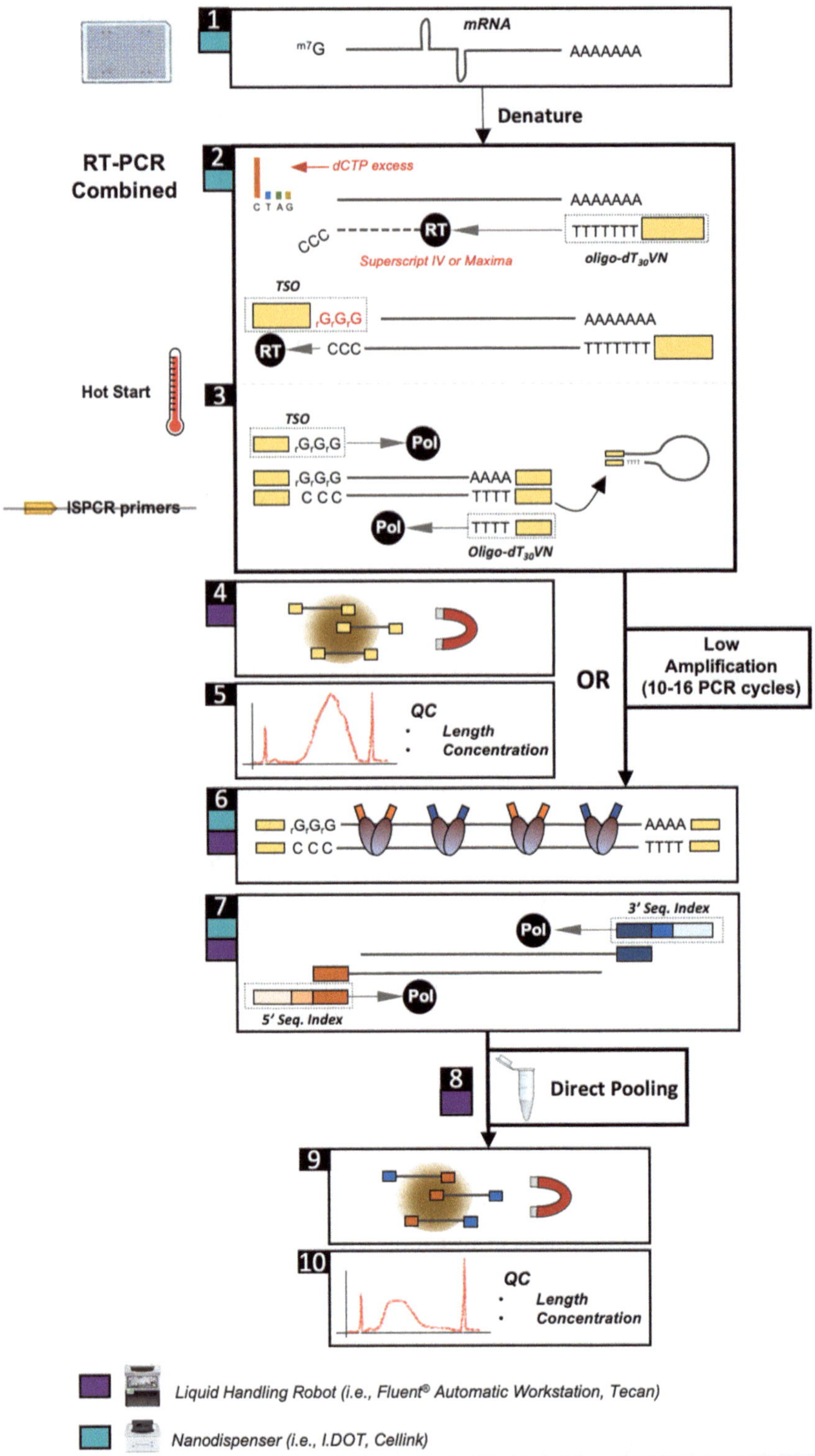

Fig. 1 Standard FLASH-seq workflow. Steps automated with a liquid-handling robot or nanodispenser are marked by purple and teal squares, respectively. **Step 1**, cells are sorted by FACS in wells of a plate containing a lysis buffer. **Step 2**, after cell lysis, the secondary structure of mRNAs is resolved by heating before priming with an oligodT primer and reverse transcription with a template-switching reverse transcriptase (RT) in the

7. FACSAria™ Fusion (BD Biosciences). Other cell dispensers can also be used.
8. Agilent 2100 Bioanalyzer.
9. Nextseq™ 550/1000/2000 or Novaseq™ 6000 Sequencing Systems, depending on sample throughput and desired read output. Other Illumina instruments can also be used.

7 Methods

7.1 Cell Lysis Mix Preparation

Follow the standard FLASH-seq protocol.

7.2 RT-PCR Mix Preparation and RT-PCR Reaction

Follow the standard FLASH-seq protocol.

IMPORTANT - Adjust the number of cycles according to the cell type. We recommend 10–12 cycles for HEK 293T cells and 14–16 cycles for hPBMC.

IMPORTANT - Do not perform QC (Bioanalyzer/PicoGreen) or sample normalization!

7.3 Library Preparation

7.3.1 Library Preparation with In-House Tn5 Transposase

Please note that the Tn5 transposase amount is a suggested starting point only. Optimizations might be necessary, depending on the specific batch activity and desired library size.

1. Remove the in-house Tn5 transposase (2 μM working dilution) and the KAPA HiFi polymerase from the −20 °C storage and place them on ice. Remove the dNTP mix, high-fidelity KAPA HiFi buffer, and the index adapter plate from the −20 °C storage and keep them at room temperature. Remove the TAPS-$MgCl_2$ buffer from the +4 °C storage and keep them at room temperature. Get the SDS and the DMF solutions.
2. For each sample, prepare 9 μL tagmentation mix: 2 μL TAPS-$MgCl_2$ buffer, 2 μL DMF, 0.025 μL in-house Tn5 transposase, nuclease-free water to volume.
3. Briefly vortex and gently spin down the tagmentation mix. Keep it on ice until needed.

Fig. 1 (continued) presence of an excess of dCTP. **Step 3**, cDNA is amplified by semi-suppressive PCR. For Steps 2 and 3, only one master mix is required. **Step 4**, cDNA is cleaned up using magnetic beads. **Step 5**, cDNA concentration and fragment size are measured and the samples are normalized to the desired concentration for the tagmentation reaction. **Step 6**, cDNA is tagmented using a Tn5, to make it amenable to Illumina sequencing. **Step 7**, tagmented cDNA is amplified by PCR and sequencing indices are added. **Step 8**, samples are pooled together to facilitate handling. **Step 9**, sequencing-ready library is cleaned up with magnetic beads. **Step 10**, concentration and average fragment size are measured in preparation for the sequencing. **Steps 4** and **5** can be skipped when using FS-LA

4. Transfer 1 μL unpurified cDNA into a new 384-well plate.
5. Add 9 μL tagmentation mix to each well.
6. Seal the plate with adhesive transparent film, vortex, spin down, place it in a thermocycler, and carry out the tagmentation reaction: 55 °C for 8 min, 4 °C hold. Once the reaction is completed, proceed immediately to the next step.
7. Add 2.5 μL 0.2% SDS solution to each well.
8. Seal the plate with adhesive transparent film, vortex, spin down, and incubate for 5 min at room temperature. Do not put the plate back on ice.
9. Add 2.5 μL N7xx + S5xx indices from the adapter index plate (5 μM each).
10. Add 10 μL enrichment PCR mix: 0.5 μL KAPA HiFi DNA polymerase, 0.75 μL dNTP mix, 5 μL high-fidelity buffer, nuclease-free water to volume.
11. Seal the plate with adhesive transparent film, vortex, spin down, place it in a thermocycler, and carry out the enrichment PCR reaction: 72 °C for 3 min, 98 °C for 30 s, then N cycles of (98 °C for 10 s, 55 °C for 30 s, 72 °C for 30 s), 72 °C for 5 min, 4 °C hold. The number of cycles "N" should be adjusted according to the number of preamplification cycles used in the experiment. As a general guideline we found that 14–16 cycles are sufficient for most applications, especially when pooling 384 cells before the library QC.
12. It is safe to stop here and store the final library in a −20 °C freezer until needed.

7.3.2 Library Preparation with the Nextera® XT Kit

1. Remove the Amplicon Tagment Mix (ATM) and the Nextera® PCR mix (NPM) from the −20 °C storage and place them on ice. Remove the Tagment DNA buffer (TD) and the adaptor index plate from the −20 °C storage and keep them at room temperature. Remove the neutralization buffer (NT) from the +4 °C storage and keep it at room temperature.
2. Take an aliquot of the preamplified and unpurified cDNA and transfer it into a new 384-well plate.
3. Add nuclease-free water to obtain a 1:10 dilution. This represents the cDNA that is going to be used for tagmentation.
4. Prepare the tagmentation mix: 1 μL TD and 0.02–0.25 μL ATM. Refer to the table below for estimating the most appropriate amount of ATM to use.

No. pre-amplification cycles	Amount ATM used (µl)	Average library size Bioanalyzer (bp)
10	0.25	460
10	0.125	498
10	0.0625	657
10	0.03125	794
10	0.015625	931
10	0.0078125	998
10	0.00390625	1069
10	0.001953125	1020
12	1	448
12	0.5	507
12	0.25	567
12	0.125	644
12	0.0625	735
12	0.03125	822
12	0.015625	1028
12	0.0078125	1085
12	0.00390625	1027

The table shows that the approximate library size is to be expected when using 1:10 diluted cDNA after 10 or 12 preamplification cycles in HEK 293T cells

5. Briefly vortex and gently spin down the mix. Keep it on ice until needed.
6. Transfer 1 µL diluted cDNA into a new 384-well plate.
7. Add 1 µL tagmentation mix to each well.
8. Seal the plate with adhesive transparent film, vortex, spin down, place it in a thermocycler, and carry out the tagmentation reaction: 55 °C for 8 min, 4 °C hold. Once the reaction is completed, proceed immediately to the next step.
9. Add 0.5 µL NT buffer into each well.

10. Seal the plate with adhesive transparent film, vortex, spin down, and incubate for 5 min at room temperature. Do not put the plate back on ice.
11. Add 1 μL N7xx + S5xx indices from the adapter index plate (5 μM each).
12. Add 1.5 μL NPM into each well.
13. Seal the plate with adhesive transparent film, vortex, spin down, place it in a thermocycler, and carry out the enrichment PCR reaction: 72 °C for 3 min, 95 °C for 30 s, then N cycles of (95 °C for 10 s, 55 °C for 30 s, 72 °C for 30 s), 72 °C for 5 min, 4 °C hold. The number of cycles "N" can vary according to several parameters (number of preamplification cycles used, RNA content, number of cells that are being processed), but we found that 12–14 cycles are generally more than sufficient, especially when pooling 384 cells before the final cleanup.
14. It is safe to stop here and store the final library in a −20 °C freezer until needed.

7.4 Pooling and Bead Cleanup of the Final Library

Follow the standard FLASH-seq protocol, regardless of which Tn5 was used for tagmentation in the steps above.

7.5 Sequencing Library QC and Sequencing

Follow the standard FLASH-seq protocol. Single-end 75 or 100 bp is generally sufficient but longer read modes or paired-end sequencing can be an option, depending on the question at hand.

8 Data Processing

Refer to the standard FLASH-seq protocol for the data processing. Particular attention must be given to the data quality checkups. As described in *Hahaut* et al. [5], performing a limited number of PCR cycles can influence data quality. We observed that an insufficient number of PCR cycles increases the percentage of intergenic and intronic reads as well as the percentage of multimapped/unmapped reads.

This is likely due to the balance between amplified cDNA, tagmented genomic DNA, and ribosomal RNAs, which are always present in the reaction. In Smart-seq2, the cDNA overamplification and the intermediate cleanup masked these contaminants. However, in FS-LA, the right number of PCR cycles must be chosen carefully, in order to obtain a sufficient amount of cDNA and avoid compromising the final data.

To ensure that the data generated are of good quality, we advise the user to compare FS-LA data with a control FS sample done with the standard number of PCR cycles.

Insufficient cDNA amplification can be identified by:

- Higher intergenic/intronic read tag percentages (ReSQC)
- Lower exonic read tag percentages (ReSQC)
- Lower uniquely mapping read percentages (STAR)
- Increased multimapped/unmapped read percentages (STAR)

As the baseline of these mapping statistics vary from one cell type to another, the thresholds may vary as well. We typically tolerate a variation of 2–5% in the percentage of uniquely mapped reads and <2% in the intergenic reads between standard FS and FS-LA values.

In addition, we recommend exploring the aligned data with a genome browser viewer such as Integrated Genome Viewer (IGV). Given the nature of the FS protocol (polyA-based), most of the reads should be confined to exons and introns of protein coding genes, with clear exon/intron boundaries. High levels of single reads in intergenic, intronic, or centromeric regions may indicate that the cDNA was not sufficiently amplified. We also observed that in lower quality samples, the intergenic reads tended to be regularly spaced, at least to some extent.

We do not recommend doing isoform reconstruction/quantification in the presence of genomic DNA contamination. However, the tagmentation is a random process which means that low levels of genomic DNA contamination are unlikely to influence differential expression results at the gene/exon level as the same DNA regions will not be systematically sequenced in every cell.

9 FLASH-seq with Unique Molecular Identifiers (FS-UMI)

To exploit the robustness and flexibility of FLASH-seq even further, we modified the original protocol by including UMIs into the TSO sequence, as originally introduced by Smart-seq3 [3], thus generating both 5′ UMI-containing reads and internal reads without UMI. However, based on existing literature, we also realized that the eight random nucleotides in the UMI followed by three ribo-G (Smart-seq3 TSO design) might increase the risk of strand invasion [24].

Strand invasion is a phenomenon occurring when the TSO anneals to an internal sequence of the cDNA or mRNA molecule that, by chance, displays sequence complementarity [5]. The end result is a molecule that is shorter and with a chimeric 5′-end. Although not always easy to exactly quantify, these artifacts should be minimized whenever possible, as they affect isoform detection and bias gene counts.

In a similar fashion to the solution implemented in the nanoCAGE protocol and the 10x Genomics Single Cell 5′ v1/v2 kits, we tested seven new TSO where the UMI and the ribo-Gs were separated by a 5-bp spacer sequence. These new TSOs were tested in combination with the STRT-seq2i-oligo-dT [25], with the pre-amplification reaction deviating from the suppression PCR used by Smart-seq2 and the standard FLASH-seq protocol.

In our experiments we observed that the cDNA yield was strongly influenced by the TSO sequence. Interestingly, most conditions displayed great mapping statistics and coverage [5]. The addition of a spacer sequence increased by up to ~15% (TSO-CATCA) the number of detected genes compared to SS3. Six out seven TSOs with a spacer even outperformed the standard FS in respect of the number of detected genes. We eventually settled on a spacer carrying the sequence "CTAAC," as it also showed excellent results and because the sequence "CTAACGGG" had the lowest likelihood to match any transcriptome or genome sequences among all tested TSO.

Our experiments indicate how robust and versatile FLASH-seq is and confirm that UMIs can also be used in our RT-PCR setup when isoform reconstruction is the primary goal of the experiment.

9.1 Cell Lysis Mix

1. 10% v/v Triton X-100. Store at +4 °C (*see* **Note 1**).
2. Premixed dNTP solution (25 mM each). Store at −20 °C.
3. Recombinant RNase inhibitor (40 U/μL). Store at −20 °C.
4. Betaine (5 M). Store at +4 °C.
5. Dithiothreitol (DTT, 100 mM; part of the SuperScript™ IV kit). Store at −20 °C.
6. dCTP (100 mM). Store at −20 °C.
7. STRT-dT_{31} oligonucleotide (5′Bio-AATGATACGGCGACCACCGATCGT_{31}-3′, 100 μM). "Bio" = biotin. Store at −20 °C.
8. Nuclease-free water. Store at room temperature.

9.2 RT-PCR Mix

1. Superscript™ IV kit: dithiothreitol (DTT, 100 mM). Store at −20 °C.
2. Superscript™ IV kit: Superscript™ IV reverse transcriptase (200 U/μL). Store at −20 °C (*see* **Note 3**).
3. Betaine (5 M). Store at +4 °C.
4. Magnesium chloride ($MgCl_2$, 1 M). Store at +4 °C.
5. Recombinant RNase inhibitor (40 U/μL). Store at −20 °C.
6. Template-switching oligonucleotide with UMIs, TSO-UMI (5′ Bio-AAGCAGTGGTATCAACGCAGAGTNNNNNNNNCTAACrGrGrG-3′, 100 μM). "Bio" = biotin; NNNNNNNN = UMI; CTAAC = spacer. Store aliquots in use at −20 °C and stock tubes at −80 °C (*see* **Note 2**).

7. Forward PCR primer, Tn5-ISPCR-F (5′-TCGTCGGCAGCG TCAGATGTGTATAAGAGACAGAAGCAGTGGTAT-CAACGCAGAGT-3′, 100 μM). Store at −20 °C.
8. Reverse PCR primer, DI-PCR-P1A-R (5′-AATGATACGGCG ACCACCGA-3′, 100 μM). Store at −20 °C.
9. KAPA HiFi HotStart ReadyMix (2×). Store at −20 °C.
10. Nuclease-free water. Store at room temperature.

9.3 Magnetic Beads Preparation

Follow the standard FLASH-seq protocol.

9.4 Library Preparation with Commercial Illumina Reagents

Follow the standard FLASH-seq protocol.

9.5 Sample QC and Sequencing

Follow the standard FLASH-seq protocol.

9.6 Instruments and Consumables

Follow the standard FLASH-seq protocol.

10 Methods

10.1 Cell Lysis Mix Preparation

1. For each sample, prepare 1 μL of the following cell lysis mix: 0.020 μL Triton X-100, 0.240 μL dNTP mix, 0.018 μL STRT-dT_{31} oligonucleotide, 0.030 μL recombinant RNase inhibitor, 0.012 μL DTT, 0.090 μL dCTP, 0.200 μL betaine, nuclease-free water to volume.
2. Mix well and gently spin down.
3. Dispense the cell lysis mix into a 384-well plate.
4. If used immediately, keep the plate on ice until needed. The cell lysis mix can be prepared and dispensed on plates several weeks or months in advance. The plates are then stored at −20 °C until needed. Avoid multiple freezing-thawing cycles.
5. Collect the cells by FACS using a FACSAria™ Fusion, preferably with a larger nozzle (85 μm or, even better, 100 μm) to minimize shearing forces upon sorting (*see* **Note 15**).
6. Once the sorting is completed, seal the plate with aluminum seal and snap-freeze it at −80 °C or by placing it on dry ice, especially if not proceeding immediately with RT-PCR. Lysed cells can be stored in these conditions for several months without appreciable decrease in RNA quality. The plate should never undergo freeze-thaw cycles for any reason.

10.2 RT-PCR Mix Preparation and RT-PCR Reaction

1. For each sample, prepare 4 μL of the following RT-PCR mix: 0.238 μL DTT, 0.800 μL betaine, 0.046 μL $MgCl_2$, 0.096 μL recombinant RNase inhibitor, 0.050 μL Superscript™ II reverse transcriptase, 2.500 μL KAPA HiFi HotStart ReadyMix, 0.092 μL TSO-UMI oligonucleotide, 0.025 μL Tn5-ISPCR- F primer, 0.005 μL DI-PCR-P1A-R primer, nuclease-free water to volume.
2. Briefly vortex and gently spin down the mix. Keep it on ice until needed.
3. Take the plate out of the −80 °C freezer and leave it for a couple of minutes at room temperature to thaw.
4. Briefly spin down to collect eventual drops that might have condensed on the lid.
5. Place the plate on a thermocycler block and perform cell lysis and mRNA denaturation at 72 °C for 3 min.
6. Remove it from the thermocycler and keep it on ice for a couple of minutes to cool. Briefly spin down to collect the lysate at the bottom of each well.
7. Dispense 4 μL RT-PCR mix in each well. The final volume is now 5 μL.
8. Seal the plate with adhesive transparent film, vortex, spin down, place it in a thermocycler, and carry out the RT-PCR reaction: 50 °C for 60 min, 98 °C for 3 min, then N cycles of (98 °C for 20 s, 65 °C for 20 s, 72 °C for 6 min), 4 °C hold. The number of cycles "N" should be adjusted according to the cell type used in the experiment. We recommend 20–21 cycles for HEK 293T cells and 23–24 cycles for hPBMC. The addition of UMI and the lack of semi-suppressive PCR decrease reaction yield. As a rule of thumb, use the same number of PCR cycles as in the Smart-seq2 protocol.
9. It is safe to stop here and store the PCR product in a −20 °C freezer until needed.

10.3 Magnetic Beads Cleanup After Preamplification

Follow the standard FLASH-seq protocol but use a 0.6:1 ratio of beads:cDNA. Check the cDNA quality on the Agilent Bioanalyzer instrument.

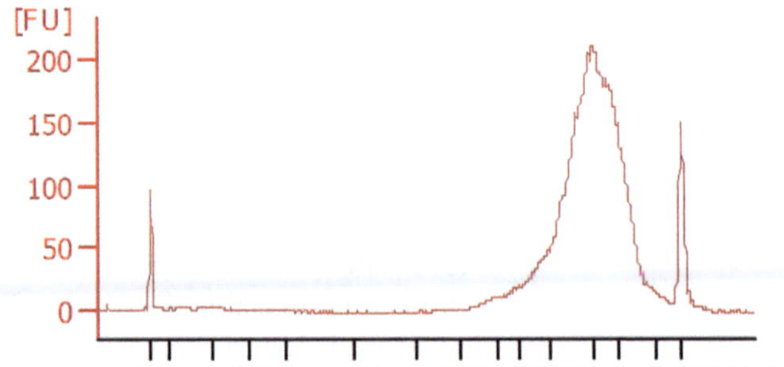

Example of amplified cDNA from a single human retinal organoid cell (24 PCR cycles)

10.4 cDNA Quantification and Sample Normalization

Follow the standard FLASH-seq protocol. In the last step, prepare a normalization plate by adding 1 μL purified cDNA and nuclease-free water to a final concentration of 100 pg/μL using the Pico-Green readings.

10.5 Library Preparation with the Nextera XT Kit

Please note that the Tn5 transposase amount indicated below is a suggested starting point for tagmenting 100 pg cDNA. Optimization might be necessary, depending on the size of the sequencing libraries that need to be obtained.

To ensure better reproducibility between experiments, we recommend using the Nextera XT kit. However, the in-house Tn5 will provide data of comparable quality.

If the cDNA quantification was accurate, then 0.1–0.2 μL ATM will give sequencing-ready libraries in the range of 700–1000 bp. In our experience, longer libraries gave higher percentages of UMI reads.

The balance between UMI- and internal reads might be complex to achieve and does not always relate to the library size. Increasing the amount of forward primer in the PCR reaction (up to 2–4-times) may sometimes help increase the proportion of UMI reads. However, it has to be taken into account that fragments of >1000 bp are not expected to efficiently bind to NextSeq or NovaSeq flow cells and should therefore be avoided.

For simplicity, here we use 0.2 μL throughout the text.

1. Remove the Amplicon Tagment mix (ATM) and the Nextera® PCR mix (NPM) from the −20 °C storage and place them on ice. Remove the Tagment DNA buffer (TD) and the adaptor index plate from the −20 °C storage and keep them at room temperature. Remove the neutralization buffer (NT) from the +4 °C storage and keep it at room temperature.
2. For each sample, prepare 1.2 μL tagmentation mix: 0.2 μL ATM and 1 μL TD.
3. Briefly vortex and gently spin down the mix. Keep it on ice until needed.
4. Transfer 1 μL normalized cDNA into a new 384-well plate.
5. Add 1.2 μL tagmentation mix to each well.
6. Seal the plate with adhesive transparent film, vortex, spin down, place it in a thermocycler, and carry out the tagmentation reaction: 55 °C for 8 min, 4 °C hold. Once the reaction is completed proceed immediately to the next step.
7. Add 0.5 μL NT buffer to each well.
8. Seal the plate with adhesive transparent film, vortex, spin down, and incubate for 5 min at room temperature. Do not put the plate back on ice.
9. Add 1 μL N7xx + S5xx indices from the adaptor index plate.

10. Add 1.5 μL NPM into each well.
11. Seal the plate with adhesive transparent film, vortex, spin down, place it in a thermocycler, and carry out the enrichment PCR reaction: 72 °C for 3 min, 95 °C for 30 s, then N cycles of (95 °C for 10 s, 55 °C for 30 s, 72 °C for 30 s), 72 °C for 5 min, 4 °C hold. The number of cycles "N" should be adjusted according to the amount of cDNA used for the tagmentation reaction. When pooling at least 32–48 cells (or more), 14 cycles are sufficient.
12. It is safe to stop here and store the final library in a −20 °C freezer until needed.

10.6 Pooling and Bead Cleanup of the Final Library

Follow the standard FLASH-seq protocol.

10.7 Sequencing Library QC and Sequencing

1. Use 1 μL of the purified sequencing pool to assess the concentration on a Qubit instrument.
2. Use 1 μL to assess the average size on a High Sensitivity DNA chip.

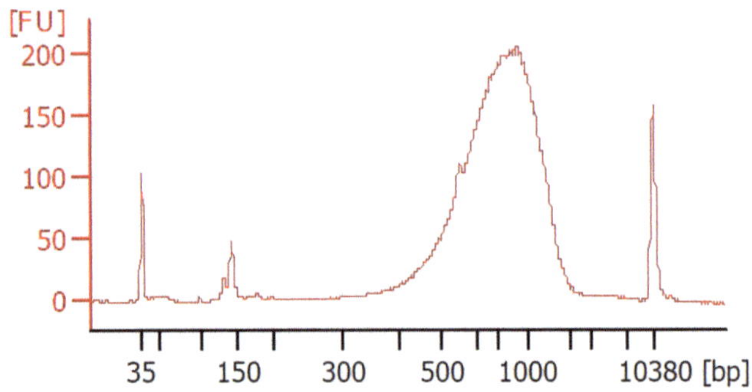

Example of a sequencing-ready library from a pool of HEK 293T cells. Average size around 800 bp

3. The purified library can be sequenced on any Illumina sequencer. Follow the specifications reported for each instrument. Paired-end sequencing is recommended to reap the most benefits in terms of isoform reconstruction. Read 1 length should not be <75 bp and preferably ≥100 bp. We regularly sequence FS-UMI libraries on a NextSeq500 using 100-8-8-50 read mode but preliminary data indicate that 90-8-8-60 or 80-8-8-70 read modes might lead to better results.

11 Data Preprocessing

The data processing will depend on the final goal. The following description provides guidelines to separate and independently preprocess UMI and internal reads.

In specific cases (e.g., isoform detection), using both internal and UMI reads together is required to properly cover the full body of the transcripts. Trimming of adapters, spacer, and UMI sequences should still be performed in the first step. Failing to properly remove these sequences can result in a decreased percentage of reads uniquely mapped/assigned to a feature.

Here we assume that the data were generated using paired-end sequencing (100 × 50) on an Illumina instrument and are processed on a unix-based machine.

11.1 Prerequisites

Download the following tools. Versions indicated are the ones tested in *Hahaut* et al. [5].

- bcl2fastq (Illumina, v2.20)
- STAR [12] (v2.7.3)
- FeatureCounts [13] (v1.6.5)
- BBMAP [14] (v38.86)
- samtools [15] (v1.9)
- umi_tools [26] (v1.1)
- IGV

When working on hundreds of samples, it is recommended to use virtual machines or cloud platforms to process the data as it will typically require >48 Gb of RAM, >12 cores, and >500 Gb of free space. The footprint of the most time and resource consuming step (= read alignment) can be reduced using pseudoalignment tools (i.e., salmon [16] or kallisto [17]).

11.2 Demultiplexing

Refer to FLASH-Seq data preprocessing.

11.3 Separate UMI/Internal Reads and Extract UMI Sequences

Umi_tools can be used to remove the TSO adapter and spacer sequences and extract the UMI sequences, which are typically located in read 1 (R1-UMI). The UMI sequence can also sometimes be found in read 2 (R2-UMI) instead of read 1 when a tagmentation event occurs upstream of the UMI in the TSO adapter sequence. For this reason, the UMI search must be performed in both R1 and R2 reads.

After obtaining the UMI reads, the internal reads can be extracted from the raw fastq file.

```
# 1. Extract R1-UMI sequences
# The barcode pattern is made of all tagmented variations of
the TSO adapter sequence
umi_tools extract --bc-pattern="^(?P<discard_1>AAGCAGTGGTAT-
CAACGCAGAGT|AGCAGTGGTATCAACGCAGAGT|GCAGTGGTATCAACGCAGAGT|
CAGTGGTATCAACGCAGAGT|AGTGGTATCAACGCAGAGT|GTGGTATCAACGCAGAGT|
TGGTATCAACGCAGAGT|GGTATCAACGCAGAGT|GTATCAACGCAGAGT|ATCAACGCA-
GAGT|CAACGCAGAGT|AACGCAGAGT|ACGCAGAGT|GCAGAGT|CAGAGT|AGAGT|
GAGT|AGT|GT)(?P<umi_1>.{8})(?P<discard_2>CTAACGG)(?P<dis-
card_3>G{0,4})" --stdin=sample.R1.fastq.gz --stdout=umi.
UMIinR1.R1.fq --read2-in=sample.R2.fastq.gz --read2-out=umi.
UMIinR1.R2.fq --extract-method=regex

# 2. Extract R2-UMI sequences
# The barcode pattern is made of the known most common
tagmented variations of the TSO adapter sequence
umi_tools extract --bc-pattern="^(?P<discard_1>GAGT|AGT|GT)(?
P<umi_1>.{8})(?P<discard_2>CTAACGG)(?P<discard_3>G{0,4})" --
stdin=sample.R2.fastq.gz --stdout=umi.UMIinR2.R2.fq --read2-
in=sample.R1.fastq.gz --read2-out=umi.UMIinR2.R1.fq --ex-
tract-method=regex

# 3. Find reads with a UMI in both R1 and R2
# In very rare cases (<0.0001%) the UMI is found in both R1 and
R2
# This can be due to short insert size (= R1 invades R2) or
retrotranscription errors when two TSO are used
# These must be removed or they will create duplicated read ID
entries

# 3.1. Get the R1-UMI / R2-UMI read IDs
cat umi.UMIinR1.R1.fq | uniq | awk 'NR%4==1{print}' | sed 's/
\@//g' > names.R1umi.txt
cat umi.UMIinR2.R1.fq | uniq | awk 'NR%4==1{print}' | sed 's/
\@//g' > names.R2umi.txt
# 3.2. Remove the UMI info from their name
# Umi_tools stores the UMI sequence in the read ID
cat names.R1umi.txt | sed 's/_........ .*$//g' > names.R1umi.
cleaned
cat names.R2umi.txt | sed 's/_........ .*$//g' > names.R2umi.
cleaned
# 3.3. Get the read IDs found in both R1-UMI and R2-UMI
comm -12 <(sort names.R1umi.cleaned) <(sort names.R2umi.
cleaned) > R1R2.toFilterOut

echo "===> Number of R1-R2 with both a UMI: $(wc -l R1R2.
toFilterOut) <==="
echo "===> Number of R1 UMI before cleanup: $(wc -l names.
```

```
R1umi.txt) <==="
echo "===> Number of R2 UMI before cleanup: $(wc -l names.
R2umi.txt) <==="
# 4. Filter out readw with dual UMI sequences

# 4.1. Get the actual read ID of dual-UMI reads
# ID + UMI sequences
grep -f R1R2.toFilterOut names.R1umi.txt > R1.toFilterOut
grep -f R1R2.toFilterOut names.R2umi.txt > R2.toFilterOut

# 4.2. Filter them out using BBMAP filterbyname.sh

filterbyname.sh -Xmx6g in=umi.UMIinR1.R1.fq in2=umi.UMIinR1.
R2.fq out=umi.UMIinR1.R1.tmp out2=umi.UMIinR1.R2.tmp
names=R1.toFilterOut include=f overwrite=t
filterbyname.sh -Xmx6g in=umi.UMIinR2.R1.fq in2=umi.UMIinR2.
R2.fq out=umi.UMIinR2.R1.tmp out2=umi.UMIinR2.R2.tmp
names=R2.toFilterOut include=f overwrite=t

# 5. Rename the files
mv umi.UMIinR1.R1.tmp umi.UMIinR1.R1.fq
mv umi.UMIinR2.R1.tmp umi.UMIinR2.R1.fq
mv umi.UMIinR1.R2.tmp umi.UMIinR1.R2.fq
mv umi.UMIinR2.R2.tmp umi.UMIinR2.R2.fq

echo "===> Number of R1 UMI after cleanup: $(grep -c \@ umi.
UMIinR1.R1.fq) <==="
echo "===> Number of R2 UMI after cleanup: $(grep -c \@ umi.
UMIinR2.R1.fq) <==="

# 6. Internal reads

# 6.1. UMI read IDs
cat umi.UMIinR1.R1.fq | uniq | awk 'NR%4==1{print}' | sed 's/
\@//g' | sed 's/_........//g' > names.R1umi.txt
cat umi.UMIinR2.R1.fq | uniq | awk 'NR%4==1{print}' | sed 's/
\@//g' | sed 's/_........//g' > names.R2umi.txt
sed -i 's/_........//g' R1.toFilterOut
sed -i 's/_........//g' R2.toFilterOut
cat names.R1umi.txt names.R2umi.txt R1.toFilterOut R2.toFil-
terOut > names.umi.txt

# 6.2. Extract internal reads
filterbyname.sh -Xmx6g in=sample.R1.fastq.gz in2=sample.R2.
fastq.gz out=internal.R1.fq out2=internal.R2.fq names=names.
umi.txt include=f overwrite=t

echo "===> Number of Internal Reads after cleanup: $(grep -c
```

```
\@ internal.R1.fq) <==="

# 7. Combine R1/R2-UMI reads
# UMI reads are stranded (= same orientation as the original
molecule)
# R2-UMI are in the opposite direction compared to the feature
due to the sequencing process
# To reconcile them use read 2 R2-UMI as read 1.
cat umi.UMIinR1.R1.fq umi.UMIinR2.R2.fq > umi.R1.fq
cat umi.UMIinR1.R2.fq umi.UMIinR2.R1.fq > umi.R2.fq

# 8. Clean-up
rm R1.toFilterOut R2.toFilterOut toFilterOut.txt names.umi.txt
rm names.* umi.UMIinR*.R*.fq
```

11.4 Optional: Remove Sequencing Adapter Leftovers

Refer to FLASH-seq data pre-processing.

11.5 Recommended: Discarding Too Short Reads

UMI extraction and trimming can result in short reads (<50 bp). Discard too short reads using trimmomatic.

This step can be skipped when using longer R2 reads (>70 bp).

```
java -jar trimmomatic-0.39.jar PE -threads 2 -phred33 FASTQ/
umi.R1.trim.fq FASTQ/umi.R2.trim.fq FASTQ/umi.R1.trim2.R1.fq
FASTQ/umi.R1.trim2.unpaired.fq FASTQ/umi.R2.trim2.R1.fq
FASTQ/umi.R2.trim2.unpaired.fq CROP:100 MINLEN:29

mv FASTQ/umi.R1.trim2.R1.fq FASTQ/umi.R1.trim.fq
mv FASTQ/umi.R2.trim2.R1.fq FASTQ/umi.R2.trim.fq

gzip -c FASTQ/umi.R1.trim.fq > FASTQ/umi.R1.fq.gz
gzip -c FASTQ/umi.R2.trim.fq > FASTQ/umi.R2.fq.gz
```

11.6 Data Mapping

11.6.1 Genome Indexing

Read aligners (e.g., STAR or HISAT2) or pseudo aligners (e.g., salmon or kallisto) can be used to map the data. The following example is demonstrated using STAR.

The first step of the data mapping process is to generate an index for the reference genome/transcriptome onto which the read will be aligned. This procedure is generally performed only once. The same index is reused for every sample.

The presence of reads of varying lengths represents an additional challenge, as it requires adapting the sjdbOverhang values accordingly. Large values should be privileged in most cases, setting sjdbOverhang to the maximal mate read length - 1. If reads <50 bp are expected after trimming, the default seedSearchStartLmax must be decreased during read alignment.

```
# 0. Define paths
OUTPUTREF="/path/to/output_reference_index/"
FASTA="/path/to/fasta_reference_genome.fa"
GTF="/path/to/genome_annotation.gtf"

# 1. Create the index repository
mkdir $OUTPUTREF
cd $OUTPUTREF

# 2. OPTIONAL: Save the fasta/GTF files to the reference folder
scp $FASTA .
scp $GTF .

# 3. Genome indexing
# runThreadN: Adapt the number of threads to your device
# sjdbOverhang: Usually set to read length -1
STAR --runThreadN 30 \
--runMode genomeGenerate \
--genomeDir $OUTPUTREF \
--genomeFastaFiles $FASTA \
--sjdbGTFfile $GTF \
--sjdbOverhang 99
```

11.6.2 Align Reads

```
mkdir STAR

# 1. UMI Reads
MYFASTQ_R1="/path/to/FASTQ/umi.R1.fq.gz"
MYFASTQ_R2="/path/to/FASTQ/umi.R2.fq.gz"

STAR --runThreadN 10 --limitBAMsortRAM 20000000000 --seed-
SearchStartLmax 30 --genomeLoad LoadAndKeep --genomeDir "$OUT-
PUTREF" --readFilesIn "$MYFASTQ_R1" "MYFASTQ_R2" --
readFilesCommand zcat --limitSjdbInsertNsj 2000000 --outFil-
terIntronMotifs RemoveNoncanonicalUnannotated --outSAMtype
BAM SortedByCoordinate --outFileNamePrefix STAR/"$ID"_UMI_

# 2. Internal Reads
MYFASTQ_R1="/path/to/FASTQ/internal.R1.fq.gz"
MYFASTQ_R2="/path/to/FASTQ/internal.R2.fq.gz"STAR --run-
ThreadN 10 --limitBAMsortRAM 20000000000 --genomeLoad LoadAnd-
Keep --genomeDir "$OUTPUTREF" --readFilesIn "$MYFASTQ_R1"
"MYFASTQ_R2" --readFilesCommand zcat --limitSjdbInsertNsj
2000000 --outFilterIntronMotifs RemoveNoncanonicalUnannotated
--outSAMtype BAM SortedByCoordinate --outFileNamePrefix STAR/
"$ID"_INTERNAL_
```

11.7 Optional: Filter Reads

Refer to FLASH-seq data preprocessing.

11.8 Optional: Remove UMI Invasion Events

As previously described [5, 24], TSO invasions can happen in SMART-seq methods and can be favored by the presence of a random sequence in the TSO (e.g., UMI). The addition of a spacer in FLASH-seq TSO significantly decreases the probability of such events. However, it can be interesting to evaluate the degree of strand-invasion and remove putative events before performing isoform reconstructions or counting UMI reads.

We created an R script that takes as input a BAM file of reads processed with umi_tools/STAR and filters out putative strand-invasion events. These events are identified based on the homology between the read's UMI sequence and the 20-bp sequence adjacent to the read starting position. Given the random nature of the UMI sequence, we do not expect an overlap between these two sequences. Partial homology can indicate that the UMI was used as an internal primer during RT/PCR and does not originate from an actual random distribution.

```
# Arguments:
# arg1: /path/to/mapped.umi_tools.bam
# arg2: sample ID
# arg3: method 'FLASH-Seq' or 'zUMIs'
# arg4: Maximum number of mismatches (recommended: 1)
# arg5: path/to/genome.fa - genome used for the mapping
# arg6: Minimum number of total reads to process the sample
(recommended: 10000)
# arg7: Minimum mapq to look for a match between UMI and
adjacent sequence (recommended: 5)
# arg8: /path/to/output/folder/"

Rscript filterINvasionEventsfromBAM.R "$ID"_"$TYPE"_"$DOW-
N"_Aligned.sorted.bam "$ID"_"$TYPE"_"$DOWN" 'FLASH-Seq'
1 $FASTA 1000 5 ./
```

This script will return two files:

/path/to/output/folder/ID_filteredInvasion.bam
/path/to/output/folder/ID_filteredInvasion.log.out

The first one is a bam file where the putative invasion events have been removed. The second one is a log file containing some filtering information.

The filterINvasionEventsfromBAM.R script can be found in the FLASH-Seq Github repository:

https://github.com/Radek91/FLASH-Seq

11.9 Assign Reads to Feature Using Featurecounts

UMI read assignment to a feature differs from the one of internal reads. First, they should be assigned to a feature in a stranded manner. Second, similar UMI sequences assigned to the same feature should be collapsed (= deduplication), as they represent PCR duplicates.

```
mkdir FEATURECOUNTS

# 1. UMI Reads

# 1.1. Assign UMI reads to a feature
# Returns a BAM file
featureCounts -T 1 -p -t exon -g gene_name -s 1 --fracOverlap
0.25 -a "$GTF" -R BAM -o FEATURECOUNTS/"$ID".featureCounts.txt
STAR/"$ID"_UMI_Aligned.sortedByCoord.filtered.bam

# 1.2. Sort and index BAM
samtools sort -@ 10 FEATURECOUNTS/"$ID"_UMI_Aligned.sortedBy-
Coord.filtered.bam.featureCounts.bam -o FEATURECOUNTS/
"$ID"_UMI_Aligned.sortedByCoord.filtered.bam.featureCounts.
sorted.bam
samtools index FEATURECOUNTS/"$ID"_UMI_Aligned.sortedByCoord.
filtered.bam.featureCounts.sorted.bam

# 1.3. Count UMI reads
umi_tools count --per-gene --paired --gene-tag=XT --chimeric-
pairs=discard --unpaired-reads=discard --assigned-status-
tag=XS -I FEATURECOUNTS/"$ID"_UMI_Aligned.sortedByCoord.fil-
tered.bam.featureCounts.sorted.bam -S FEATURECOUNTS/"$ID".
umi.counts.tsv.gz

# 2. Internal Reads
featureCounts -T 1 -p -t exon -g gene_name --fracOverlap 0.25
-a "$GTF" -o FEATURECOUNTS/"$ID"_INTERNAL_featureCounts.txt
STAR/"$ID"_INTERNAL_Aligned.sortedByCoord.filtered.bam
```

11.10 Recommended: Quality Checkups

Refer to FLASH-seq preprocessing.

12 Data Postprocessing

The postprocessing steps will greatly vary depending on the question at hand. In our experience, most tools working for Smart-seq2/Smart-seq3 can be used as well with FLASH-seq-UMI data.

13 Notes

1. The concentration of Triton X-100 is important and should not be significantly increased. Although concentrations of 1% or higher still result in cDNA that is indistinguishable from that obtained with 0.2%, the amount of multi-mapped and unmapped reads increases significantly [5]. Other non-ionic surfactants such as Tween-20, Igepal® CA-630, NP-40, or similar could be used but we did not test them extensively. The use of anionic detergents such as SDS results in the complete inactivation of enzymes even at very low concentrations and should be avoided.
2. Primers should be biotinylated and, whenever possible, HPLC-purified. In the RT reaction, the biotinylation prevents secondary strand-switch events and creation of concatemers.
3. In our hands, there was no difference when replacing Superscript™ IV with Maxima H-RT while using the same reaction buffer and reaction conditions.
4. Commercial alternatives like Ampure XP beads or SPRI beads can be used without any negative impact on the reaction.
5. It is recommended to add Tween-20 at the end, as it tends to foam.
6. Due to its toxicity, it is recommended to add sodium azide at the very end.
7. It is a good practice to titrate every new batch of Sera-Mag SpeedBeads™ side by side against commercial products, such as AMPure XP or SPRI magnetic beads.
8. For all tagmentation protocols described here, indices from the Nextera® index kit can be diluted 1:5 with nuclease-free water or low-EDTA TE buffer (10 mM Tris–HCl, 0.1 mM EDTA, pH 8.0) before combining them on a plate. Each well will contain a unique combination of N7xx and S5xx (each 10 times diluted compared to their original concentration).
9. Tn5MErev primer needs to be pre-annealed in separate vials with Tn5ME-A and Tn5ME-B before loading the two partly double-stranded oligonucleotides on the Tn5 transposase. See [10] for a detailed protocol.
10. Dimethylformamide is a potent liver toxin that is readily absorbed through the skin. Avoid inhalation of vapors and always handle the solution under a fume hood.
11. SDS concentration is extremely important for the inactivation of the Tn5 transposase and for the enrichment of PCR reaction to take place afterward. Best results are obtained when the concentration of SDS is in the range 0.1–0.2%. Do not increase

the concentration further, as already 0.3% leads to a complete failure of the following PCR.

12. Additional sets of index adaptors guarantee higher multiplexing capabilities, thus maximizing the number of cells that can be pooled and sequenced in each run. All index adaptors should carry a 5′-biotin to minimize artifacts as well as a phosphorothioate bond between the last and second last nucleotide at the 3′-end to make them more resistant to exonucleases. IDT is our preferred vendor but several others can manufacture oligos of comparable quality. Please see [5] for an extensive list of additional index adaptors.
13. Always choose fully skirted plates when carrying out the protocol with liquid handling robots, which also have the additional advantage of offering increased rigidity, thus reducing warping during thermal cycling. Twin.tec® (Eppendorf) is our preferred choice.
14. The choice of a good seal that can resist long-term −80 °C storage is rather limited. We recommend AlumaSeal® 384 film (VWR International) for optimal results. For storage at −20 °C of preamplified cDNA or final libraries, use the adhesive plastic foil of your choice.
15. Whenever possible, use both a positive and a negative control on every plate. The negative control well contains only 1 μL cell lysis mix but no cell. The positive control well contains either high-quality total RNA (10–100 pg) or a pool of 10–20 cells.
16. For long-term storage of preamplified cDNA or sequencing-ready libraries, Lobind tubes and/or Lobind Twin.tec® plates should be preferred. This ensures a minimal adsorption of the DNA to the plastic and guarantees that little DNA is lost upon storage.

References

1. Tang F et al (2009) mRNA-Seq whole-transcriptome analysis of a single cell. Nat Methods 6:377–382
2. Zheng GXY et al (2017) Massively parallel digital transcriptional profiling of single cells. Nat Commun 8:14049
3. Hagemann-Jensen M et al (2020) Single-cell RNA counting at allele and isoform resolution using Smart-seq3. Nat Biotechnol 38:708–714
4. Picelli S et al (2013) Smart-seq2 for sensitive full-length transcriptome profiling in single cells. Nat Methods 10:1096–1098
5. Hahaut V, Pavlinic D, Cowan C, Picelli S (2021) Lightning Fast and Highly Sensitive Full-Length Single-cell sequencing using FLASH-Seq. http://biorxiv.org/lookup/doi/10.1101/2021.07.14.452217. https://doi.org/10.1101/2021.07.14.452217
6. Deng Q, Ramsköld D, Reinius B, Sandberg R (2014) Single-cell RNA-seq reveals dynamic, random monoallelic gene expression in mammalian cells. Science 343:193–196
7. Petropoulos S et al (2016) Single-cell RNA-Seq reveals lineage and X chromosome dynamics in human preimplantation embryos. Cell 165:1012–1026

8. Reinius B et al (2016) Analysis of allelic expression patterns in clonal somatic cells by single-cell RNA-seq. Nat Genet 48:1430–1435
9. Zhu YY, Machleder EM, Chenchik A, Li R, Siebert PD (2001) Reverse transcriptase template switching: a SMART™ approach for full-length cDNA library construction. BioTechniques 30:892–897
10. Picelli S et al (2014) Tn5 transposase and tagmentation procedures for massively scaled sequencing projects. Genome Res 24:2033–2040
11. Chen W, Gardeux V, Meireles-Filho A, Deplancke B (2017) Profiling of single-cell transcriptomes. Curr Protoc Mouse Biol 7: 145–175
12. Dobin A et al (2013) STAR: ultrafast universal RNA-seq aligner. Bioinformatics 29:15–21
13. Liao Y, Smyth GK, Shi W (2014) featureCounts: an efficient general purpose program for assigning sequence reads to genomic features. Bioinformatics 30:923–930
14. Bushnell B BBMap – sourceforge.net/projects/bbmap/
15. Li H et al (2009) The sequence alignment/map format and SAMtools. Bioinformatics 25: 2078–2079
16. Patro R, Duggal G, Love MI, Irizarry RA, Kingsford C (2017) Salmon provides fast and bias-aware quantification of transcript expression. Nat Methods 14:417–419
17. Bray NL, Pimentel H, Melsted P, Pachter L (2016) Near-optimal probabilistic RNA-seq quantification. Nat Biotechnol 34:525–527
18. Bolger AM, Lohse M, Usadel B (2014) Trimmomatic: a flexible trimmer for Illumina sequence data. Bioinformatics 30:2114–2120
19. Kim D, Langmead B, Salzberg SL (2015) HISAT: a fast spliced aligner with low memory requirements. Nat Methods 12:357–360
20. Hao Y et al (2021) Integrated analysis of multimodal single-cell data. Cell 184:3573–3587.e29
21. Wolf FA, Angerer P, Theis FJ (2018) SCANPY: large-scale single-cell gene expression data analysis. Genome Biol 19:15
22. Lun ATL, McCarthy DJ, Marioni JC (2016) A step-by-step workflow for low-level analysis of singlecell RNA-seq data with Bioconductor. F1000 Res 5:1–71
23. Hagemann-Jensen M, Ziegenhain C, Sandberg R (2021) Scalable full-transcript coverage single cell RNA sequencing with Smart-seq3xpress. http://biorxiv.org/lookup/doi/10.1101/2021.07.10.451889, https://doi.org/10.1101/2021.07.10.451889
24. Tang DTP et al (2013) Suppression of artifacts and barcode bias in high-throughput transcriptome analyses utilizing template switching. Nucleic Acids Res 41:e44
25. Hochgerner H et al (2017) STRT-seq-2i: dual-index 5′ single cell and nucleus RNA-seq on an addressable microwell array. Sci Rep 7:16327
26. Smith T, Heger A, Sudbery I (2016) UMI-tools: modelling sequencing errors in Unique Molecular Identifiers to improve quantification accuracy. http://biorxiv.org/lookup/doi/10.1101/051755 https://doi.org/10.1101/051755

Chapter 6

Plant Single-Cell/Nucleus RNA-seq Workflow

Sandra Thibivilliers, Andrew Farmer, Susan Schroeder, and Marc Libault

Abstract

Single-cell transcriptomics technologies allow researchers to investigate how individual cells, in complex multicellular organisms, differentially use their common genomic DNA. In plant biology, these technologies were recently applied to reveal the transcriptomes of various plant cells isolated from different organs and different species and in response to environmental stresses. These first studies support the potential of single-cell transcriptomics technology to decipher the biological function of plant cells, their developmental programs, cell-type-specific gene networks, programs controlling plant cell response to environmental stresses, etc. In this chapter, we provide information regarding the critical steps and important information to consider when developing an experimental design in plant single-cell biology. We also describe the current status of bioinformatics tools used to analyze single-cell RNA-seq datasets and how additional emerging technologies such as spatial transcriptomics and long-read sequencing technologies will provide additional information on the differential use of the genome by plant cells.

Key words Plant, Single-cell transcriptomics, Single-cell RNA-seq, Single-nucleus RNA-seq, Protoplasts

1 Introduction

Single-cell –omics analysis emerged as a driving force in biological science upon the development of microfluidic technologies. Plant biologists successfully applied microfluidic-based single-cell transcriptomics for the first time in 2019 [1–5]. The results generated from these studies enhanced our understanding of the differential regulation of gene expression between the cell types composing the *Arabidopsis* root and revealed how powerful the use of such technology could be to better characterize gene function, predict gene networks, and reveal the differential use of the genomic information between the different cells and cell types composing an organ [6, 7]. Subsequently, this technology has been successfully applied on various plant species (i.e., poplar [8], rice [9, 10], maize [11, 12], *Chlamydomonas reinhardtii* [13]) and on different organs (i.e., leaf [10, 14–16], gametophytes [17], the embryo [18], maize

Raffaele A. Calogero and Vladimir Benes (eds.), *Single Cell Transcriptomics: Methods and Protocols*, Methods in Molecular Biology, vol. 2584, https://doi.org/10.1007/978-1-0716-2756-3_6,

ears [11], and the vegetative shoot apex [12, 19]). Taken together, these studies support the potential for the broad application and impact of single-cell transcriptomics in plant biology [20]. However, this presupposes the ability to isolate biological entities from various species and organs.

Before conducting a single-cell RNA-seq experiment, the biological sample used for analysis, the access to isolated biological entities (i.e., isolated plant protoplasts or nuclei; see below for more details), the accessibility to the genome sequence of the species, and the quality of the annotation of the genome must be carefully considered to ensure the generation of high-quality datasets and their proper analysis using bioinformatics packages. In this chapter, we discuss the best practices to manipulate plant biological samples before conducting single-cell/nucleus RNA-seq experiments, provide information about the current bioinformatics packages available to analyze the sequencing results, and share exciting new directions in the analysis of these libraries to develop a more in-depth knowledge of the differential use of the genomic information in plants.

In this book chapter, we highlight the pros and cons associated with the use of plant protoplasts and isolated nuclei when conducting single-cell/nucleus RNA-seq experiments, provide information regarding the critical steps associated with the isolation of plant protoplasts and nuclei, and share important indicators to evaluate the quality of protoplasts or nuclei suspensions.

Biosynthesis of Single-Cell and Single-Nucleus RNA-seq Libraries Using Microfluidic Technologies

The isolation of high-quality plant protoplasts or nuclei is required to generate biologically meaningful single-cell/nuclei complementary DNA (cDNA) libraries. The use of low-quality protoplasts or nuclei (e.g., damaged or burst cells/nuclei) will result in a decrease in the number of biological entities analyzed, and in a lower number of sequenced reads per cell/nucleus, a consequence of the loss of polyadenylated transcripts. While increasing the depth of sequencing might partially help compensate for the latter problem, we consider that the optimization of the isolation, manipulation, and processing of protoplasts and nuclei is required before considering conducting single-cell/nucleus transcriptomic analyses. Here, we provide information about the critical steps associated with the isolation of plant nuclei and protoplasts (Fig. 1).

When comparing the information collected from both entities, it is important to consider that the nuclear transcriptome better reflects the set of actively transcribed genes at the time of the isolation (i.e., a snapshot of the transcriptional activity of the genes), while the protoplasts' transcriptome, a product of the integration over time of the mRNA considering the half-life of the mRNA, better reflects the pool of mRNA molecules present for translation.

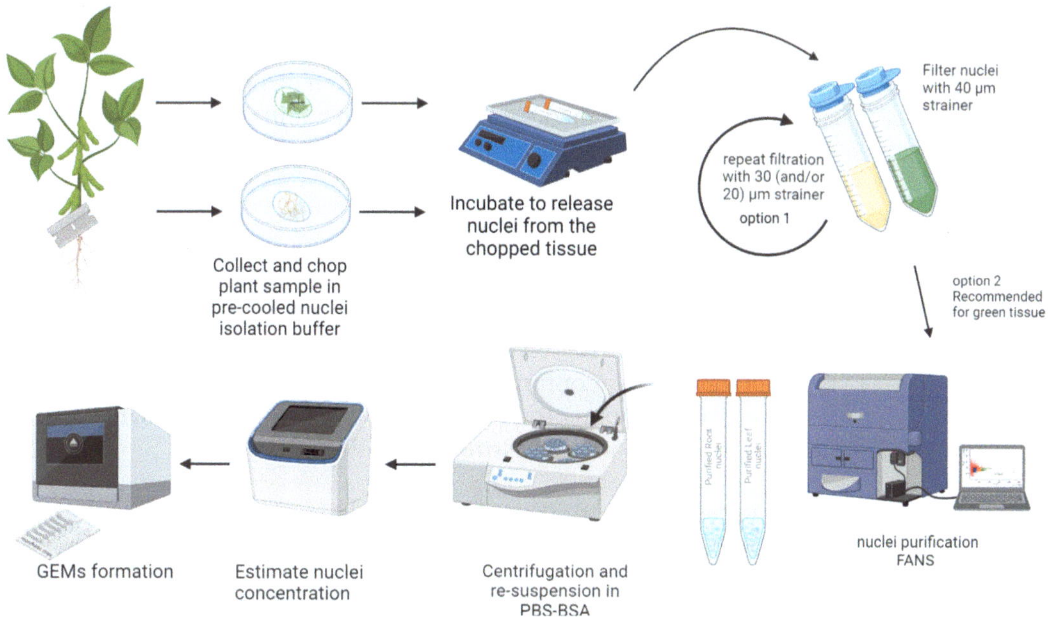

Fig. 1 Schematic representation of the methodology used to isolate plant nuclei or protoplasts before biosynthesis of Illumina-compatible single-cell/nucleus RNA-seq libraries. (This figure was created by using Biorender)

2 Materials

We strongly recommend working with high-quality chemical reagents; preferentially those suggested by single-cell biology suppliers. Using other chemicals may severely negatively influence the quality of the library. If the alternative product(s) is to be used, they should be validated to confirm their quality when conducting the single-cell gene expression assay. Special attention needs to be given to the cleanup and size selection of the magnetic beads (e.g., SPRIselect Reagent Kit and Dynabeads reagents) used to purify the cDNA fragments. These reagents require a thorough resuspension of the beads, which pellet easily, to perform an accurate size selection of the cloned fragments (*see* **Note 1**). When considering nuclei isolation, high-quality DNase- and RNase-free Bovine Serum Albumin (BSA) is recommended (*see* **Note 2**).

3 Methods

3.1 Preparation of Protoplast Suspensions

To date, most of the plant single-cell transcriptomic analyses were conducted on isolated protoplasts. Access to protoplasts requires the optimization of the cocktails of enzymes used to digest the cell wall and the resuspension media (e.g., the osmotic pressure of the resuspension media strongly influences the viability of the plant protoplasts) [21]. As part of the criteria for optimization, the enzymatic composition of these cocktails must consider the biochemical composition of the cell wall. For instance, the cell wall composition significantly differs between plant species (e.g., monocotyledons and dicotyledons [22]). In addition, the digestibility of the cell wall might vary between cell layers due to the differential accessibility of the cell layers to the enzymes and the unique composition in cell wall polymers. For instance, we noticed that the suberization and deposition of cutin in a subset of the cells composing the *Arabidopsis* root prevented the characterization of their transcriptome [23]. Taken together, it is important to evaluate the heterogeneous representation of the different cell types composing the organ when optimizing protoplastization.

Critical points to consider when isolating high-quality protoplasts are reported in **Notes 3–6**.

3.2 Isolation of Plant Nuclei

One alternative to the use of protoplasts to establish the transcriptome of plant single cells consists in measuring the abundance of transcripts from isolated nuclei. Compared to protoplasts, isolated nuclei have several advantages but also some limitations [24]. Considering their advantages, the plant nuclei isolation method required only minor changes to be broadly applied to various plant species and organs [25]. The method described below will also lead to a more homogeneous release of the nuclei from the different cell layers and cell types composing the organ. Therefore, the population of nuclei processed better represents the cellular complexity of the organ analyzed. In addition, the smaller size in the nuclei compared to protoplasts allow their use in any microfluidic technology. Finally, plant nuclei can be used to access multiple molecular features enhancing correlation analyses (e.g., correlation analysis of single-nuclei RNA-seq and ATAC-seq datasets [23, 26]). Despite these advantages, it is important to be aware of several limitations associated with the use of isolated nuclei in transcriptomic analysis. First, the more limited pool of polyadenylated transcripts in a nucleus versus a protoplast would lead to the detection of a lower number of transcripts and expressed genes per biological entities. Therefore, the disruption of the nuclear membrane during the isolation of the nuclei leading to RNA leakage might have even more important consequences on the depth of the transcriptomic information generated. Hence, nuclei isolation must be conducted with great care as described below.

Published isolation methods rely on the release of the plant nuclei upon chopping of the plant samples in a buffer and the purification of the nuclei by combining low-speed centrifugations and Fluorescent-Activated Nuclei Sorting (FANS). The use of this method on *Arabidopsis*, poplar, and maize led to the generation of single-nuclei RNA-seq and ATAC-seq libraries and the characterization of the transcriptome and/or epigenome of selected organs from these species at the single-cell level [23, 26, 27]. We expect that the broader use of this method on various crop species and its continuous enhancement will expand our knowledge of single-cell transcriptome and will enable comparative analysis of the use of plant genomes both between cell types and between homologous cell types of different species.

Critical points to consider when isolating high-quality plant nuclei are reported in **Notes** from **7** to **16**.

3.3 Bioinformatics Analysis of Plant Single-Cell Transcriptomes

Compared with the challenges encountered by plant researchers in creating suitable libraries for single-cell sequencing, it has so far been relatively straightforward to adopt workflows previously developed for the downstream analysis of such data by the community of computational researchers working on single-cell methods focused primarily on data from animal systems. With few exceptions (e.g., [18, 28]), published single-cell transcriptome studies in plants to date have used droplet-based methods (10× genomics Chromium, DropSeq), lending themselves to perform primary analysis of read data using standard pipelines for these platforms (https://support.10xgenomics.com/single-cell-gene-expression/software/pipelines/latest/what-is-cell-ranger, https://github.com/broadinstitute/Drop-seq). These take care of read alignment and UMI-based counting against annotated reference genomes, as well as estimation of which barcodes belong to empty droplets and which are associated with true cells, to produce cell-by-gene counts matrices. This primary analysis has generally been followed by the application of toolkits such as Seurat [29] which implements a wide range of secondary analysis methods including batch effects correction and integration, dimensionality reduction (PCA, tSNE, UMAP), unsupervised clustering, and between-cluster differential expression characterization for marker genes. In addition, more specialized techniques such as trajectory inference have been applied, as implemented in a wide variety of tools [30], generally to explore developmental changes or stress-induced changes. Since best practices associated with these techniques have been well covered elsewhere [31], we will focus attention here on aspects of such analyses that merit particular consideration due either to special characteristics of plant genomes or to the nature of the library preparation methods used in plants.

3.3.1 Bioinformatics Analysis of Protoplast- and Nuclei-Based Single-Cell Transcriptomes

Most single-cell transcriptome studies in plants have utilized protoplasting as the means of dissociating cells from the rigid cell walls that surround them in living plant tissues into independent units for encapsulation in droplet-based or plate well-based platforms. As is well known, the enzymatic digestion of the cell wall performed during protoplasting can alter the expression of hundreds of genes, and such treatment-induced changes should be accounted for when performing downstream analyses; for example, correlation of pseudo-bulked protoplast-derived transcriptome data against bulk RNA-seq data have been shown to increase when genes known to be influenced by protoplasting are excluded [3, 23]. Similarly, protoplasting may yield populations of dissociated cells biased in composition toward cell types more accessible to the action of the enzymatic treatment or less prone to subsequent bursting; for example, in comparisons of protoplast-derived single-cell transcriptome data and single-nucleus data from *Arabidopsis* roots, several clusters were found whose composition was heavily biased toward or exclusively formed from the single-nucleus data, presumably due to characteristics of the cell types from which they were derived. Nuclear isolation is a technique that has been used in animal tissues such as those found in the brain [32] and liver [33] where cell morphological characteristics can prevent the successful dissociation of whole cells. This technique has also now been applied in plants to contexts like early *Arabidopsis* embryo development where protoplasting would be extremely challenging [18], as well as plant tissues such as the syncytial early developing endosperm where single-nucleus sequencing may be the only option to resolve within-tissue heterogeneity [34]. However, analysis of nuclear-derived transcript data comes with its own set of challenges, many of which are due to the reduced quantities of RNA available to library construction, leading generally to sparser representation of the transcriptome for any individual data point. However, a lack of systematic bias in this regard leads to comparable representation of genes and their relative abundances when correlation is made across values aggregated at the level of clusters obtained from unsupervised methods. Quality control metrics such as the abundance of organelle-derived reads can also be expected to have significantly different profiles in libraries derived from nuclei. Single-nucleus data has also been observed often to contain higher levels of ambient RNA than in corresponding single-cell data [33], although methods for subtracting out this background based on analysis of the average contents of non-cell droplets have been developed [35].

3.3.2 Considering Introns When Mapping Single-Nuclei RNA-seq Transcripts

Another major difference expected between nuclear-derived transcripts and those characteristics of the cell as a whole is the degree to which the population of transcripts contains fully spliced mature transcripts versus transcripts that contain introns due to incomplete processing by the spliceosome. Earlier versions of the 10× Genomics Cell Ranger pipeline required that annotation files be specially processed for single-nucleus analysis, in order to allow intron-matching reads to be counted as properly belonging to their parent transcripts, while current versions have deprecated this approach and provided an "include-introns" option to the mkref command to obtain similar behavior. Given that the 10× 3′ Gene Expression protocol produces reads anchored at the polyA tail of processed transcripts, it may be surprising that intron-containing reads would be present at all, regardless of whether the origin is nuclear or cytoplasmic; however, it has been shown that these reads can be generated from priming off of polyA/polyT stretches contained within intron sequences [36] which tend to be an order of magnitude larger in animal species than in plants [37]. Intron-derived reads could simply be viewed as a nuisance factor, complicating normalization procedures for comparisons of relative abundances of genes. However, a specialized analysis methodology known as RNA velocity has been developed that utilizes estimated ratios of spliced and unspliced transcripts in order to determine whether a given cell is in the process of up- or downregulating a particular gene, or whether the transcriptional dynamics are in a steady state in which the rate of production of new transcripts is equal to the rate of degradation of processed transcripts [36, 38]. Of plant single-cell transcriptome data sets published to date, only a few have reported results for RNA velocity analysis, and these have generally indicated that the proportion of intron-containing reads is only a few percent and that the confidence of velocity estimates are therefore likely to be lower than when applied to human data, where the proportion of informative reads is around an order of magnitude higher [5, 39]. This may be an area where plant systems will benefit from the application of long-read sequencing techniques to single-cell/nucleus libraries [40]. It should be noted too that the application of RNA velocity to single-nucleus data is so far relatively uncommon even in animal systems [41, 42], and questions may remain as to whether different dynamical models ought to be applied in these cases [43].

3.3.3 Cross-Species Integration of Single-Cell Transcriptome

One aspect of single-cell analysis in plants that remains challenging, especially as the diversity of species assayed with these technologies continues to grow, is the assignment of cell-type identities to the clusters produced by unsupervised methods. The earliest single-cell transcriptomic studies in plants were all applied to the developing roots of *Arabidopsis*, where much previous work in the model system resulted in marker genes for the expected cell types and developmental stages [44, 45] in addition to different cell states such as the progression through phases of the cell cycle and entry into endoreduplication [46] followed by the acquisition of different endoploidy levels [47]. Although curated resources have recently been developed to mine literature for marker genes across *Arabidopsis* and a handful of crop species [48], and formalized methods such as the Index of Cell Identity (ICI, [49]) have been applied to some of the *Arabidopsis* single-cell datasets, the determination of the meaning of clusters obtained and the interpretation of the structuring of the outputs of nonlinear dimensionality reduction methods like tSNE and UMAP continue to be time-consuming and subjective procedures. Ideally, some conservation of cell-type specificity across evolutionary time would allow label transfer; however, the complex many-to-many relationships seen in many gene families due to the prevalence of whole-genome duplication and other more localized duplication events in plant genomes often lead to subfunctionalization that can partition expression into cell-type specificity between paralogs that would be expected to give different patterns for duplications occurring independently in different species [50]. Further, more recently duplicated genes within a plant genome can also be quite difficult to discriminate due to high sequence similarity, leading to issues of interpretation of non-uniquely mapped reads, especially when read information is limited to a subset of the full transcript. Thus, methods developed for cross-species integration that could enable sharing of information about the type-identity of differing expression profiles must be able to take into account these more complex family relationships in order to be relevant to application in plant genomic contexts [51]. Another approach that shows much promise for helping to resolve the identities of transcriptional states is the integration of single-cell data with spatial transcriptomic data [52].

3.3.4 Future Directions in the Field of Plant Single-Cell Transcriptomic

Spatial Transcriptomic Analysis Using Visium and Probe-Based Technologies

Single-cell transcriptomic leads to the characterization of the transcriptome of each cell and cell types composing a cellularly complex organ. However, per se, this knowledge is biologically limited if we cannot functionally annotate the cells of the organs based on their transcriptomic profiles. When considering model species (e.g., *A. thaliana* in plants), decades of molecular studies reveal the genes specifically expressed in various organs and cell types [44, 45, 53–55]. This knowledge has been utilized to functionally annotate single-cell clusters according to their transcriptomic profiles [1–5, 17, 23, 56]. When considering non-model plant species, where foundational knowledge on cell-type marker genes is more limited, alternative strategies need to be implemented to better bridge the transcriptome of the cells with their biological functions. Emerging strategies are the use of spatial transcriptomic and molecular cartography of gene expression.

Spatial transcriptomic (e.g., 10× Genomics Visium Spatial Gene Expression) technology uses ≈5000 spatially barcoded oligonucleotides that are printed on a 6.5 mm × 6.5 mm glass microscopic slide. Upon their release from the cells, the polyadenylated transcripts are attached to the printed oligonucleotides. Reverse transcription is performed, creating a library of barcoded cDNAs. The unique location of each barcoded oligonucleotide on the capture area allows the mapping of the sequenced cDNA in the context of the morphology of the tissue. In plants, this technology has been applied to *Populus tremula, A. thaliana, and Picea abies* [57, 58]. One current limitation of this technology is its resolution which does not support the analysis of gene expression at the single-cell level. The use of bioinformatics deconvolution tools could help in the analysis of the application of this technology, which currently suffers from its moderately low resolution (i.e., 55 μm in diameter). However, a recent press release suggests that a high-definition version of this technology will be soon accessible to biologists (https://www.bio-itworld.com/news/2021/01/14/10x-genomics-million-cell-platform-high-def-spatial-transcriptomics-more). Considering that 10× Genomics Visium Spatial Gene Expression technology does not require prior knowledge of cell-type marker genes allowing their de novo discovery, this technology might become an attractive solution to characterize at the single-cell-level resolution the activity of plant genes and bridge this information with the functional annotation of plant cells.

Molecular cartography (e.g., from Resolve Biosciences) nicely complements spatial transcriptomic technology. This technology allows the analysis of gene activity at the highest resolution (i.e., subcellular localization). However, it requires the design of fluorescent probes to interrogate the transcriptional activity of hundreds of genes. Therefore, this technology requires prior knowledge of the expression of plant genes to support their selection to establish their molecular cartography.

Nanopore Sequencing for More In-Depth Discoveries of Plant Gene Regulatory Mechanisms

Oxford Nanopore Technologies has developed nanopore-sequencing devices that are capable of reading single molecules of nucleic acid rapidly and cost-effectively. Nanopore sequencing directly reads DNA and RNA, detects nucleotide methylations, and achieves read lengths up to 4.2 Mb [59–61] (https://nanoporetech.com/about us/news/blog-kilobases-whales-short-history-ultra-long-reads-and-high-throughput-genome). The nanopore-sequencing technology uses a motor protein in a pore in a synthetic bilayer to transfer a single molecule of DNA or RNA across the membrane. Sample preparation involves attaching an adapter sequence that is specific for the motor protein that hydrolyzes ATP while translocating the nucleic acid strand across the synthetic membrane. The voltage in picoamperes is measured as a function of time and sampled at 4 kHz. The voltage varies depending on the sequence of 5–6 nucleotides that is located in the channel of the pore. Base calling can be done from analysis of the voltage directly or squiggle patterns in the data [62].

The advantages of nanopore sequencing are long-read single-molecule data in a scalable, cost-effective format. The single-molecule, long reads enable unambiguous splicing isoform identification, which can be challenging in plant genomes with complex alternative splicing gene regulation [63, 64]. The main disadvantage of the Oxford Nanopore sequencing approach has been lower accuracy than the gold-standard Illumina sequencing. However, technology developments in motor protein engineering and bioinformatic analysis with machine learning algorithms have steadily improved sequencing accuracy. The most recently released Q20+ technology and R10.2 flow cells achieve 99.5% base calling accuracy [65]. Nanopore technology is a potentially disruptive innovation for traditional next-generation sequencing with Illumina. Currently, many approaches combine the high accuracy of Illumina sequencing with nanopore long-read sequencing to achieve both long-reads and high accuracy.

The R2C2 method uses rolling circle amplification to further improve the base calling accuracy for barcodes in single-cell 10× Genomics approaches. High-accuracy base calling for short barcodes is essential for single-cell transcriptome studies with 10× Genomics workflows. The rolling circle amplification of a single barcoded transcript generates multiple copies of the same transcript and barcode sequence on a single molecule. Taking advantage of nanopore long-read sequencing, a consensus sequence can be generated from the analysis of the multiple copies of a transcript on a single molecule. When applied to human T cells, R2C2 methods improved the base calling accuracy from 95% to 98% using the earlier R9.4 minion flow cells [66]. Even as the base calling accuracy steadily improves, methods such as R2C2 for reducing errors in barcode sequencing will be important for fully optimizing efficient single transcriptome sequencing.

The input for the R2C2 protocol is the reverse-transcribed cDNA library from a 10× Genomics workflow. The steps in the R2C2 protocol include the following:

1. Generate cDNA and reamplify if necessary.
2. Generate DNA splint.
3. Circularize the cDNA with a DNA splint.
4. Rolling circle amplification with phi29 polymerase.
5. DNA repair and adapter ligation for nanopore sequencing.

At each step, digestion of unreacted DNA and side products, such as branched DNA, are digested, and the product is bead purified before the next step. The DNA splint contains UMI for multiplexing and ends that are compatible with cDNA ends. The R2C2 protocol includes 16 DNA splint sequences that can be purchased as IDT Ultramer oligos that are compatible with standard Smart-seq2 cDNA [66].

R2C2 methods and 10× Genomics methods have been combined for studies in human cells [66, 67], but not yet in plant cells. Because plants have a complex alternative splicing and plant transcription factors use alternative splicing to regulate responses to abiotic stress [63], the R2C2 approach with 10× Genomics has the potential to generate new insights into plant gene regulatory networks. Traditional Illumina short-read sequencing may provide only ambiguous isoform identification, and PacBio single-molecule sequencing methods do not yet have cost-efficient high throughput for single-cell and perturbation screening applications. This R2C2 approach is modular and will enable more information about alternative splicing regulation to be collected in concert with other omics methods.

Alternative approaches to solving the challenges of nanopore sequencing accuracy for bar codes in single-cell transcriptomes have been applied to plant transcriptomics. The method FlsnRNAseq (full-length single-nucleus RNA-Seq) uses a combination of Illumina high-accuracy short-read sequencing with nanopore long-read sequencing for transcriptomics in *Arabidopsis* roots and endosperm [40]. After the 10× barcoding step, the library was split and then nanopore and Illumina sequencing were conducted in parallel. The authors developed a bioinformatic pipeline, snuppy (single-nucleus utility in python), to use the results from the Illumina sequencing to guide the alignments of the nanopore data. Similar yields of single-nucleus transcriptomes were generated from both branches in the flsnRNAseq protocol. Interestingly, 54.3% of the plant nuclear RNA is incompletely spliced, which makes the long-read sequencing from nanopore very useful in identifying new cell types. In *Arabidopsis* root cells, new cellular subtypes were identified in mature non-hair and cortex cell types based on different splicing patterns. For example, transcripts at AT3G19010 which includes JAZ7, a splicing regulator in response to jasmonate, were largely unspliced in the second subcell type of cluster 2 mature nonhair cells. Thus, nanopore long-read sequencing has the potential to reveal cell-type-specific regulation of alternative splicing. A variety of bioinformatic approaches to the challenges of bar code sequencing accuracy with nanopore data methods being developed in human and mouse cells, such as scCOLORSeq [68], ScNaUMI-seq [69], may also be applied to plant studies in the future. Bioinformatics workflows such as FLAMES demonstrate the potential for long-read nanopore sequencing and Illumina data to provide a multi-omics perspective with the integration of 10× scRNASeq and scATACSeq [70] to probe in exquisite detail the synergy between transcriptional start sites and isoform and its varied regulation in different cell types.

4 Notes

1. It is recommended to use SPRIselect Reagent Kit instead of AMPure XP Beads, as the former leads to more accurate size selection. Inconsistent size selection of the DNA fragment between samples would create technological and not biological variations between replicates and/or conditions.

2. The addition of BSA will facilitate the resuspension of the nuclei upon centrifugation. As a note, using over 1% of BSA in the nuclei buffer does not interfere with the quality of the cDNA library.
3. The release of protoplasts requires the use of optimized enzymatic cocktails. It is important to consider that cell wall digestibility also differs between cell layers based on the biochemical composition of their cell wall and the accessibility of the enzymes to the cell wall.
4. Upon generation, the protoplasts are extremely fragile and prone to bursting if not manipulated properly. Their bursting should also be considered when manipulating large size protoplasts in microfluidic devices due to the size exclusion of the microfluidic system.
5. The protoplast transcriptome will provide deeper information than the nuclear transcriptome, a consequence of the larger population of polyadenylated transcripts.
6. Protoplastization leads to the activation of stress-related genes that can interfere with the interpretation of the transcriptomic data.
7. The plant sample should be processed at 4 °C to prevent RNA degradation, and maximize the integrity of the nuclear membrane.
8. Isolating high-quality nuclei is key to obtaining a meaningful single-nuclei RNA-seq library and relies on mechanical dissociation of the plant sample to release the nuclei. One strategy consists in finely chopping the plant sample in an optimized nuclei isolation buffer to release the nuclei from different cell types and protect their integrity. Alternatively, gently grinding flash-frozen *A. thaliana* tissue into mortar and pestle can also lead to the release of nuclei [71].
9. Cell strainer should be used to remove debris. Depending on the contamination level of the nuclei by plant debris, or chloroplasts, the nuclei may be purified using Fluorescent-Activated Nuclei Sorting (FANS) [23, 27]. We recommend the use of propidium iodide or SYTOX green. As a note, the spectra of excitation and emission of the SYTOX green help differentiate the stained nuclei versus the chloroplasts.
10. When washing the nuclei, use low-speed centrifugation to pellet them. Gently resuspend the pellet of nuclei in PBS-BSA (0.1–1%). All pipetting needs to be done with wide-opening tips and at a slow rate to not shear or damage the nuclei. Keeping the integrity of the nuclei is critical, as a low-quality input into the microfluidic system will lead to a low-quality library (lack of diversity in the mRNA detected).

11. Confirm that the biological entities are not forming clumps. Clumps might clog the microfluidic system leading to a limited yield in the cDNA library.
12. Carefully estimate the density of protoplasts/nuclei resuspension before initiating the creation of the cDNA libraries. An underestimation of this density would require additional sequencing to properly saturate the sequencing of the transcriptome of each nucleus/protoplast and will increase the loss of usable data due to the formation of multiplets. Contrarily, an overestimation of this density will lead to the analysis of a smaller population of nuclei/protoplasts than desired. This would reduce the amount of information collected especially when considering samples with high cellular complexity. We recommend the use of an automated cell counter to minimize the time between the initiation of the isolation of nuclei and their use to generate single-nuclei RNA-seq libraries.
13. The isolation of plant nuclei is relatively rapid (<1 h in total) and requires no or very limited optimization. The absence of enzymatic treatment and incubation reduces the occurrence of transcriptional regulation; notably in response to stress [e.g., decrease in the transcriptional response of the genes differentially expressed upon protoplastization in *Arabidopsis* [23]].
14. Estimate the number of biological entities to analyze based on the cellular complexity of the samples and the estimated percentage of each cell type. This information can be estimated using predictive online tools (e.g., https://satijalab.org/howmanycells/).
15. Equipment: In addition to accessing a microfluidic system and its associated chips, several additional items are needed to successfully generate plant single-cell/nucleus RNA-seq libraries. It ranges from DNase/RNAse free and low-binding/retention pipette tips, gutter, and conical tubes. We also recommend the use of a cold block on ice to maintain the sample at 4 °C during the entire preparation of the cDNA library.
16. Once the nuclei are isolated, purified, and their concentration is established, the construction of the Single Cell Gene Expression Assay will start with the generation of the GEMs when using 10× Genomics technology. An accurate pipetting is essential. This accuracy is reached by not using wide-bore tips. Hence, at this stage, the pipetting needs to be especially slow to not damage the nuclei in this step.

References

1. Denyer T, Ma X, Klesen S, Scacchi E, Nieselt K, Timmermans MCP (2019) Spatiotemporal developmental trajectories in the Arabidopsis root revealed using high-throughput single-cell RNA sequencing. Dev Cell 48(6): 840–852.e845. https://doi.org/10.1016/j.devcel.2019.02.022
2. Ryu KH, Huang L, Kang HM, Schiefelbein J (2019) Single-cell RNA sequencing resolves molecular relationships among individual plant cells. Plant Physiol 179(4):1444–1456. https://doi.org/10.1104/pp.18.01482
3. Shulse CN, Cole BJ, Ciobanu D, Lin J, Yoshinaga Y, Gouran M, Turco GM, Zhu Y, O'Malley RC, Brady SM, Dickel DE (2019) High-throughput single-cell transcriptome profiling of plant cell types. Cell Rep 27(7): 2241–2247.e2244. https://doi.org/10.1016/j.celrep.2019.04.054
4. Zhang TQ, Xu ZG, Shang GD, Wang JW (2019) A single-cell RNA sequencing profiles the developmental landscape of Arabidopsis root. Mol Plant 12(5):648–660. https://doi.org/10.1016/j.molp.2019.04.004
5. Jean-Baptiste K, McFaline-Figueroa JL, Alexandre CM, Dorrity MW, Saunders L, Bubb KL, Trapnell C, Fields S, Queitsch C, Cuperus JT (2019) Dynamics of gene expression in single root cells of Arabidopsis thaliana. Plant Cell 31(5):993–1011. https://doi.org/10.1105/tpc.18.00785
6. Xie Y, Jiang S, Li L, Yu X, Wang Y, Luo C, Cai Q, He W, Xie H, Zheng Y, Xie H, Zhang J (2020) Single-cell RNA sequencing efficiently predicts transcription factor targets in plants. Front Plant Sci 11:603302. https://doi.org/10.3389/fpls.2020.603302
7. Li S, Yan H, Lee J (2021) Identification of gene regulatory networks from single-cell expression data. Methods Mol Biol 2328:153–170. https://doi.org/10.1007/978-1-0716-1534-8_9
8. Li H, Dai X, Huang X, Xu M, Wang Q, Yan X, Sederoff RR, Li Q (2021) Single-cell RNA sequencing reveals a high-resolution cell atlas of xylem in Populus. J Integr Plant Biol 63(11):1906–1921. https://doi.org/10.1111/jipb.13159
9. Liu Q, Liang Z, Feng D, Jiang S, Wang Y, Du Z, Li R, Hu G, Zhang P, Ma Y, Lohmann JU, Gu X (2021) Transcriptional landscape of rice roots at the single-cell resolution. Mol Plant 14(3):384–394. https://doi.org/10.1016/j.molp.2020.12.014
10. Wang Y, Huan Q, Li K, Qian W (2021) Single-cell transcriptome atlas of the leaf and root of rice seedlings. J Genet Genomics 48(10): 881–898. https://doi.org/10.1016/j.jgg.2021.06.001
11. Xu X, Crow M, Rice BR, Li F, Harris B, Liu L, Demesa-Arevalo E, Lu Z, Wang L, Fox N, Wang X, Drenkow J, Luo A, Char SN, Yang B, Sylvester AW, Gingeras TR, Schmitz RJ, Ware D, Lipka AE, Gillis J, Jackson D (2021) Single-cell RNA sequencing of developing maize ears facilitates functional analysis and trait candidate gene discovery. Dev Cell 56(4):557–568.e556. https://doi.org/10.1016/j.devcel.2020.12.015
12. Satterlee JW, Strable J, Scanlon MJ (2020) Plant stem-cell organization and differentiation at single-cell resolution. Proc Natl Acad Sci U S A 117(52):33689–33699. https://doi.org/10.1073/pnas.2018788117
13. Ma F, Salomé PA, Merchant SS, Pellegrini M (2021) Single-cell RNA sequencing of batch Chlamydomonas cultures reveals heterogeneity in their diurnal cycle phase. Plant Cell 33(4): 1042–1057. https://doi.org/10.1093/plcell/koab025
14. Lee LR, Wengier DL, Bergmann DC (2019) Cell-type-specific transcriptome and histone modification dynamics during cellular reprogramming in the Arabidopsis stomatal lineage. Proc Natl Acad Sci U S A 116(43): 21914–21924. https://doi.org/10.1073/pnas.1911400116
15. Liu Z, Zhou Y, Guo J, Li J, Tian Z, Zhu Z, Wang J, Wu R, Zhang B, Hu Y, Sun Y, Shangguan Y, Li W, Li T, Hu Y, Guo C, Rochaix JD, Miao Y, Sun X (2020) Global dynamic molecular profiling of stomatal lineage cell development by single-cell RNA sequencing. Mol Plant 13(8):1178–1193. https://doi.org/10.1016/j.molp.2020.06.010
16. Lopez-Anido CB, Vatén A, Smoot NK, Sharma N, Guo V, Gong Y, Anleu Gil MX, Weimer AK, Bergmann DC (2021) Single-cell resolution of lineage trajectories in the Arabidopsis stomatal lineage and developing leaf. Dev Cell 56(7):1043–1055.e1044. https://doi.org/10.1016/j.devcel.2021.03.014
17. Song Q, Ando A, Jiang N, Ikeda Y, Chen ZJ (2020) Single-cell RNA-seq analysis reveals ploidy-dependent and cell-specific transcriptome changes in Arabidopsis female gametophytes. Genome Biol 21(1):178. https://doi.org/10.1186/s13059-020-02094-0

18. Kao P, Schon MA, Mosiolek M, Enugutti B, Nodine MD (2021) Gene expression variation in Arabidopsis embryos at single-nucleus resolution. Development 148(13):dev199589. https://doi.org/10.1242/dev.199589
19. Zhang TQ, Chen Y, Wang JW (2021) A single-cell analysis of the Arabidopsis vegetative shoot apex. Dev Cell 56(7):1056–1074.e1058. https://doi.org/10.1016/j.devcel.2021.02.021
20. Jha SG, Borowsky AT, Cole BJ, Fahlgren N, Farmer A, Huang SSC, Karia P, Libault M, Provart NJ, Rice SL, Saura-Sanchez M, Agarwal P, Ahkami AH, Anderton CR, Briggs SP, Brophy JA, Denolf P, Di Costanzo LF, Exposito-Alonso M, Giacomello S, Gomez-Cano F, Kaufmann K, Ko DK, Kumar S, Malkovskiy AV, Nakayama N, Obata T, Otegui MS, Palfalvi G, Quezada-Rodriguez EH, Singh R, Uhrig RG, Waese J, Van Wijk K, Wright RC, Ehrhardt DW, Birnbaum KD, Rhee SY, Consortium PCA (2021) Vision, challenges and opportunities for a plant cell atlas. Elife 10: e66877. https://doi.org/10.7554/eLife.66877
21. Reed KM, Bargmann BOR (2021) Protoplast regeneration and its use in new plant breeding technologies. Front Genome Ed 3:734951. https://doi.org/10.3389/fgeed.2021.734951
22. Carré B, Brillouet J-M (1986) Yield and composition of cell wall residues isolated from variuos feedstuffs used for non-ruminant farm animal. J Sci Food Agric 37:341–351. https://doi.org/10.1002/jsfa.2740370403
23. Farmer A, Thibivilliers S, Ryu KH, Schiefelbein J, Libault M (2021) Single-nucleus RNA and ATAC sequencing reveals the impact of chromatin accessibility on gene expression in Arabidopsis roots at the single-cell level. Mol Plant 14(3):372–383. https://doi.org/10.1016/j.molp.2021.01.001
24. Thibivilliers S, Libault M (2021) Plant single-cell multiomics: cracking the molecular profiles of plant cells. Trends Plant Sci 26(6):662–663. https://doi.org/10.1016/j.tplants.2021.03.001
25. Thibivilliers S, Anderson D, Libault M (2020) Isolation of plant root nuclei for single cell RNA sequencing. Curr Protoc Plant Biol 5(4):e20120. https://doi.org/10.1002/cppb.20120
26. Marand AP, Chen Z, Gallavotti A, Schmitz RJ (2021) A cis-regulatory atlas in maize at single-cell resolution. Cell 184(11):3041–3055.e21. https://doi.org/10.1016/j.cell.2021.04.014
27. Conde D, Triozzi PM, Balmant KM, Doty AL, Miranda M, Boullosa A, Schmidt HW, Pereira WJ, Dervinis C, Kirst M (2021) A robust method of nuclei isolation for single-cell RNA sequencing of solid tissues from the plant genus Populus. PLoS One 16(5):e0251149. https://doi.org/10.1371/journal.pone.0251149
28. Kubo M, Nishiyama T, Tamada Y, Sano R, Ishikawa M, Murata T, Imai A, Lang D, Demura T, Reski R, Hasebe M (2019) Single-cell transcriptome analysis of Physcomitrella leaf cells during reprogramming using microcapillary manipulation. Nucleic Acids Res 47(9):4539–4553. https://doi.org/10.1093/nar/gkz181
29. Stuart T, Butler A, Hoffman P, Hafemeister C, Papalexi E, Mauck WM, Hao YH, Stoeckius M, Smibert P, Satija R (2019) Comprehensive integration of single-cell data. Cell 177(7): 1888–1902.e21. https://doi.org/10.1016/j.cell.2019.05.031
30. Saelens W, Cannoodt R, Todorov H, Saeys Y (2019) A comparison of single-cell trajectory inference methods. Nat Biotechnol 37(5): 547–554. https://doi.org/10.1038/s41587-019-0071-9
31. Luecken MD, Theis FJ (2019) Current best practices in single-cell RNA-seq analysis: a tutorial. Mol Syst Biol 15(6):e8746. https://doi.org/10.15252/msb.20188746
32. Grindberg RV, Yee-Greenbaum JL, McConnell MJ, Novotny M, O'Shaughnessy AL, Lambert GM, Arauzo-Bravo MJ, Lee J, Fishman M, Robbins GE, Lin XY, Venepally P, Badger JH, Galbraith DW, Gage FH, Lasken RS (2013) RNA-sequencing from single nuclei. Proc Natl Acad Sci U S A 110(49):19802–19807. https://doi.org/10.1073/pnas.1319700110
33. Andrews TS, Atif J, Liu JC, Perciani CT, Ma XZ, Thoeni C, Slyper M, Eraslan G, Segerstolpe A, Manuel J, Chung S, Winter E, Cirlan I, Khuu N, Fischer S, Rozenblatt-Rosen O, Regev A, McGilvray ID, Bader GD, MacParland SA (2021) Single-cell, single-nucleus, and spatial RNA sequencing of the human liver identifies cholangiocyte and mesenchymal heterogeneity. Hepatol Commun. https://doi.org/10.1002/hep4.1854
34. Picard CL, Povilus RA, Williams B, Gehring M (2021) Transcriptional and imprinting complexity in Arabidopsis seeds at single-nucleus resolution. Nat Plants 7(6):730–738. https://doi.org/10.1038/s41477-021-00922-0
35. Young MD, Behjati S (2020) SoupX removes ambient RNA contamination from droplet-based single-cell RNA sequencing data. Gigascience 9(12):giaa151. https://doi.org/10.1093/gigascience/giaa151

36. La Manno G, Soldatov R, Zeisel A, Braun E, Hochgerner H, Petukhov V, Lidschreiber K, Kastriti ME, Lönnerberg P, Furlan A, Fan J, Borm LE, Liu Z, van Bruggen D, Guo J, He X, Barker R, Sundström E, Castelo-Branco G, Cramer P, Adameyko I, Linnarsson S, Kharchenko PV (2018) RNA velocity of single cells. Nature 560(7719):494–498. https://doi.org/10.1038/s41586-018-0414-6
37. Zhu L, Zhang Y, Zhang W, Yang S, Chen J-Q, Tian D (2009) Patterns of exon-intron architecture variation of genes in eukaryotic genomes. BMC Genomics 10(1):47. https://doi.org/10.1186/1471-2164-10-47
38. Bergen V, Lange M, Peidli S, Wolf FA, Theis FJ (2020) Generalizing RNA velocity to transient cell states through dynamical modeling. Nat Biotechnol 38(12):1408–1414. https://doi.org/10.1038/s41587-020-0591-3
39. Zhang T-Q, Chen Y, Liu Y, Lin W-H, Wang J-W (2021) Single-cell transcriptome atlas and chromatin accessibility landscape reveal differentiation trajectories in the rice root. Nat Commun 12(1):2053. https://doi.org/10.1038/s41467-021-22352-4
40. Long Y, Liu Z, Jia J, Mo W, Fang L, Lu D, Liu B, Zhang H, Chen W, Zhai J (2021) FlsnRNA-seq: protoplasting-free full-length single-nucleus RNA profiling in plants. Genome Biol 22(1):66. https://doi.org/10.1186/s13059-021-02288-0
41. Wen Y, Englund DA, Peck BD, Murach KA, McCarthy JJ, Peterson CA (2021) Myonuclear transcriptional dynamics in response to exercise following satellite cell depletion. iScience 24(8):102838. https://doi.org/10.1016/j.isci.2021.102838
42. Marsh B, Blelloch R (2020) Single nuclei RNA-seq of mouse placental labyrinth development. Elife 9. https://doi.org/10.7554/eLife.60266
43. Bergen V, Soldatov RA, Kharchenko PV, Theis FJ (2021) RNA velocity-current challenges and future perspectives. Mol Syst Biol 17(8): e10282. https://doi.org/10.15252/msb.202110282
44. Birnbaum K, Shasha DE, Wang JY, Jung JW, Lambert GM, Galbraith DW, Benfey PN (2003) A gene expression map of the Arabidopsis root. Science 302(5652):1956–1960. https://doi.org/10.1126/science.1090022
45. Brady SM, Orlando DA, Lee JY, Wang JY, Koch J, Dinneny JR, Mace D, Ohler U, Benfey PN (2007) A high-resolution root spatiotemporal map reveals dominant expression patterns. Science 318(5851):801–806. https://doi.org/10.1126/science.1146265
46. Torii K, Kubota A, Araki T, Endo M (2019) Time-series single-cell RNA-Seq data reveal auxin fluctuation during endocycle. Plant Cell Physiol 61(2):243–254. https://doi.org/10.1093/pcp/pcz228
47. Bhosale R, Boudolf V, Cuevas F, Lu R, Eekhout T, Hu Z, Van Isterdael G, Lambert GM, Xu F, Nowack MK, Smith RS, Vercauteren I, De Rycke R, Storme V, Beeckman T, Larkin JC, Kremer A, Höfte H, Galbraith DW, Kumpf RP, Maere S, De Veylder L (2018) A spatiotemporal DNA endoploidy map of the Arabidopsis root reveals roles for the endocycle in root development and stress adaptation. Plant Cell 30(10):2330–2351. https://doi.org/10.1105/tpc.17.00983
48. Jin J, Lu P, Xu Y, Tao J, Li Z, Wang S, Yu S, Wang C, Xie X, Gao J, Chen Q, Wang L, Pu W, Cao P (2021) PCMDB: a curated and comprehensive resource of plant cell markers. Nucleic Acids Res 50(D1):D1448–D1455. https://doi.org/10.1093/nar/gkab949
49. Efroni I, Ip P-L, Nawy T, Mello A, Birnbaum KD (2015) Quantification of cell identity from single-cell gene expression profiles. Genome Biol 16(1):9. https://doi.org/10.1186/s13059-015-0580-x
50. Coate JE, Farmer AD, Schiefelbein JW, Doyle JJ (2020) Expression partitioning of duplicate genes at single cell resolution in Arabidopsis roots. Front Genet 11. https://doi.org/10.3389/fgene.2020.596150
51. Tarashansky AJ, Musser JM, Khariton M, Li P, Arendt D, Quake SR, Wang B (2021) Mapping single-cell atlases throughout Metazoa unravels cell type evolution. Elife 10. https://doi.org/10.7554/eLife.66747
52. Andersson A, Bergenstråhle J, Asp M, Bergenstråhle L, Jurek A, Fernández Navarro J, Lundeberg J (2020) Single-cell and spatial transcriptomics enables probabilistic inference of cell type topography. Commun Biol 3(1):565. https://doi.org/10.1038/s42003-020-01247-y
53. Dinneny JR, Long TA, Wang JY, Jung JW, Mace D, Pointer S, Barron C, Brady SM, Schiefelbein J, Benfey PN (2008) Cell identity mediates the response of Arabidopsis roots to abiotic stress. Science 320(5878):942–945. https://doi.org/10.1126/science.1153795
54. Gifford ML, Dean A, Gutierrez RA, Coruzzi GM, Birnbaum KD (2008) Cell-specific nitrogen responses mediate developmental plasticity. Proc Natl Acad Sci U S A 105(2):803–808. https://doi.org/10.1073/pnas.0709559105
55. Walker L, Boddington C, Jenkins D, Wang Y, Gronlund JT, Hulsmans J, Kumar S, Patel D, Moore JD, Carter A, Samavedam S,

Bonomo G, Hersh DS, Coruzzi GM, Burroughs NJ, Gifford ML (2017) Changes in gene expression in space and time orchestrate environmentally mediated shaping of root architecture. Plant Cell 29(10):2393–2412. https://doi.org/10.1105/tpc.16.00961

56. Shahan R, Hsu C-W, Nolan TM, Cole BJ, Taylor IW, Vlot AHC, Benfey PN, Ohler U (2020) A single cell *Arabidopsis* root atlas reveals developmental trajectories in wild type and cell identity mutants. bioRxiv:2020.2006.2029.178863. https://doi.org/10.1101/2020.06.29.178863

57. Giacomello S (2021) A new era for plant science: spatial single-cell transcriptomics. Curr Opin Plant Biol 60:102041. https://doi.org/10.1016/j.pbi.2021.102041

58. Giacomello S, Salmén F, Terebieniec BK, Vickovic S, Navarro JF, Alexeyenko A, Reimegård J, McKee LS, Mannapperuma C, Bulone V, Ståhl PL, Sundström JF, Street NR, Lundeberg J (2017) Spatially resolved transcriptome profiling in model plant species. Nat Plants 3(6):17061. https://doi.org/10.1038/nplants.2017.61

59. Xu Y, Luo H, Wang Z, Lam HM, Huang C (2021) Oxford nanopore technology: revolutionizing genomics research in plants. Trends Plant Sci. https://doi.org/10.1016/j.tplants.2021.11.004

60. Jain M, Koren S, Miga KH, Quick J, Rand AC, Sasani TA, Tyson JR, Beggs AD, Dilthey AT, Fiddes IT, Malla S, Marriott H, Nieto T, O'Grady J, Olsen HE, Pedersen BS, Rhie A, Richardson H, Quinlan AR, Snutch TP, Tee L, Paten B, Phillippy AM, Simpson JT, Loman NJ, Loose M (2018) Nanopore sequencing and assembly of a human genome with ultra-long reads. Nat Biotechnol 36(4):338–345. https://doi.org/10.1038/nbt.4060

61. Shafin K, Pesout T, Chang P-C, Nattestad M, Kolesnikov A, Goel S, Baid G, Kolmogorov M, Eizenga JM, Miga KH, Carnevali P, Jain M, Carroll A, Paten B (2021) Haplotype-aware variant calling with PEPPER-Margin-DeepVariant enables high accuracy in nanopore long-reads. Nat Methods 18(11):1322–1332. https://doi.org/10.1038/s41592-021-01299-w

62. McIntyre ABR, Alexander N, Grigorev K, Bezdan D, Sichtig H, Chiu CY, Mason CE (2019) Single-molecule sequencing detection of N6-methyladenine in microbial reference materials. Nat Commun 10(1):579–579. https://doi.org/10.1038/s41467-019-08289-9

63. Chaudhary S, Khokhar W, Jabre I, Reddy ASN, Byrne LJ, Wilson CM, Syed NH (2019) Alternative splicing and protein diversity: plants versus animals. Front Plant Sci 10. https://doi.org/10.3389/fpls.2019.00708

64. Reddy ASN, Marquez Y, Kalyna M, Barta A (2013) Complexity of the alternative splicing landscape in plants. Plant Cell 25(10): 3657–3683. https://doi.org/10.1105/tpc.113.117523

65. Brown C (2021) Nanopore community meeting update. Paper presented at the Nanopore Community Meeting 2021, December 1, 2021

66. Volden R, Vollmers C (2021) Highly multiplexed single-cell full-length cDNA sequencing of human immune cells with 10X Genomics and R2C2. bioRxiv:2020.2001.2010.902361. https://doi.org/10.1101/2020.01.10.902361

67. Volden R, Palmer T, Byrne A, Cole C, Schmitz RJ, Green RE, Vollmers C (2018) Improving nanopore read accuracy with the R2C2 method enables the sequencing of highly multiplexed full-length single-cell cDNA. Proc Natl Acad Sci 115(39):9726–9731. https://doi.org/10.1073/pnas.1806447115

68. Philpott M, Watson J, Thakurta A, Brown T, Brown T, Oppermann U, Cribbs AP (2021) Nanopore sequencing of single-cell transcriptomes with scCOLOR-seq. Nat Biotechnol 39(12):1517–1520. https://doi.org/10.1038/s41587-021-00965-w

69. Lebrigand K, Magnone V, Barbry P, Waldmann R (2020) High throughput error corrected Nanopore single cell transcriptome sequencing. Nat Commun 11(1):4025. https://doi.org/10.1038/s41467-020-17800-6

70. Tian L, Jabbari JS, Thijssen R, Gouil Q, Amarasinghe SL, Voogd O, Kariyawasam H, Du MRM, Schuster J, Wang C, Su S, Dong X, Law CW, Lucattini A, Prawer YDJ, Collar-Fernández C, Chung JD, Naim T, Chan A, Ly CH, Lynch GS, Ryall JG, Anttila CJA, Peng H, Anderson MA, Flensburg C, Majewski I, Roberts AW, Huang DCS, Clark MB, Ritchie ME (2021) Comprehensive characterization of single-cell full-length isoforms in human and mouse with long-read sequencing. Genome Biol 22(1):310. https://doi.org/10.1186/s13059-021-02525-6

71. Sunaga-Franze DY, Muino JM, Braeuning C, Xu XC, Zong ML, Smaczniak C, Yan WH, Fischer C, Vidal R, Kliem M, Kaufmann K, Sauer S (2021) Single-nucleus RNA sequencing of plant tissues using a nanowell-based system. Plant J 108(3):859–869. https://doi.org/10.1111/tpj.15458

Chapter 7

Ensuring Quality Cell Input for Single Cell Sequencing Experiments by Viability and Singlet Enrichment Using Cell Sorting

Malte Paulsen

Abstract

As with every sensitive analysis technology, the golden principle of "input quality rules output quality" also applies to single cell sequencing methods. Given the sensitivity of the current methods in single cell sequencing and the minuscule amounts of RNA present within a single cell, any extrinsic source of variability should be reduced by ensuring a homogenous input right at the start. Not every tissue is as readily handled as a single cell suspension like blood and most tissues will have to undergo digestions to free the cells from their spatial organization to undergo single cell transcriptomics workflows. This chapter provides working protocols for two simple, but very precise and powerful methods to ensure only the most viable cells are introduced into single cell assays.

Key words Single cell RNA sequencing, Cell sorting, Apoptosis, Metabolic labeling, Cell viability, Enrichment

1 Introduction

The key part of any nonspatially resolved single cell sequencing experiment is to obtain cells as single cell suspension. It does not matter if the cells are destined to undergo 10×Genomics [1] or BD-Rhapsody [2] or traditional SmartSeq2 [3] and MARseq [4], for example. Whenever a tissue is digested using enzymatic and chemically assisted tissue dissociation protocols, many cells will be physical destroyed, partially damaged, or might undergo triggering of apoptosis since they are not anymore residing in a solid tissue. Even the best protocol and experienced laboratory technician will never be able to fully avoid this loss. Therefore, it is advisable to use readily available methods to remove debris, dying or dead cells from the feed suspension that is destined to be introduced into a single cell sequencing experiment. Depending on the type of downstream analysis, even debris containing residual RNAs could be a source of

Raffaele A. Calogero and Vladimir Benes (eds.), *Single Cell Transcriptomics: Methods and Protocols*, Methods in Molecular Biology, vol. 2584, https://doi.org/10.1007/978-1-0716-2756-3_7,

noise within a single cell RNA-seq. Of course, there are methods utilizing computational removal of "nonconforming cells" after the wet laboratory work has been completed. The methods rely on assumptions that the increase in the occurrence of certain markers, like mitochondrial RNAs for example, represents a hallmark of an apoptotic or dead cell [5]. Although there is usability of such filters, they are not perfect and should be only used in combination with proper biochemical staining and cell purification methods [6].

As a matter of fact, the difference in the lipid components of the 10×Genomics and BD-Rhapsody lysis buffers tends to show a higher release of mitochondrial RNAs paired with an improved mRNA detection in BD-Rhapsody runs, although the cells were of fantastic viability just prior to the lysis. Were the cells less viable? No, because metabolic viability staining revealed their health. One has to be very careful when relying only on post-assay filtering tools if one is to shy of investing the work of enriching viable cells for single cell RNA-seq studies.

Cell sorting today is a readily available technology allowing for simple purification of cells. Almost every available sorter will be able to perform very simple two-parametric viability cell sorts based on the presence of active Caspase-3/7 probes or the metabolic viability stain Calcein-AM (7) in combination with either PI or DAPI. These two color viability assays will be able to provide a very broad range of usability across many model organisms and cell types.

2 Materials

2.1 Reagent and Buffers

1. The best Caspase-3/7 probes are sold by Sartorius, called Incucyte Caspase 3/7 dyes, either as a blue laser (488 nm) excitable green label (FITC or GFP filter, 530-30 nm bandpass) or as red laser (640 nm) red label (APC or AlexaFluor 647 filter, 670-14 nm bandpass). Catalog numbers are 4440 or 4704, respectively (https://www.sartorius.com/en/applications/life-science-research/cell-analysis/live-cell-assays).
2. The most suitable Calcein probes can be obtained from Invitrogen under the CellTrace Calcein AM label either as a blue laser (488 nm) excitable green label (FITC or GFP filter, 530-30 nm bandpass) or as a violet laser (405 nm) blue label (V450 or BV421 filter, 450-40 nm bandpass). Catalog numbers are C34852 or C34858, respectively.

3. DAPI can be obtained from any source as it is a simple compound. It should be carefully diluted to a 1 mM stock solution. The final concentration needs to be tested but should be around 0.5–2 μM. Sometimes it can be even further diluted.
4. Use 5% FBS or 1% BSA in PBS, pH 7.4, supplemented with 5 mM EDTA as cell suspension buffer for all stains.

2.2 Cell Sorter

Any cell sorter providing a 405 nm, 488 nm, and 640 nm laser excitation will be suitable to enable the entire bandwidth of dye combinations. Typical cell sorters are BD ARIA-type cell sorters, SONY SH800's, BDMelody, Beckman Astrios, or MoFlo. A large nozzle should be used, typically 100 μm or 130 μm, to enhance post sort viability. The lower the pressure used, the better for the cells and the downstream assay quality.

It is advisable to refer to spectral viewers to find the best possible combination according to the experimental conditions (presence of fluorescent proteins, etc.) and the available cell sorter. Generally, if a GFP stain is present in any of the cells of interest, one should use a red Caspase 3/7 or violet Calcein probe combined with PI as live-dead counter stain. If Calcein violet is not an option, then the Calcein violet channel can be used for DAPI live-dead counter staining.

3 Methods

3.1 Caspase-Based Viability Enrichment

Tissue or cells should be dissociated with the most suitable method to obtain a suspension containing as many as possible single cells. The suspension should contain 0.5–1.0 million cells per ml using the 1% FBS, 0.5 mM EDTA PBS buffer. The cells can be stored on ice or at RT depending on their sensitivity (must be explored independently). As a typical, 10×Genomics or BD-Rhapsody experiment only utilizes 10,000–50,000 cells per reaction, one will not need to purify more than 1/2 million cells if viabilities are within 20–30% of all cells.

Add 0.5 μl per ml cell suspension of the Incucyte Caspase 3/7 stain and carefully mix the suspension. Incubate the cells for 5–10 min either on ice or at RT and then add the necessary amount of either PI or DAPI live-dead counter stain to obtain 1 μg/ml PI or 0.5–2 μM DAPI staining concentration. PI and DAPI stainings are very rapid and can be analyzed within 2–3 min after staining.

Load the sample on to the cell sorter and establish proper FSC/SSC gating to select singlet cells—any experience or otherwise trained cell sorter operator will be able to assist you with this step. On the final bivariate dot plot displaying live-dead versus Caspase 3/7 signals (see Fig. 1), draw a final sorting gate around the double negative population. This population represents the

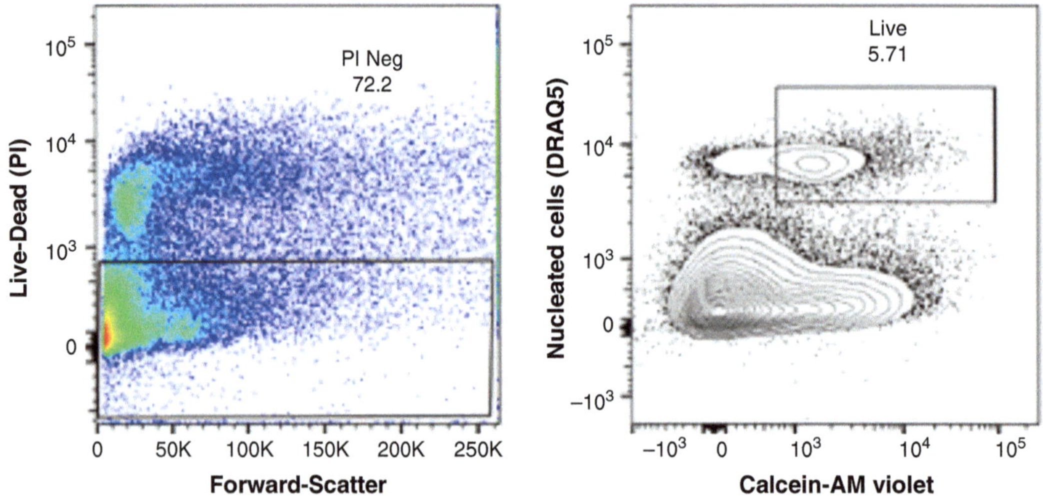

Fig. 1 Bivariate dot plot: live-dead versus Caspase 3/7 signals. Cells are gated first against the live-dead marker PI and then further analyzed on a plot displaying the Calcein-AM. Violet signal indicates health and the presence of strong signal from the viable DNA label DRAQ5 indicates proper cells (*see* **Note 2**). Viable cells are identified by a strong Calcein and DRAQ5 signal

healthiest single cells that will hopefully provide you with the best cellular material for the single cell RNA-seq experiment. The cells should be sorted into a tube containing 300 μl of the 1% FBS, 5 mM EDTA PBS buffer—preferably chilled at 4 °C (*see* **Note 1**).

3.2 Calcein Metabolic Labeling Enrichment

Using Calcein as metabolic viability indicator requires incubation of the cells with the stain for 10–20 min at RT or even 37 °C. Calcein-AM is converted by esterases to be freed from the lipid modification it carries to get across membranes. Once it is cleaved, Calcein is membrane impermeable, and it is retained inside healthy living cells with active esterases. In comparison with the Caspase 3/7 labels, the live cells will have a bright signal when probed with Calcein.

The procedure itself is simple and straightforward: bring your cells into single cell suspension, spin them down using 5 min 90xrcf, and take up in 1 ml 1% FBS, 0.5 mM EDTA PBS buffer. The Calcein-AM substrate is generally dissolved in DMSO and stored as 1 mM stock solution if not otherwise specified by the supplying company. Add 1–2 μl of the stock into the 1 ml of cell suspension and incubate the cells at RT or even better 37 °C for 10 min. If your cells are sensitive to 37 °C incubation, revert to RT and optimize the duration of the incubation. Please note that 4 °C incubation will significantly slow the reaction and you will see very little signal. Should there be a lot of background signal from Calcein, for example, a lack for good separation of low/unstained cells from bright cells, one can implement a washing step prior to sorting to reduce the amount of free Calcein that could be bound to the outside of cells or debris.

Load the sample on to the cell sorter and establish proper FSC/SSC gating to select singlet cells—any experience or otherwise trained cell sorter operator will be able to assist you with this step (*see* **Note 2**). On the final bivariate dot plot displaying live-dead versus Calcein signals. Be mindful to gate the brightest Calcein-labeled population that is negative for the live-dead counter stain (see Fig. 1) (*see* **Note 3**).

4 Notes

1. After the sort, when the cells need to be spun down to remove the EDTA (it has to be removed to enable proper functionality of 10×Genomics or BD-Rhapsody assays), one should only spin with 90-rcf and rather extend the time to pellet the cells. Using classic polypropylene 5 ml FACS tubes paired with swingout rotors will provide the best results with the supernatant is carefully decanted. After spinning and washing the cells, the cell suspension should be immediately introduced into the single cell sequencing assay. In case of plate-based single cell methods, the sorting step is part of the plate preparation if single cell sorting is used.
2. If you deal with very-difficult-to-gate cells because of large amounts of debris, for example, then it is advisable to change the suggested simple setup and include the membrane-permeable DNA stain DRAQ5 (best obtained by BioStatus, http://www.biostatus.com/DRAQ5/). Particles with a bright DRAQ5 signal (640 nm excitation, 670/14 emission) will be particles containing high amounts of DNA, which should be the cells. Low signal for the live-dead dye and corresponding signal for the apoptosis or metabolic label will provide additional security to select the best cells (for more on this, [7]).
3. 96, 384, or 1536 plates should be immediately shock-frosted after completion of the sort using a mixture of dry-ice and 80% EtOH. Be careful to always wear proper PPE.

References

1. Zheng GX, Terry JM, Belgrader P, Ryvkin P, Bent ZW, Wilson R, Ziraldo SB, Wheeler TD, McDermott GP, Zhu J, Gregory MT, Shuga J, Montesclaros L, Underwood JG, Masquelier DA, Nishimura SY, Schnall-Levin M, Wyatt PW, Hindson CM, Bharadwaj R, Wong A, Ness KD, Beppu LW, Deeg HJ, McFarland C, Loeb KR, Valente WJ, Ericson NG, Stevens EA, Radich JP, Mikkelsen TS, Hindson BJ, Bielas JH (2017) Massively parallel digital transcriptional profiling of single cells. Nat Commun 8: 14049. https://doi.org/10.1038/ncomms14049
2. Triana S, Vonficht D, Jopp-Saile L, Raffel S, Lutz R, Leonce D, Antes M, Hernandez-Malmierca P, Ordonez-Rueda D, Ramasz B, Boch T, Jann JC, Nowak D, Hofmann WK, Muller-Tidow C, Hubschmann D, Alexandrov T, Benes V, Trumpp A, Paulsen M, Velten L, Haas S (2021) Single-cell proteo-genomic reference maps of the hematopoietic system enable the purification and massive

profiling of precisely defined cell states. Nat Immunol 22(12):1577–1589. https://doi.org/10.1038/s41590-021-01059-0

3. Picelli S, Faridani OR, Bjorklund AK, Winberg G, Sagasser S, Sandberg R (2014) Full-length RNA-seq from single cells using Smart-seq2. Nat Protoc 9(1):171–181. https://doi.org/10.1038/nprot.2014.006
4. Keren-Shaul H, Kenigsberg E, Jaitin DA, David E, Paul F, Tanay A, Amit I (2019) MARS-seq2.0: an experimental and analytical pipeline for indexed sorting combined with single-cell RNA sequencing. Nat Protoc 14(6): 1841–1862. https://doi.org/10.1038/s41596-019-0164-4
5. Butler A, Hoffman P, Smibert P, Papalexi E, Satija R (2018) Integrating single-cell transcriptomic data across different conditions, technologies, and species. Nat Biotechnol 36(5): 411–420. https://doi.org/10.1038/nbt.4096
6. Ordonez-Rueda D, Baying B, Pavlinic D, Alessandri L, Yeboah Y, Landry JJM, Calogero R, Benes V, Paulsen M (2020) Apoptotic cell exclusion and bias-free single-cell selection are important quality control requirements for successful single-cell sequencing applications. Cytometry A 97(2):156–167. https://doi.org/10.1002/cyto.a.23898
7. Cossarizza A, Chang HD, Radbruch A, Acs A, Adam D, Adam-Klages S, Agace WW, Aghaeepour N, Akdis M, Allez M, Almeida LN, Alvisi G, Anderson G, Andra I, Annunziato F, Anselmo A, Bacher P, Baldari CT, Bari S, Barnaba V, Barros-Martins J, Battistini L, Bauer W, Baumgart S, Baumgarth N, Baumjohann D, Baying B, Bebawy M, Becher B, Beisker W, Benes V, Beyaert R, Blanco A, Boardman DA, Bogdan C, Borger JG, Borsellino G, Boulais PE, Bradford JA, Brenner D, Brinkman RR, Brooks AES, Busch DH, Buscher M, Bushnell TP, Calzetti F, Cameron G, Cammarata I, Cao X, Cardell SL, Casola S, Cassatella MA, Cavani A, Celada A, Chatenoud L, Chattopadhyay PK, Chow S, Christakou E, Cicin-Sain L, Clerici M, Colombo FS, Cook L, Cooke A, Cooper AM, Corbett AJ, Cosma A, Cosmi L, Coulie PG, Cumano A, Cvetkovic L, Dang VD, Dang-Heine C, Davey MS, Davies D, De Biasi S, Del Zotto G, Dela Cruz GV, Delacher M, Della Bella S, Dellabona P, Deniz G, Dessing M, Di Santo JP, Diefenbach A, Dieli F, Dolf A, Dorner T, Dress RJ, Dudziak D, Dustin M, Dutertre CA, Ebner F, Eckle SBG, Edinger M, Eede P, Ehrhardt GRA, Eich M, Engel P, Engelhardt B, Erdei A, Esser C, Everts B, Evrard M, Falk CS, Fehniger TA, Felipo-Benavent M, Ferry H, Feuerer M, Filby A, Filkor K, Fillatreau S, Follo M, Forster I, Foster J, Foulds GA, Frehse B, Frenette PS, Frischbutter S, Fritzsche W, Galbraith DW, Gangaev A, Garbi N, Gaudilliere B, Gazzinelli RT, Geginat J, Gerner W, Gherardin NA, Ghoreschi K, Gibellini L, Ginhoux F, Goda K, Godfrey DI, Goettlinger C, Gonzalez-Navajas JM, Goodyear CS, Gori A, Grogan JL, Grummitt D, Grutzkau A, Haftmann C, Hahn J, Hammad H, Hammerling G, Hansmann L, Hansson G, Harpur CM, Hartmann S, Hauser A, Hauser AE, Haviland DL, Hedley D, Hernandez DC, Herrera G, Herrmann M, Hess C, Hofer T, Hoffmann P, Hogquist K, Holland T, Hollt T, Holmdahl R, Hombrink P, Houston JP, Hoyer BF, Huang B, Huang FP, Huber JE, Huehn J, Hundemer M, Hunter CA, Hwang WYK, Iannone A, Ingelfinger F, Ivison SM, Jack HM, Jani PK, Javega B, Jonjic S, Kaiser T, Kalina T, Kamradt T, Kaufmann SHE, Keller B, Ketelaars SLC, Khalilnezhad A, Khan S, Kisielow J, Klenerman P, Knopf J, Koay HF, Kobow K, Kolls JK, Kong WT, Kopf M, Korn T, Kriegsmann K, Kristyanto H, Kroneis T, Krueger A, Kuhne J, Kukat C, Kunkel D, Kunze-Schumacher H, Kurosaki T, Kurts C, Kvistborg P, Kwok I, Landry J, Lantz O, Lanuti P, LaRosa F, Lehuen A, LeibundGut-Landmann S, Leipold MD, Leung LYT, Levings MK, Lino AC, Liotta F, Litwin V, Liu Y, Ljunggren HG, Lohoff M, Lombardi G, Lopez L, Lopez-Botet M, Lovett-Racke AE, Lubberts E, Luche H, Ludewig B, Lugli E, Lunemann S, Maecker HT, Maggi L, Maguire O, Mair F, Mair KH, Mantovani A, Manz RA, Marshall AJ, Martinez-Romero A, Martrus G, Marventano I, Maslinski W, Matarese G, Mattioli AV, Maueroder C, Mazzoni A, McCluskey J, McGrath M, McGuire HM, McInnes IB, Mei HE, Melchers F, Melzer S, Mielenz D, Miller SD, Mills KHG, Minderman H, Mjosberg J, Moore J, Moran B, Moretta L, Mosmann TR, Muller S, Multhoff G, Munoz LE, Munz C, Nakayama T, Nasi M, Neumann K, Ng LG, Niedobitek A, Nourshargh S, Nunez G, O'Connor JE, Ochel A, Oja A, Ordonez D, Orfao A, Orlowski-Oliver E, Ouyang W, Oxenius A, Palankar R, Panse I, Pattanapanyasat K, Paulsen M, Pavlinic D, Penter L, Peterson P, Peth C, Petriz J, Piancone F, Pickl WF, Piconese S, Pinti M, Pockley AG, Podolska MJ, Poon Z, Pracht K, Prinz I, Pucillo CEM, Quataert SA, Quatrini L, Quinn KM, Radbruch H, Radstake T, Rahmig S, Rahn HP, Rajwa B, Ravichandran G, Raz Y, Rebhahn JA, Recktenwald D, Reimer D, Reis e Sousa C, Remmerswaal EBM, Richter L, Rico LG,

Riddell A, Rieger AM, Robinson JP, Romagnani C, Rubartelli A, Ruland J, Saalmuller A, Saeys Y, Saito T, Sakaguchi S, Sala-de-Oyanguren F, Samstag Y, Sanderson S, Sandrock I, Santoni A, Sanz RB, Saresella M, Sautes-Fridman C, Sawitzki B, Schadt L, Scheffold A, Scherer HU, Schiemann M, Schildberg FA, Schimisky E, Schlitzer A, Schlosser J, Schmid S, Schmitt S, Schober K, Schraivogel D, Schuh W, Schuler T, Schulte R, Schulz AR, Schulz SR, Scotta C, Scott-Algara D, Sester DP, Shankey TV, Silva-Santos B, Simon AK, Sitnik KM, Sozzani S, Speiser DE, Spidlen J, Stahlberg A, Stall AM, Stanley N, Stark R, Stehle C, Steinmetz T, Stockinger H, Takahama Y, Takeda K, Tan L, Tarnok A, Tiegs G, Toldi G, Tornack J, Traggiai E, Trebak M, Tree TIM, Trotter J, Trowsdale J, Tsoumakidou M, Ulrich H, Urbanczyk S, van de Veen W, van den Broek M, van der Pol E, Van Gassen S, Van Isterdael G, van Lier RAW, Veldhoen M, Vento-Asturias S, Vieira P, Voehringer D, Volk HD, von Borstel A, von Volkmann K, Waisman A, Walker RV, Wallace PK, Wang SA, Wang XM, Ward MD, Ward-Hartstonge KA, Warnatz K, Warnes G, Warth S, Waskow C, Watson JV, Watzl C, Wegener L, Weisenburger T, Wiedemann A, Wienands J, Wilharm A, Wilkinson RJ, Willimsky G, Wing JB, Winkelmann R, Winkler TH, Wirz OF, Wong A, Wurst P, Yang JHM, Yang J, Yazdanbakhsh M, Yu L, Yue A, Zhang H, Zhao Y, Ziegler SM, Zielinski C, Zimmermann J, Zychlinsky A (2019) Guidelines for the use of flow cytometry and cell sorting in immunological studies (second edition). Eur J Immunol 49(10):1457–1973. https://doi.org/10.1002/eji.201970107

Chapter 8

Tissue RNA Integrity in Visium Spatial Protocol (Fresh Frozen Samples)

Federica Antico, Marta Gai, and Maddalena Arigoni

Abstract

The transcriptome of a tissue can be acquired both by single-cell RNAseq (scRNA-seq) and by spatial transcriptomics (ST). The dissociation step, which is mandatory in scRNA-seq methods, might lead to the loss of fragile cells and of spatial information, thus limiting the acquisition of the tissue cellular organization. Spatial transcriptomics methods moderate the above-mentioned issues and provide single-cell transcripts detection over an intact fresh frozen tissue section. Visium platform, commercialized from 10× Genomics, provides a whole transcriptome spatial transcriptomics platform, which does not require dedicated instruments, other than those available in any pathology laboratory. In spatial transcriptomics, proper tissue handling is mandatory to preserve the morphological quality of the tissue sections and the integrity of mRNA transcripts. Proper tissue handling is critical for downstream library preparation and sequencing performance. In this chapter, we describe the most critical steps of Visium protocol on fresh frozen tissues and we provide indications on how to interpret the data obtained from the quality control analysis recommended during the workflow.

Key words Spatial Transcriptomics, Visium, RNA quality control, Tissue preparation

1 Introduction

In 2020, *Nature Methods* crowned spatially resolved transcriptomics the method of the year [1]. The conventional bulk RNA sequencing (RNA-seq) requires tissue homogenization, which results in average transcripts quantification given by bulk sequencing. In population composed of different cell types, bulk RNA-seq might obscure critical differences among cell types. For example, in bulk RNA-seq of a cancer tissue, transcripts from different types of cells, including tumor cells, immune cells, fibroblasts, and endothelial cells cannot be differentiated. Population heterogeneity can be instead depicted by single-cell RNA-sequencing technique (scRNA-seq). Since tissue dissociation is required before scRNA-seq analysis, scRNA-seq, as bulk RNA-seq, is characterized by the

Raffaele A. Calogero and Vladimir Benes (eds.), *Single Cell Transcriptomics: Methods and Protocols*,
Methods in Molecular Biology, vol. 2584, https://doi.org/10.1007/978-1-0716-2756-3_8,

loss of spatial information, which is very important for tissue characterization [2].

The Visium Spatial Gene Expression Solution from 10× Genomics (www.10xgenomics.com) allows measuring the amount of polyA$^+$ mRNAs in intact tissue sections, mapping at the same time the locations where mRNAs are located. It is the most straightforward and widely used whole transcriptome method for spatial biology. Visium platform measures the mRNA expression profile of cells, within a tissue section of 6.5 × 6.5 mm, using 5000 spots (each spot is 55 μm diameter, with a 100 μm center-to-center distance among spots).

A frozen tissue section, 10–20 μm thick, is prepared and positioned on the Visium slide with oligos containing spatial barcodes and UMIs. By sequencing the synthesized cDNA libraries, RNA-seq data from each spot, with its spatial information, is obtained. By comparing these data with an hematoxylin/eosin (H&E)-stained image, gene expression patterns can be correlated with histopathological information. Spatial transcriptomics requires few extra steps compared with the RNA-seq analysis of homogenized tissue, but it has the advantage of retaining spatial information. A main limitation of Visium spatial transcriptomics tools is its lack of single-cell resolution: each spot captures the transcriptome of few cells, where the cell number in each spot depends on the tissue organization and cell size. Techniques, providing at the same time whole transcriptome and single cell resolution [3, 4] are not mature, yet. Moreover, current spatial transcriptome techniques focus exclusively on the poly-A tagged transcriptome.

In this chapter, we describe the five steps of the Visium spatial gene expression protocol, with a particular focus on the tissue preparation and RNA extraction. Tissue preparation and RNA extraction are critical steps for downstream library preparation and generation of high-quality sequencing data. We also highlight the relationship between input RNA integrity status and sequencing quality.

2 Materials

All solution must be prepared with nuclease-free water. Reagents must be fully thawed and thoroughly mixed before use. All reagents and enzymes must be kept on ice during the experiment setup. Retrieve the latest version of the Visium protocol and diligently follow all the instructions. The tissue preparation procedure is a critical step, since only proper tissue handling preserves the integrity of mRNA transcripts and the morphological quality of the sections.

2.1 Tissue Preparation and Section Placement

1. A bath of isopentane (2-methylbutane) and liquid nitrogen is used to freeze the freshly obtained tissue. Alternatively use dry ice instead of liquid nitrogen.
2. Prior cryosectioning, frozen tissue samples are embedded in Tissue Tek O.C.T compound.
3. Visium slide (10× Genomics, choose appropriate kit according to the number of samples).
4. Dewar for liquid nitrogen.
5. Razor blades.
6. Cryostat.
7. Sealed container.
8. Cryomolds.

2.2 RNA Quality Assessment

1. RNA quality of tissue block is estimated before the placement of the tissue sections on the slide. RNA isolation is done using Qiagen RNeasy mini kit.
2. RNA quality is evaluated using Agilent bioanalyzer with Agilent RNA 6000 Nano or Pico kits.
3. RNAse-free tubes.
4. Ethanol.
5. Nuclease-free water.
6. Tissue homogenizer.
7. RNAse ZAP (Invitrogen).
8. 2-Mercaptoethanol.
9. DNA LoBind Tubes, 1.5 ml.

2.3 Tissue Fixation, H&E Staining, and Tissue Imaging

1. Tissue is fixed in cold methanol (methanol ≥99.9%).
2. 2-Propanol (isopropanol ≥99.5%).
3. Hematoxylin solution, Mayer's (Lillie's modification).
4. Eosin Y solution (aqueous) 0.5% (w/v) in water.
5. Tris–acetic acid buffer (0.45 M pH 6.0).
6. Bluing buffer Dako (Agilent).
7. pH meter.
8. Thermal Cycler.
9. Dry ice.
10. Any imaging systems with both brightfield and fluorescent capacity (images have a minimum size requirement of 2000 pixels in at least one direction. For more information, check the Visium Spatial Gene Expression Imaging Guidelines Technical Note).

2.4 Permeabilization and Reverse Transcription

1. Appropriate Visium gene expression reagent kit.
2. Visium Accessories kit (10× Genomics).
3. Nuclease-free water.
4. 20× SSC.
5. Low TE buffer (10 mM Tris–HCl pH 8.0, 0.1 mM EDTA).
6. Hydrochloric acid solution (HCl 0.1 N).
7. Thermal Cycler.

2.5 Second Strand Synthesis and Denaturation

1. Visium gene expression kit.
2. Qiagen buffer EB.
3. Tris 1 M pH 7.0 (Tris–HCl) RNAse free.
4. Potassium hydroxide solution 8 M.
5. Thermal Cycler.
6. PCR tubes 0.2 ml 8-tube strips.
7. DNA low-bind tubes (1.5 ml, 2.0 ml).

2.6 cDNA Amplification and Quality Control

1. Visium gene expression kit.
2. SPRIselect reagent (Beckman Coulter).
3. Bioanalyzer (Agilent).
4. DNA high sensitivity kit (Agilent).
5. KAPA SYBR fast qPCR Master Mix 2× (KAPA Biosystem).
6. Ethanol.
7. Magnetic separator.
8. Real-time qPCR system.
9. qPCR plates.
10. Thermal Cycler.

2.7 Libraries Preparation and Sequencing

1. Visium gene expression kit.
2. Dual index Plate TT set A (10× Genomics).
3. DNA high-sensitivity kit for Bioanalyser (Agilent).
4. Sequencer (All Illumina sequencers, but miSeq and miniSeq).
5. Appropriate kit for sequencing (a standard Illumina paired-end dual index sequencing, at least 150 cycles, 400 million reads are strongly suggested for a Visium slide).
6. Thermal Cycler.
7. SPRIselect reagent (Beckman Coulter).

3 Methods

3.1 Tissue Preparation, RNA Quality Assessment, and Section Placement on Visium Slide

Here, we describe the Visium procedure from 10× Genomics and highlight the relevant critical steps using a challenging tissue: pancreas. The general principles of Visium protocol are compatible with many tissues (listed in the 10× Genomics supported website). Pancreatic tissue is not included in this list, and it requires further precautions being a "difficult" tissue, due to the high enzymatic content.

In this section, we use, as exemplary models, RNAs of different quality (high, medium, and low) extracted from the following:

1. Pancreatic ductal adenocarcinoma murine cell line orthotopically injected in FBV/n mice [5].
2. Mouse model of pancreatic cancer (PDXCre/LSL-*Kras*^{G12D} mouse model [6]).

Here, we describe the procedure to:

1. Remove and freeze pancreatic tissue from mice.
2. To assess RNA quality.
3. To place tissue on Visium slide (*see* **Note 1**).

The following steps were performed on animals:

1. Anesthetize mouse.
2. Leave mouse at least 10 min on ice.
3. Sacrifice mouse. From this step on, proceed very quickly.
4. Remove pancreas.
5. Wash the organ in precooled PBS (phosphate buffer saline) and briefly blot dry.
6. Embed the organ in precooled OCT compound (precooled at least for 2 h at +4 °C).
7. Put the cryomold container (containing OCT+ organ) on a bath of cold isopentane on liquid nitrogen and wait until OCT becomes white (few minutes). Alternatively, put a box with precooled isopentane (precooled at least for 2 h at −80 °C) on dry ice and wait until OCT becomes white.
8. Close the cryomold with a tinfoil and put in a plastic envelope in dry ice.
9. Close the plastic envelope and store at −80 °C.

Keep the tissue in dry ice to avoid any damage. Snap-frozen tissue sample must be stored in a sealed container at −80 °C for long-term storage.

Remove OCT-embedded tissue blocks from the −80 °C storage and cryosection in a cryostat to generate 15 sections (12 μm thickness) for total RNA extraction and RNA quality estimation and 1 section (12 μm thickness) for H&E staining (*see* **Note 2**).

Put the 15 tissue sections in a precooled nuclease-free LoBind Eppendorf (2 ml). Extract RNA using RNeasy miniKit (Qiagen) following manufacturer's instruction (*see* **Note 3**). Quantify total RNA and perform quality control using Agilent bioanalyzer (*see* **Note 4**). RNA quality must be estimated before placing the tissue on Visium slide and RIN (RNA integrity number) should be ≥7. Place tissue sections on appropriate Visium slide following 10× Genomics protocol (*see* **Note 5**). Size of tissue section must fit the capture area of the Visium slide (*see* **Note 6**). Store slide at −80 °C for up to a week or immediately proceed to tissue fixation.

3.2 Tissue Fixation, H&E Staining, and Tissue Imaging

Remove slide from −80 °C and place it for 1 min at 37 °C in a thermal cycler following Visium protocol instructions. Completely submerge the slide in prechilled methanol for 30 min at −20 °C. Add 500 μl isopropanol to cover all tissue sections on the slide and incubate for 1 min at room temperature. Discard reagent and air dry the slide for 5 to 10 min.

Add 1 ml hematoxylin to cover uniformly all tissue sections on the slide for 7 min at room temperature. Immerse the slide in the water following diligently Visium protocol recommendations. Add 1 ml of bluing buffer for 2 min at room temperature. Immerse the slide in water and add 1 ml eosin mix for 1 min at room temperature. Wash the slide in water following Visium protocol and incubate it on thermal cycler with lid open for 5 min at 37 °C. Proceed to tissue imaging (*see* **Note 7**).

3.3 Permeabilization, Reverse Transcription, Second Strand Synthesis and Denaturation

Permeabilize the fixed and stained tissue sections on the slide using the permeabilization enzyme from Visium Gene expression kit. The poly-adenylate mRNA released from the cells is captured by the primers on the spots. Place the slide in the slide cassette following Visium protocol. Add 70 μl permeabilization enzymes and place the slide cassette at 37 °C on the thermocycler adaptor in thermal cycler with lid close. Incubate for the pre-determined permeabilization time (*see* **Note 8**).

Add RT Master Mix, containing reverse transcription reagent, to the permeabilizated tissue section to obtain spatially barcoded, full-length cDNA from poly-adenylated mRNA on the slide. Add second strand mix to the tissue sections on the slide to synthetize second strand cDNA. Denature and transfer the cDNA from each capture area in a new tube.

3.4 cDNA Amplification and Quality Control

Amplify cDNA via PCR to generate enough material for library construction using a predetermined number of cycles (*see* **Note 9**). After amplification, cDNA is purified through SPRI-select beads and can be stored at 4 °C for 4 weeks. Run 1 μl of cDNA on an Agilent Bioanalyzer High Sensitivity chip for quality control (*see* **Note 10**).

3.5 Libraries Preparation and Sequencing

Perform library preparation following Visium protocol. Enzymatic fragmentation, end repair, A-tailing, adaptors ligation, and PCR are done using reagents from Visium gene expression kit. Size selection and clean-up are performed using SPRIselect reagent. Final Visium spatial gene expression library is composed of standard Illumina P5, i5 index, spatial barcode, 12 bp UMI, cDNA fragment, i7 index, and standard Illumina P7. Run 1 μl of library on Agilent DNA High sensitivity chip (*see* **Note 11**). Prepare the sequencing libraries following Visium protocol (*see* **Note 12**).

3.6 Sequencing Quality Control

10× Genomics offers the possibility to analyze data exploiting SpaceRanger pipeline. Even if analysis details are beyond the aim of this chapter, we used SpaceRange QC data to correlate tissue RNA integrity status with sequencing data quality.

4 Notes

Before proceeding with the tissue removal, clean all surface and forceps with RNAse Zap and nuclease-free water.

1. Visium gene expression slide (10×Genomics) contains capture areas (6.5 × 6.5 mm), each defined by a fiducial frame (fiducial frame + capture area is 8 × 8 mm) with gene expression spots (5000 spots in each capture area). Tissue sections are placed within the frames of capture areas. For this reason, it is important to identify a priori the area of interest that fits the capture area (6.5 × 6.5 mm). 12 μm thickness tissue section is placed on a slide and stained with H&E. Select the area of interest and highlight it with a marker. Make a deep cut within tissue block with a precooled razor blade, proceed to the cryosection of the area of interest, and place tissue sections on the appropriate Visium slide, following diligently Visium protocols (Fig. 1). 10× Genomics recommends the assessment of RNA quality before placing the tissue sections on Visium spatial slides.
2. The use of other kits is not recommended.
3. RNA quality is assessed with bioanalyzer (Agilent). Here, we report four representative images of electropherograms of different quality RNA extracted from cancer pancreatic tissues (Fig. 2). RNA integrity number (RIN) is reported for each sample. RIN value is between 1 and 10, where 1 indicates the most degraded and 10 the most intact RNA profile.

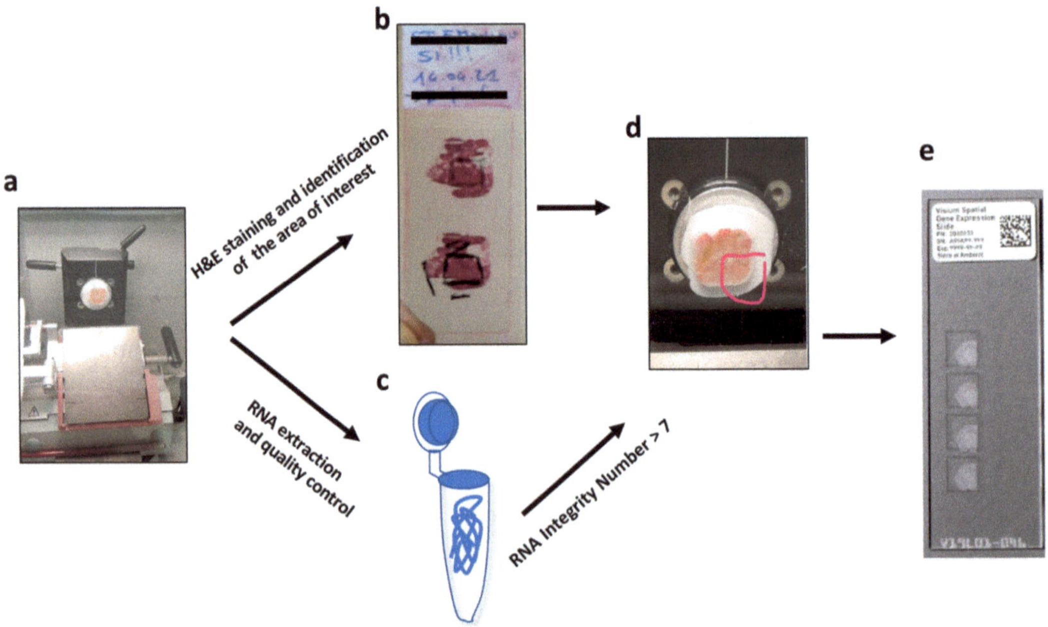

Fig. 1 Definition of the area of interest for spatial transcriptomics analysis and tissue section placement on Visium slide. (**a**) Cryosection of OCT embedding tissue; (**b**) placement of 1 section on a slide for H&E staining; (**c**) tissue RNA extraction from 12 sections for RNA quality assessment; (**d**) identification and marking of the area of interest.; (**e**) cryosectioning of the area of interest and placement on appropriate Visium slide for gene expression analysis (only for tissue with RIN $\geq$ 7 RNA)

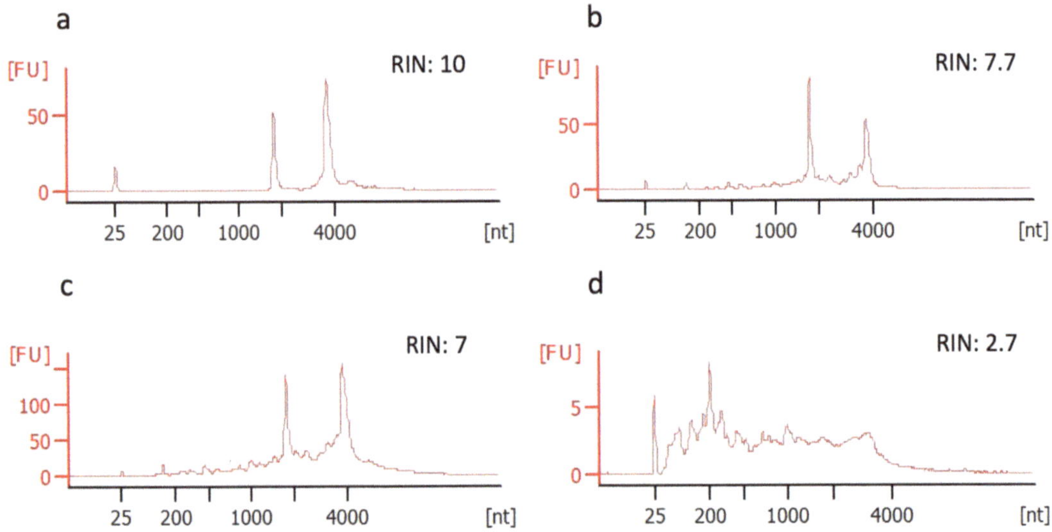

Fig. 2 RNA quality assessment. (**a**) High-quality RNA extracted from pancreatic ductal adenocarcinoma murine cell line orthotopically injected in FBV/n mice (RNA integrity number = RIN); (**b**) and (**c**) medium-quality and (**d**) low-quality RNA extracted from pancreatic ductal adenocarcinoma in a PDXCre/LSL-*Kras*^{G12D} mouse model

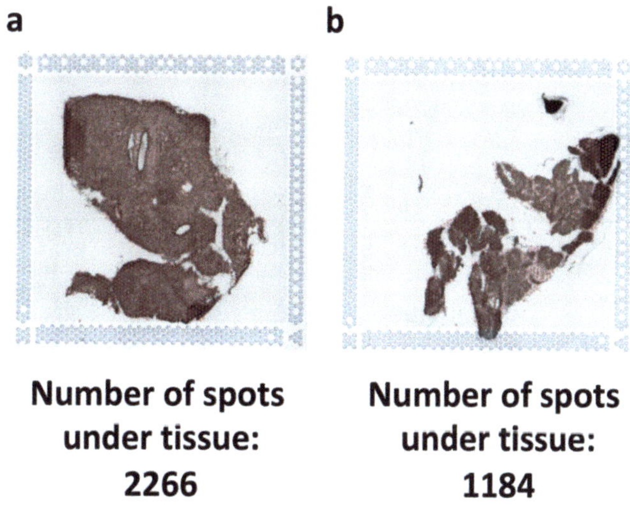

Fig. 3 Tissue placement on Visium slide. Different pancreatic tissue is positioned on capture area of Visium slide. The number of spots under tissue varies according to the tissue size. H&E images are aligned to fiducial frames of Visium slide using Loupe browser 5.0 (10×Genomics). (**a**) Pancreatic ductal adenocarcinoma murine cell line orthotopically injected in FBV/n mice (**b**) Mouse model of pancreatic cancer (PDXCre/LSL-Kras^{G12D} mouse model)

4. Slide from Visium spatial gene expression kit has four areas to permit the positioning of four tissue sections. 10× Genomics recommends placing four sections from the same tissue on each slide. However, if needed, four sections from four different tissues per slide can be loaded. In this case, all the procedures (cryosection and tissue placement) must be performed very quickly.
5. It is important that the tissue section remains within the slide capture area, and it is large enough to cover most of the capture area to make a good use of the whole Visium slide (Fig. 3). In case it would not be possible to cover completely the capture area, the number of cDNA amplification PCR cycles should be increased accordingly with the Visium protocol.
6. After tissue imaging, proceed immediately to tissue permeabilization.
7. Permeabilization time is optimized prior to use of Visium gene expression slide using Visium tissue optimization slide and following manufacturers' instructions.
8. Prepare real-time PCR using KAPA SYBR FAST qPCR Master Mix and cDNA primers provided by gene expression Visium kit following Visium protocol. The number of PCR cycles for cDNA amplification is determined based on Cq value for each sample recorded after real-time PCR (as reported in Visium protocol).

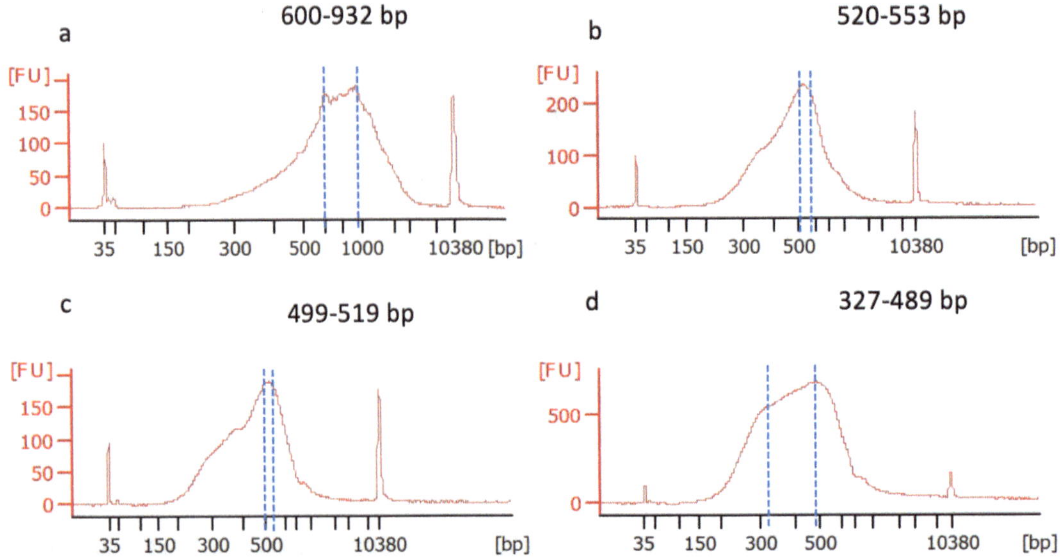

Fig. 4 cDNA quality assessment. (**a**) cDNA fragment size distribution obtained from high-quality, (**b**, **c**) medium-, and (**d**) low-quality RNA from pancreatic ductal adenocarcinoma murine cell line orthotopically injected in FBV/n mice (**a**) and pancreatic ductal adenocarcinoma in a PDXCre/LSL-*Kras*^{G12D} mouse model (**b**, **c**, and **d**). The dashed blue lines show peak fragment size (numbers reported size in bp)

9. cDNA profile may vary depending on tissue type and quality. Even if RNA input shows a high level of degradation, an amount of cDNA sufficient for subsequent steps is always available. It is very important to check fragment size distribution since it impacts on libraries constructions and sequencing results. cDNA optimal fragment size distribution must span from 350 to 1200 bp, with fragment enrichment from 600 to 900 bp, variation around 100 bp is acceptable, but lower fragment size distribution, which is typical sign of high RNA degradation, is a clear warning of low-quality cDNA and it could dramatically affect sequencing performance. Here, we report cDNA (Fig. 4) profiles obtained from RNA showing different degradation level (RIN from 10 to 2.8, *see* Fig. 1) using 13 PCR cycles. Figure 4a shows cDNA with an optimal fragment size distribution (from 350 to 1200 bp) with most of the fragments larger than 600 bp according to the high quality of RNA input (RIN 10). Figure 4b, c shows cDNA profiles from medium-quality RNA (RIN 7.7 or 7). cDNAs fragments distribution is different compared to that in Fig. 4a, since most of the fragments are shorter, even if most of them are larger than 400 bp. Figure 4d shows cDNA profile from RNA with a high degradation level (RIN 2.8). Trend's distribution is significantly different compared to those from Fig. 4a–c. Fragments are shorter and most of them are below 500 bp. cDNA quality control is a very important step in the Visium workflow,

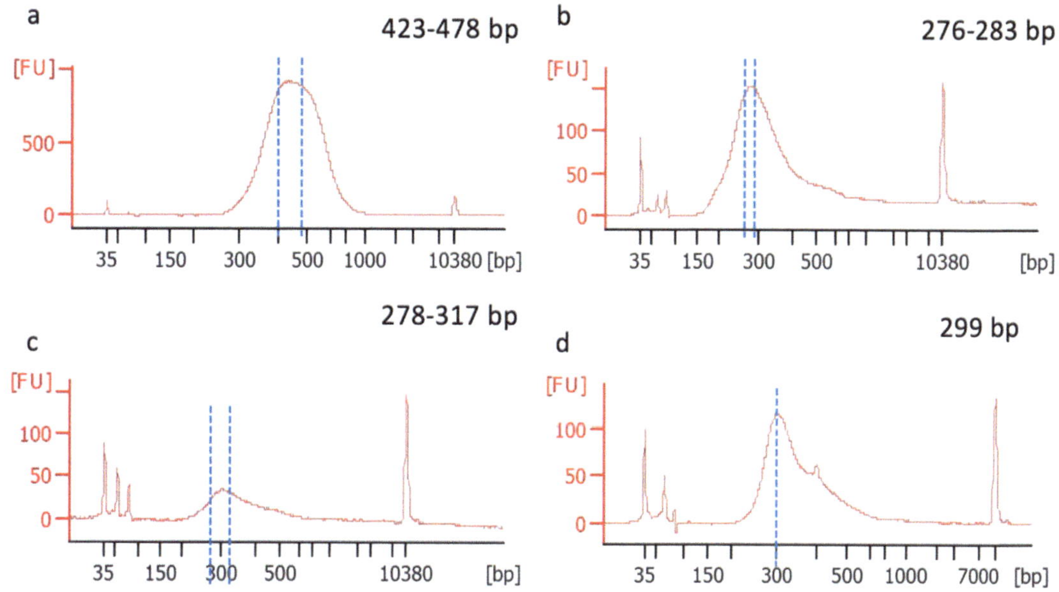

Fig. 5 Library quality assessment. (**a**) Size distribution of libraries obtained from high-quality, (**b**, **c**) medium-quality, and (**d**) low-quality RNA from pancreatic ductal adenocarcinoma murine cell line orthotopically injected in FBV/n mice (**a**) and pancreatic ductal adenocarcinoma in a PDXCre/LSL-*Kras*^{G12D} mouse model (**b**, **c** and **d**). The dashed blue lines show peak fragment size (numbers reported size in bp)

and it could help to predict the success of the library construction and consequently the sequencing data performance. Even if 10× Genomics recommends the use of RNA input with a RIN ≥ 7, for certain difficult tissues, e.g. pancreas, the use of a medium-quality RNA does not guarantee good-quality libraries and high sequencing data achievement (*see* **Note 11**).

10. Here, we show (Fig. 5) the profiles from libraries obtained using the cDNAs described in Fig. 4. Figure 5a shows fragment size distribution from high-quality RNA: most of the fragments span from 300 to 1000 bp with an enrichment around 423–478 bp. The use of high-quality RNA as input permits the construction of high-quality library, which guarantees high sequencing performance. Figure 5b, c shows profiles from medium-quality RNAs, while Fig. 5d shows the results obtained using a very low-quality RNA. Fragments are shorter and their distribution appears different compared to Fig. 5a. Sequencing of libraries in Fig. 5b, c, d resulted in low-quality data, according to cDNA profiles described previously (*see* **Note10**).
11. Sequencing mode is described in Visium protocol, and loading concentration is defined according to sequencer type.

Tissue	RNA (RIN)	Number of spots under tissue	Mean reads per spot	Median genes per spot	Median UMI count per spot	Reads mapped confidentially to transcriptome (%)
	10	2436	40015	2836	7378	82.4%
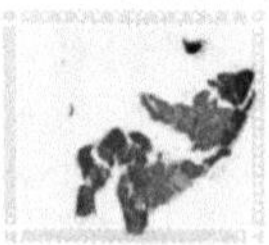	7.7	1184	14066	340	434	18.3%
	7	1963	28618	325	436	21.5%
	2.8	2044	23742	128	154	13.4%

Fig. 6 Sequencing data quality control. Summary table of sequencing data quality control starting from four different pancreatic tissue sections with different RNA quality (previously described). RNA integrity number, number of spots under tissue, mean reads per spot, median genes per spot, median UMI count per spot, and % of reads mapped confidentially to transcriptome are reported. Sequencing mode is the same for all tissues

12. Libraries described in **Note 11** were sequenced following Visium protocol, loading 1.8 pM libraries pool on NextSeq 500 sequencer. Data was analyzed with SpaceRanger pipeline from 10× Genomics [7–9]. Here, we show (Fig. 6) the quality control analysis of the samples. Overall, high-quality RNA (RIN = 10) gives rise to good sequencing data, with a median UMI count per spot >7000 and a percentage of reads mapped confidentially to transcriptome >80, allowing further clustering analysis. Low-quality RNA (RIN = 2.8) negatively affects sequencing performance: median UMI count per spot is <200 and the percentage of reads mapped to transcriptome is too low for further clustering analysis. Medium-quality RNA (RIN from 7.7 to 7.0) represents a very critical issue, since even if some data is recovered (median UMI count per spot is >400), it is not possible to perform any effective analysis, since the percentage of reads mapped to transcriptome remains very low, even increasing sequencing coverage (data not shown). Although Visium protocol from 10× Genomics recommends using RNA input with an RIN value ≥7, it does not ensure to obtain good data for further analysis.

References

1. Marx V (2021) Method of the year: spatially resolved transcriptomics. Nat Methods 18(1): 9–14. https://doi.org/10.1038/s41592-020-01033-y
2. Kashima Y, Sakamoto Y, Kaneko K, Seki M, Suzuki Y, Suzuki A (2020) Single-cell sequencing techniques from individual to multiomics analyses. Exp Mol Med 52(9):1419–1427. https://doi.org/10.1038/s12276-020-00499-2
3. Cho CS, Xi J, Si Y, Park SR, Hsu JE, Kim M, Jun G, Kang HM, Lee JH (2021) Microscopic examination of spatial transcriptome using Seq-Scope. Cell 184(13):3559–3572.e22. https://doi.org/10.1016/j.cell.2021.05.010
4. Moncada R, Barkley D, Wagner F, Chiodin M, Devlin JC, Baron M, Hajdu CH, Simeone DM, Yanai I (2020) Integrating microarray-based spatial transcriptomics and single-cell RNA-seq reveals tissue architecture in pancreatic ductal adenocarcinomas. Nat Biotechnol 38(3): 333–342. https://doi.org/10.1038/s41587-019-0392-8
5. Gilles ME, Maione F, Cossutta M, Carpentier G, Caruana L, Di Maria S, Houppe C, Destouches D, Shchors K, Prochasson C, Mongelard F, Lamba S, Bardelli A, Bouvet P, Couvelard A, Courty J, Giraudo E, Cascone I (2016) Nucleolin targeting impairs the progression of pancreatic cancer and promotes the normalization of tumor vasculature. Cancer Res 76(24):7181–7193. https://doi.org/10.1158/0008-5472.CAN-16-0300
6. Hingorani SR, Petricoin EF, Maitra A, Rajapakse V, King C, Jacobetz MA, Ross S, Conrads TP, Veenstra TD, Hitt BA, Kawaguchi Y, Johann D, Liotta LA, Crawford HC, Putt ME, Jacks T, Wright CV, Hruban RH, Lowy AM, Tuveson DA (2003) Preinvasive and invasive ductal pancreatic cancer and its early detection in the mouse. Cancer Cell 4(6):437–450. https://doi.org/10.1016/s1535-6108(03)00309-x
7. Alessandri L, Cordero F, Beccuti M, Arigoni M, Olivero M, Romano G, Rabellino S, Licheri N, De Libero G, Pace L, Calogero RA (2019) rCASC: reproducible classification analysis of single-cell sequencing data. Gigascience 8(9). https://doi.org/10.1093/gigascience/giz105
8. Alessandri L, Cordero F, Beccuti M, Licheri N, Arigoni M, Olivero M, Di Renzo MF, Sapino A, Calogero R (2021) Sparsely-connected autoencoder (SCA) for single cell RNAseq data mining. NPJ Syst Biol Appl 7(1):1. https://doi.org/10.1038/s41540-020-00162-6
9. Alessandri L, Ratto ML, Contaldo SG, Beccuti M, Cordero F, Arigoni M, Calogero RA (2021) Sparsely connected autoencoders: a multi-purpose tool for single cell omics analysis. Int J Mol Sci 22(23). https://doi.org/10.3390/ijms222312755

Chapter 9

Single-Cell RNAseq Data QC and Preprocessing

Martina Olivero and Raffaele A. Calogero

Abstract

The first step in single-cell RNAseq data analysis is the evaluation of the overall quality of the cell transcriptome and the preparation of the single-cell transcription data for clustering. In this chapter, we describe one of the possible approaches to perform single-cell data preprocessing for 3′ end single-cell RNAseq transcriptomics data.

Key words Quality control, Data preprocessing, Single-cell transcriptomics

1 Introduction

Ribosomal RNA and ribosomal proteins represent a significant part of cell content. Large cells and actively proliferating cells will have, respectively, more ribosomes and more active ribosome synthesis [1]. Thus, ribosomal proteins expression might represent one of the major confounding factors in cluster formation between active and quiescent cells.

The main function of mitochondria is to produce energy through aerobic respiration. The number of mitochondria depends on its metabolic demands [2]. This might also affect single-cell clustering, favoring the separation between metabolic active and resting cells, instead of other functional differences among subpopulations. Usually, mitochondrial DNA proportion is used as a threshold to filter out low-quality cells. Here, the UMI proportion of ribosomal and mitochondrial proteins [3] is used as surrogate measurement of ribosomes and mitochondria cell abundance. Then, ribosomes/mitochondria content is used, together with the estimation of the minimal number of genes called "present" in each cell, to select cells which are not stressed, and contain a user defined number of informative genes.

Another potential confounding factor is the cell cycle status of each cell. Here, we used Seurat [4] for cell cycle status assignment,

Raffaele A. Calogero and Vladimir Benes (eds.), *Single Cell Transcriptomics: Methods and Protocols*,
Methods in Molecular Biology, vol. 2584, https://doi.org/10.1007/978-1-0716-2756-3_9,

as further preprocessing step, to select cells in the same cell cycle stage, e.g., G1. A critical issue in single-cell RNAseq (scRNAseq) is the presence of dropout events leading to zero expression measurements (zero-inflated data) [5], that is, sparsity, and systematic technical noise [6]. Imputation methods aim to address the sparsity observed in single-cell data to recover biologically meaningful transcriptome dynamics in single cells. Hou and coworkers published a comparison among 18 different imputation methods and showed that MAGIC [7] and SAVER [8, 9] outperform the other methods [6]. Here, we describe the steps to impute data using SAVER [8].

2 Materials

2.1 Minimal Hardware/Software Requirements

For the analyses described in this chapter, a workstation with at least 32 Gb RAM, 1Tb SSD, and 4 CPUs is required. All analytical steps require a UNIX-like environment (*see* **Notes 1** and **2**).

3 Methods

3.1 rCASC Installation

rCASC [3, 10, 11] is a single-cell analysis suite providing a computation reproducible infrastructure based on docker containers. The installation of rCASC is very simple. In a R session, type the code indicated below (*see* **Note 3**).

```
# Open an R session and type the following commands
library(devtools)
install_github("kendomaniac/rCASC", ref="master")
library(rCASC)
downloadContainers()# this step is not mandatory, if the
computer is always connected to internet. Any time a new docker
is required, it is downloaded from the docker.com repository.
```

3.2 Converting a Sparse Matrix in a Dense Matrix

10×Genomics counts matrix output is provided in a compressed format made of three files: barcodes.tsv.gz, features.tsv.gz, and matrix.mtx.gz. The rCASC function *h5tocsv* allows the conversion of 10×Genomics sparse format (*see* **Notes 4** and **5**) to dense comma separated format.

```
library(rCASC)
WD = "/path_to_cell-ranger_output_data"
INPUT = "matrix.mtx.gz"
SEPARATOR = ","
setwd(WD)
```

```
h5tocsv(
 group = "docker",
 file = paste(getwd(), INPUT, sep="/"),
 type = "10xgenomics",
 version = "5"
)
```

The output is a comma separated file named matrix.csv.

3.3 Plotting Cells Based on Their Ribosomal/ Mitochondrial Content

The following code is required to run *mitoRiboUmi* function (*see* **Notes 6** and **7**), which provides information on the overall quality of the cells RNAseq:

```
WD = "/path_to_the_matrix_generated_by_h5tocsv"
INPUT = "matrix.csv"
SEPARATOR = ","
GTF = "the_name_of_ENSEMBL.gtf"
SCRATCH = "/somewhere/scratch_folder"
setwd(WD)
library(rCASC)
mitoRiboUmi(
 group = "docker",
 scratch.folder = SCRATCH,
 file = paste(getwd(), INPUT, sep="/"),
 separator = ",",
 gtf.name = GTF,
 bio.type = "protein_coding",
 umiXgene = 3
)
```

The output is a pdf file named Ribo_mito.pdf (*see* **Note 8**).

3.4 Filtering the Count Matrix Based on the Information Retrieved from the Inspection of mitoRiboUmi Output

The function *scannobyGtf* allows the user to filter cells characterized by a set of parameters related to the output of *mitoRiboUmi* function (*see* **Notes 9** and **10**).

```
WD = "/path_to_the_matrix_generated_by_h5tocsv"
INPUT = "matrix.csv"
SEPARATOR = ","
GTF = "the_name_of_ENSEMBL.gtf"
SCRATCH = "/somewhere/scratch_folder"
setwd(WD)
library(rCASC)
scannobyGtf(
 group = "docker",
 file = paste(getwd(), INPUT, sep="/"),
 gtf.name = GTF,
 biotype = "protein_coding",
```

```
 mt = FALSE,
 ribo.proteins = FALSE,
 umiXgene = 3,
 riboStart.percentage = 1,
 riboEnd.percentage = 50,
 mitoStart.percentage = 1,
 mitoEnd.percentage = 10,
 thresholdGenes = 101
)
```

The outputs of the *scannobyGtf* annotation function are as follows:

1. filteredStatistics.txt: a file that provides a summary of the genes and cells retained/removed.
2. annotated_XXXX.csv: identical to the count matrix, but with including only genes belonging to the ENSEMBL biotype.
3. filtered_annotated_XXXX.csv: the count matrix where are only retained the cells fulfilling the threshold indicated in the scann-byGtf parameters.
4. XXXX_annotated_genes.pdf: a plot showing the UMI counts changes upon removal of mitochondrial and ribosomal protein genes.

3.5 Imputing Missing Values

scRNA-seq data often contains a high percentage of missing values, also known as zero-inflation, due to the low capture efficiency and stochastic gene expression. Here, we show how to use SAVER [8] to model accurate expression for genes in unique molecule index (UMI)-based scRNA-seq data.

```
install.packages("SAVER")
library(SAVER)
WD = "/somewhere/in_your_computer/path_to_raw_data"
INPUT = "matrix.csv"
SEPARATOR = ","
THREADS = 1 # the number of cores dedicated to run such
analysis
setwd(WD)
raw.data <- read.table(INPUT, sep=SEPARATOR,
 header = T, row.names=1)
dataset <- as.matrix(raw.data)
dataset.saver <- saver(dataset, ncores = THREADS,
 estimates.only = TRUE)
write.table(dataset.saver, paste("saver", INPUT,
 sep="_"), sep=SEPARATOR, col.names=NA)
```

The output of the above-described procedure is a file embedding the normalized expression data to be used for subpopulation partitioning (*see* **Note 11**).

Low-quality cell filtering must be applied on SAVER imputed data on the results provided on the *mitoRiboUmi* analysis on the raw counts data (*see* Subheading 3.4).

```
WD = "/path_to_the_matrix_generated_by_SAVER"
INPUT = "matrix.csv"
SEPARATOR = ","
GTF = "the_name_of_ENSEMBL.gtf"
SCRATCH = "/somewhere/scratch_folder"
setwd(WD)
library(rCASC)

scannobyGtf(
 group = "docker",
 file=paste(getwd(), paste("saver", INPUT,
 sep="_"), sep="/"),
 gtf.name = GTF,
 biotype = "protein_coding",
 mt = FALSE,
 ribo.proteins = FALSE,
 umiXgene = 3,
 riboStart.percentage = 0,
 riboEnd.percentage = 100,
 mitoStart.percentage = 0,
 mitoEnd.percentage = 100,
 thresholdGenes = 1)

raw <- read.table(paste("filtered_annotated", INPUT,
 sep="_"),
 sep=SEPARATOR, header=T, row.names=1)
saver <- read.table(paste("annotated_saver", INPUT,
 sep="_"),
 sep=SEPARATOR, header=T, row.names=1)
saver_ribomito <- saver[,which(names(saver)%in%names(raw))]
write.table(saver_ribomito, paste("filtered_annotated_saver_-
ribomito", INPUT, sep="_"), sep=",", col.names=NA)
```

The output is the file filtered_annotated_saver_ribomito_matrix.csv, which will be used for cell cycle estimation.

3.6 Estimating the Cell Cycle State

Cell cycle state can be another confounding factor affecting subpopulation partitioning. Here, we show how to calculate the cell cycle state using the Seurat [12] method implemented in rCASC.

```
library(rCASC)
WD = "/somewhere/in_your_computer/path_to_SAVER_output"
INPUT = "filtered_annotated_saver_ribomito_matrix.csv"
SEPARATOR = ","
SCRATCH = "/somewhere/in_your_computer/scratch"
setwd(WD)
seurat_ccycle(group = "docker",
 scratch.folder = SCRATCH,
 file = paste(getwd(), INPUT, sep = "/"),
 separator = SEPARATOR,
 seed=111)
```

The output is a file with _cellCycle.csv extension. It contains two columns, the first one is cell ids and the second is the cell state. It could be useful to track of the cell cycle state. Thus, it might be useful to associate it to the cell names.

```
# Open an R session and type the following commands

WD = "/somewhere/in_your_computer/path_to_SAVER_output"
setwd(WD)
INPUT = "filtered_annotated_saver_ribomito_matrix.csv"
SEPARATOR = ","
CC = "filtered_annotated_saver_ribomito_matrix_cellCycle.csv"
cc <- read.table(CC, sep = SEPARATOR, header = F, string-
sAsFactors = F)
saver <- read.table(INPUT,sep=SEPARATOR, header=T, row.
names=1)
if(identical(names(saver), cc[,1])){
newnames <- paste(cc[,1],cc[,2], sep=":")
names(saver) <- newnames}
write.table(saver, "filtered_annotated_saver_ribomito_cc_ma-
trix.csv", sep = SEPARATOR, col.names=NA)
```

4 Notes

1. The installation of docker is required. Docker is available for Windows, Mac, and UNIX infrastructures at https://docs.docker.com/get-docker/.
2. R installation is required (https://cran.r-project.org/). Specifically, it is required the installation of devtools library.
3. The description of all rCASC functions is available at https://kendomaniac.github.io/rCASC/reference/index.html.

4. The compressed sparse matrix is a format in which zeros are not indicated. It provides a way to distribute very large data characterized by being zero inflated minimizing the file size, since only the position and the value of non-zeros values are stored.
5. An exemplary data set to test the function *h5tocsv* is available at https://www.ncbi.nlm.nih.gov/geo/query/acc.cgi?acc=GSM4679532 [13]. The file GSM4679532_K16733_barcodes.tsv.gz must be renamed in barcodes.tsv.gz; the file GSM4679532_K16733_feature.tsv.gz must be renamed in features.tsv.gz; the file GSM4679532_K16733_matrix.mtx.gz must be renamed in matrix.mtx.gz. Please note that if, instead of the file features.tsv.gz, it is present the file genes.tsv.gz, this is an old version of the features file, which lacks the third column indicating the data type. To use *h5tocsv* function, the user needs to edit the gene.tsv adding a third column containing in all rows the words Gene Expression and rename the edited file in features.tsv. The features.tsv file can then be compressed using the UNIX command gzip features.tsv.
6. The description of the *mitoRiboUmi* parameters is available at https://kendomaniac.github.io/rCASC/reference/mitoRiboUmi.html. The ENSEMBL GTF file can be retrieved at: https://www.ensembl.org/info/data/ftp/index.html. It must be in the same folder where the count matrix is located.
7. The identification of stressed cells, on the basis of ribosome and mitochondria cargo, is dependent by cell types under analysis [14]. In Fig. 1, there are four examples of the outputs produced on various datasets using the *mitoRiboUmi* function. The bone marrow sample, Fig. 1a, is characterized by a limited number of cells with high mitochondria content, which are also characterized by being very little informative by the number of the detected genes, i.e., less than 100 genes called present. The sample in Fig. 1b, is an example where the majority of the cells is characterized by an apoptotic profile [15]. In Fig. 1c, the activated T cells are of good quality, although few of them containing less than 100 genes. In Fig. 1d, there is an example of a spatial transcriptomics sample. Each point in the figure represents one of the 5000 spots of 55 μm of diameter presents in the acquisition area of visium chip (https://www.10xgenomics.com/products/spatial-gene-expression). In each spot, there might be more than one cell; therefore, the overall number of detectable genes is much higher of those depictable using the 10×genomics platform, which is instead single-cell specific.

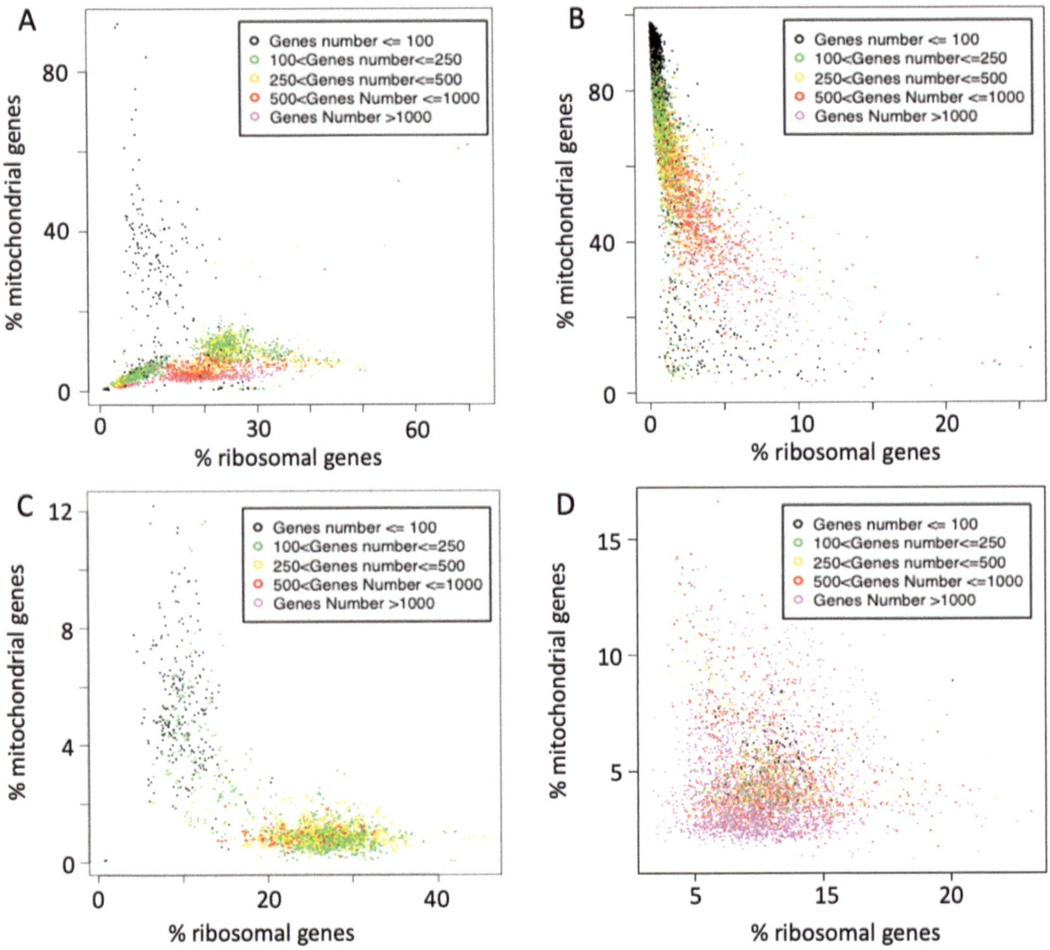

Fig. 1 mitoRiboUmi output. Percentage of mitochondrial protein genes counts is plotted, for each cell, with respect to the percentage of ribosomal protein genes counts. Cells are colored on the basis of total number detected genes, as default a gene a called present if supported by at least three UMI. The rationale of this metric is that: (i) Stressed cells are characterized by high percentage of mitochondrial genes. (ii) Active cells are characterized by the presence of high percentage of ribosomes. (iii) Low informative cells are characterized by few genes called present. (**a**) Mouse bone marrow cells, 10×Genomics platform v2 chemistry. (**b**) Human ovarian cancer cells from ascite, 10×Genomics platform v3 chemistry. (**c**) Mouse activated T cells, 10×Genomics platform v1 chemistry. (**d**) Breast cancer cells, visium spatial transcriptomics platform v2 chemistry

8. In Fig. 2, the results of the analysis of the sample indicated in **Note 5** are shown. Most of the cells show a percentage of mitochondrial genes below 10%. The cells characterized by less than 100 genes called (Fig. 2) black dots are also those prevalently showing a percentage of mitochondrial genes greater than 10%. Few cells are characterized by more than 50% ribosomal genes, which is in an indication of potential doublets. Thus, it would be ideal to retain cells with percentage

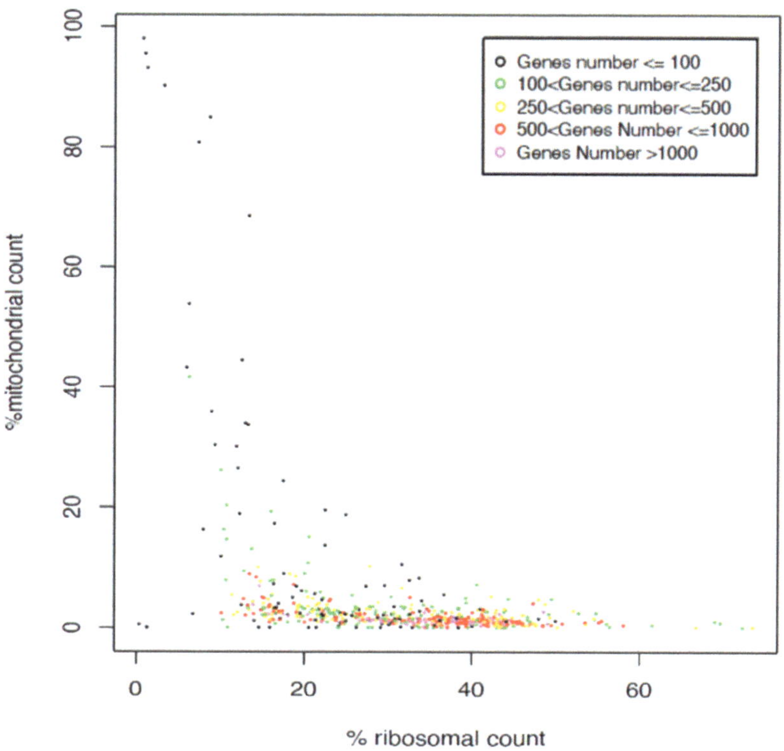

Fig. 2 mitoRiboUmi plot for the sample GSM4679532 (*see* **Note 5**)

mitochondrial genes between 1 and 10% and with percentage of ribosomal genes between 1 and 50%. It would be ideal to retain the cells with more than 100 genes called present. Furthermore, we suggest removing mitochondrial and ribosomal genes from the dataset before further analysis.

9. Defining the optimal thresholds to remove noninformative cells requires a good knowledge of cellular diversity in the tissue under study. Removal of cells characterized by more than 5% of mitochondrial proportion (mtDNA%, i.e., the ratio of reads mapped to mitochondrial DNA-encoded genes to the total number of reads mapped) was used in the past. Recently, Osorio and coworkers [14] compared the mitochondrial content between human and mouse data, stratified by the type of scRNA-seq technologies. Their data indicate that early defined 5% mtDNA is not appropriate for human healthy donor cells. Their data suggest a standardized mtDNA threshold of 10% for scRNA-seq QC of human samples. For mouse samples, the threshold of 5% accurately discriminates between healthy and low-quality cells. It has to be noted that these thresholds might not be suitable for some types of pathological cells, e.g., tumor cells.

10. According to Osorio [14], the removal of mitochondrial proteins genes does not affect the overall structure of cell subpopulations partitioning. Notably, the use of patterns of ribosomal proteins could be of particular interest to discriminate among normal and malignant human cells [16].
11. Data imputing should be done on the raw count matrix before performing any filtering. Since SAVER borrows information from genes, it is important to use the unfiltered raw counts data set for the imputation. Subsequently the low-quality cells in the SAVER imputed dataset can be removed based on the *mitoRiboUmi* analysis performed on the raw dataset.

References

1. Montanaro L, Trere D, Derenzini M (2008) Nucleolus, ribosomes, and cancer. Am J Pathol 173(2):301–310. https://doi.org/10.2353/ajpath.2008.070752
2. Nasrallah CM, Horvath TL (2014) Mitochondrial dynamics in the central regulation of metabolism. Nat Rev Endocrinol 10(11): 650–658. https://doi.org/10.1038/nrendo.2014.160
3. Alessandri L, Cordero F, Beccuti M, Arigoni M, Olivero M, Romano G, Rabellino S, Licheri N, De Libero G, Pace L, Calogero RA (2019) rCASC: reproducible classification analysis of single-cell sequencing data. Gigascience 8(9). https://doi.org/10.1093/gigascience/giz105
4. Stuart T, Butler A, Hoffman P, Hafemeister C, Papalexi E, Mauck WM 3rd, Hao Y, Stoeckius M, Smibert P, Satija R (2019) Comprehensive integration of single-cell data. Cell 177(7):1888–1902.e21. https://doi.org/10.1016/j.cell.2019.05.031
5. Pierson E, Yau C (2015) ZIFA: dimensionality reduction for zero-inflated single-cell gene expression analysis. Genome Biol 16:241. https://doi.org/10.1186/s13059-015-0805-z
6. Hou W, Ji Z, Ji H, Hicks SC (2020) A systematic evaluation of single-cell RNA-sequencing imputation methods. Genome Biol 21(1):218. https://doi.org/10.1186/s13059-020-02132-x
7. van Dijk D, Sharma R, Nainys J, Yim K, Kathail P, Carr AJ, Burdziak C, Moon KR, Chaffer CL, Pattabiraman D, Bierie B, Mazutis L, Wolf G, Krishnaswamy S, Pe'er D (2018) Recovering gene interactions from single-cell data using data diffusion. Cell 174(3):716–729.e27. https://doi.org/10.1016/j.cell.2018.05.061
8. Huang M, Wang J, Torre E, Dueck H, Shaffer S, Bonasio R, Murray JI, Raj A, Li M, Zhang NR (2018) SAVER: gene expression recovery for single-cell RNA sequencing. Nat Methods 15(7):539–542. https://doi.org/10.1038/s41592-018-0033-z
9. Wang J, Agarwal D, Huang M, Hu G, Zhou Z, Ye C, Zhang NR (2019) Data denoising with transfer learning in single-cell transcriptomics. Nat Methods 16(9):875–878. https://doi.org/10.1038/s41592-019-0537-1
10. Alessandri L, Cordero F, Beccuti M, Licheri N, Arigoni M, Olivero M, Di Renzo MF, Sapino A, Calogero R (2021) Sparsely-connected autoencoder (SCA) for single cell RNAseq data mining. NPJ Syst Biol Appl 7(1):1. https://doi.org/10.1038/s41540-020-00162-6
11. Alessandri L, Ratto ML, Contaldo SG, Beccuti M, Cordero F, Arigoni M, Calogero RA (2021) Sparsely connected autoencoders: a multi-purpose tool for single cell omics analysis. Int J Mol Sci 22(23). https://doi.org/10.3390/ijms222312755
12. Hao Y, Hao S, Andersen-Nissen E, Mauck WM 3rd, Zheng S, Butler A, Lee MJ, Wilk AJ, Darby C, Zager M, Hoffman P, Stoeckius M, Papalexi E, Mimitou EP, Jain J, Srivastava A, Stuart T, Fleming LM, Yeung B, Rogers AJ, McElrath JM, Blish CA, Gottardo R, Smibert P, Satija R (2021) Integrated analysis of multimodal single-cell data. Cell 184(13): 3573–3587.e29. https://doi.org/10.1016/j.cell.2021.04.048
13. Lin W, Noel P, Borazanci EH, Lee J, Amini A, Han IW, Heo JS, Jameson GS, Fraser C, Steinbach M, Woo Y, Fong Y, Cridebring D, Von Hoff DD, Park JO, Han H (2020) Single-cell transcriptome analysis of tumor and stromal compartments of pancreatic ductal

adenocarcinoma primary tumors and metastatic lesions. Genome Med 12(1):80. https://doi.org/10.1186/s13073-020-00776-9

14. Osorio D, Cai JJ (2021) Systematic determination of the mitochondrial proportion in human and mice tissues for single-cell RNA-sequencing data quality control. Bioinformatics 37(7):963–967. https://doi.org/10.1093/bioinformatics/btaa751

15. Ordonez-Rueda D, Baying B, Pavlinic D, Alessandri L, Yeboah Y, Landry JJM, Calogero R, Benes V, Paulsen M (2020) Apoptotic cell exclusion and bias-free single-cell selection are important quality control requirements for successful single-cell sequencing applications. Cytometry A 97(2):156–167. https://doi.org/10.1002/cyto.a.23898

16. Carrara M, Beccuti M, Cavallo F, Donatelli S, Lazzarato F, Cordero F, Calogero RA (2013) State of art fusion-finder algorithms are suitable to detect transcription-induced chimeras in normal tissues? BMC Bioinformatics 14 (Suppl 7):S2. https://doi.org/10.1186/1471-2105-14-S7-S2

Chapter 10

Single-Cell RNAseq Complexity Reduction

Francesca Cordero and Raffaele A. Calogero

Abstract

An important step in single-cell RNAseq data analysis is the preparation of the single cell transcription data for cell sub-population partitioning. In this chapter, we describe how to perform complexity reduction for 3′ end single-cell RNAseq transcriptomics data.

Key words Dimensionality reduction, Data preprocessing, Single cell transcriptomics

1 Introduction

Single-cell transcriptomics could provide a characterization of the transcriptional status of individual cells enabling to gain full insight into the interplay of transcripts within single cells. Single-cell RNA-seq (scRNA-seq) is highly dimensional and noisy. Some of the noise comes from the large number of single cell transcripts quantified with a value equal to zero [1]. Kim and coworkers [2] observed that most dropout events are associated with cell type heterogeneity. Because of the high dimensionality of scRNA-seq data, the complexity reduction is a crucial step in downstream analysis of scRNA-seq. For sake of clarification, dimensionality reduction is an approach to transform numerical data in a simplified form aiming to provide a condensed description of the original data maintaining their overall quality. In scRNA-seq, dimensionality reduction is a preliminary step to sub-population partitioning, i.e., clustering.

Xiang and coworkers [3] performed a comparison among various dimensionality reduction methods, including deep learning, linear, nonlinear, model-based, and ensemble methods. Based on their analysis, it came out that t-distributed stochastic neighbor embedding (t-SNE) [4] and uniform manifold approximation and projection (UMAP) [5] exhibited respectively the highest accuracy and the highest stability. However, neither tSNE nor UMAP are free of imperfections, the t-SNE method is powerful in preserving

Raffaele A. Calogero and Vladimir Benes (eds.), *Single Cell Transcriptomics: Methods and Protocols*, Methods in Molecular Biology, vol. 2584, https://doi.org/10.1007/978-1-0716-2756-3_10,

local data structure, but it can lose the global geometry property of the data [6]. On the other side, the UMAP method can depict global geometry patterns [5], but it can be inefficient to identify rare cell types, when dominant/common cell types exist. EDGE [7] is an ensemble data reduction method, which might be useful to compensate for some of the UMAP limits to detect rare cell types. Here, we describe the steps required to perform t-SNE, UMAP, and EDGE data reduction.

2 Materials

2.1 Minimal Hardware/Software Requirements

For the analyses described in this chapter, a workstation with at least 32 GB RAM, 2 TB SSD, and 2 CPUs is required. All analytical steps require a UNIX-like environment.

2.2 Exemplary Dataset

Two exemplary datasets (RNA-5c, RNA-5c-lowH1975) are available at the GitHub repository https://github.com/kendomaniac/single_cell_transcriptomics.git. Both datasets include the human lung adenocarcinoma cell lines A549, H1975, H2228, H838, and HCC827. RNA-5c, RNA-5c-lowH1975 are derived from GEO repository, ID: GSM3618014 [8]. In both dataset count tables, cell identifiers are associated with cell line names, e.g. Lib90_00000.HCC827, Lib90_00002.H838. RNA-5c includes 1242 cells of A549 cell line, 436 cells belonging to H1975, 749 cells from H2228 cell line, 879 cells belonging to H838, and 598 cells from HCC827 cell line. To draw attention to a dataset characterized by an under-represented cell population, the dataset RNA-5c-lowH1975 is generated. RNA-5c-lowH1975 includes all the RNA-5c cells, but with only 50 cells from H1975 cell line (*see* **Note 1**).

Homo sapiens ENSEMBL GTF can be retrieved at https://www.ensembl.org/info/data/ftp/index.html.

All data are also available at the figshare repository: https://doi.org/10.6084/m9.figshare.19390676.

3 Methods

3.1 Data Normalization

The RNA-5c and RNA-5c-lowH1975 datasets are imputed using SAVER [9–12] (*see* **Note 2**). Then, SAVER expression is converted in $\log_2$CPM (*see* **Note 3**).

```
# Open an R session an type the following commands
install.packages("SAVER")
library(SAVER)
#RNA-5c
WD = "/somewhere/in_your_computer/path_to_raw_data"
INPUT = "RNA-5c.csv"
```

```
SEPARATOR = ","
THREADS = 1 # the number of cores dedicated to run such
analysis
setwd(WD)
raw.data <- read.table(INPUT, sep=SEPARATOR,
 header = T, row.names=1)
dataset <- as.matrix(raw.data)
dataset.saver <- saver(dataset, ncores = THREADS,
 estimates.only = TRUE)
write.table(dataset.saver, paste("saver", INPUT,
 sep="_"), sep=SEPARATOR, col.names=NA)

counts2cpm <- function(file, sep = ","){
 tmp <- read.table(file, sep=sep, header=T, row.names=1)
 col.sum <- apply(tmp, 2, sum)
 tmp1 <- t(tmp)/col.sum
 tmp1 <- t(tmp1)
 tmp1 <- tmp1 * 1000000
 write.table(tmp1, "cpm.csv", sep=",",
 col.names=NA)
 write.table(log2(tmp1 + 1), "log2cpm.csv",
 sep=",", col.names=NA)
}

counts2cpm(file="saver_RNA-5c.csv", sep=",")
file.rename(from="log2CPM.csv", to="RNA-5c_log2CPM.csv")

#RNA-5c-lowH1975
INPUT = "RNA-5c-lowH1975.csv"
SEPARATOR = ","
THREADS = 1 # the number of cores dedicated to run such
analysis
setwd(WD)
raw.data <- read.table(INPUT, sep=SEPARATOR,
 header = T, row.names=1)
dataset <- as.matrix(raw.data)
dataset.saver <- saver(dataset, ncores = THREADS,
 estimates.only = TRUE)
write.table(dataset.saver, paste("saver", INPUT,
 sep="_"), sep=SEPARATOR, col.names=NA)

counts2cpm(file="saver_RNA-5c-lowH1975.csv", sep=",")
file.rename(from="log2CPM.csv", to="RNA-5c-lowH1975_log2CPM.
csv")
```

The output of the above-described procedure is the file RNA-5c-lowH1975_log2cpm.csv, embedding the normalized expression data to be used for dimensionality reduction.

3.2 Dimensionality Reduction with UMAP

Uniform Manifold Approximation and Projection (UMAP) is available in the umap R package. The use of umap requires the editing of a configuration parameters table available in umap.defaults variable (*see* **Notes 4** and **5**).

```
# Open an R session an type the following commands install.
packages("umap")
install.packages("ggplot2")
library(umap)
library(ggplot2)
#RNA-5c
rna5c <- read.table("RNA-5c_log2CPM.csv", sep=",", header=T,
row.names=1)
rna5c.labels <- sapply(strsplit(names(rna5c), '\cr.'), func-
tion(x)x[2])
cell_line <- as.factor(rna5c.labels)
rna5c.umap <- umap(t(rna5c), random_state=111, n_epochs =
1000)
f=data.frame(x=as.numeric(rna5c.umap$layout[,1]),y=as.numeric
(rna5c.umap$layout[,2]))
#plotting UMAP
sp &lt;- ggplot(f, aes(x=x,y=y)) + geom_point(pch=19, cex=0.3,
aes(color = cell_line))
pdf("rna5c.pdf")
 print(sp)
dev.off()

#RNA-5c-lowH1975
rna5clow <- read.table("RNA-5c-lowH1975_log2CPM.csv",
sep=",", header=T, row.names=1)
rna5clow.labels <- sapply(strsplit(names(rna5clow), '\cr.'),
function(x)x[2])
cell_line <- as.factor(rna5clow.labels)
rna5clow.umap <- umap(t(rna5clow), random_state=111, n_epochs
= 1000)
f=data.frame(x=as.numeric(rna5clow.umap$layout[,1]),y=as.nu-
meric(rna5clow.umap$layout[,2]))
#plotting UMAP
sp &lt;- ggplot(f, aes(x=x,y=y)) + geom_point(pch=19, cex=0.3,
aes(color = cell_line))
pdf("rna5clow.pdf")
print(sp)
dev.off()
```

3.3 Dimensionality Reduction with t-SNE

t-distributed stochastic neighbor embedding (t-SNE) is available in the Rtsne R package (*see* **Note 6**).

```
# Open an R session an type the following commands install.
packages("Rtsne")
library(Rtsne)
library(ggplot2)
rna5c <- read.table("RNA-5c_log2CPM.csv", sep=",", header=T,
row.names=1)
rna5c.labels <- sapply(strsplit(names(rna5c), '\cr.'), func-
tion(x)x[2])
cell_line <- as.factor(rna5c.labels)

set.seed(111)
#no PCA
tsne_out <- Rtsne(as.matrix(t(rna5c)), pca=FALSE, perplex-
ity=30, theta=0.0)
f=data.frame(x = as.numeric(tsne_out$Y[,1]),y = tsne_out$Y
[,2])
#plotting tSNE
sp &lt;- ggplot(f, aes(x=x,y=y)) + geom_point(pch=19, cex=0.3,
aes(color = cell_line))
pdf("rna5ctsne_noPCA.pdf")
print(sp)
dev.off()

#PCA 5 dims & max_iter 5K
tsne_out <- Rtsne(as.matrix(t(rna5c)), pca=TRUE, perplex-
ity=30, theta=0.0, num_threads=8, dims=2, initial_dims=5,
max_iter=5000
f=data.frame(x = as.numeric(tsne_out$Y[,1]),y = tsne_out$Y
[,2])
#plotting tSNE
sp &lt;- ggplot(f, aes(x=x,y=y)) + geom_point(pch=19, cex=0.3,
aes(color = cell_line))
pdf("rna5ctsne_PCA-5_maxiter-5k.pdf")
print(sp)
dev.off()

#PCA 5 dims & max_iter 5K
tsne_out2_10k <- Rtsne(as.matrix(t(rna5c)), pca=TRUE, per-
plexity=30, theta=0.0, num_threads=8, dims=2, initial_dims=5,
max_iter=10000)
f=data.frame(x = as.numeric(tsne_out$Y[,1]),y = tsne_out$Y
[,2])
#plotting tSNE
sp &lt;- ggplot(f, aes(x=x,y=y)) + geom_point(pch=19, cex=0.3,
aes(color = cell_line))
```

```
pdf("rna5ctsne_PCA-5_maxiter-10k.pdf")
print(sp)
dev.off()

#RNA-5c-lowH1975 PCA 5 dims & max_iter 5K
tsne_outlow_5k <- Rtsne(as.matrix(t(rna5clow)), pca=FALSE,
perplexity=30, theta=0.0, num_threads=8, initial_dims=5,
max_iter=5000)
f=data.frame(x = as.numeric(tsne_outlow_5k$Y[,1]),y = tsne_ou-
tlow_5k$Y[,2])
#plotting tSNE
sp &lt;- ggplot(f, aes(x=x,y=y)) + geom_point(pch=19, cex=0.3,
aes(color = cell_line))
pdf("rna5clow_tsne_PCA5_maxiter5k.pdf")
print(sp)
dev.off()
```

The output is an R structure including various information (*see* **Note 7**). In the output R structure, the Y element contains the plotting coordinates in the low dimension space for the cells. Exemplary plots of the RNA-5c and RNA-5c-lowH1975 are given in **Note 8**.

3.4 Dimensionality Reduction with EDGE

Ensemble Dimensionality reduction and feature Gene Extraction (EDGE) is available as github R package (*see* **Note 9**, https://github.com/shawnstat/EDGE).

```
# Open an R session an type the following commands library
("devtools")
install_github("shawnstat/EDGE")
library(EDGE)
install.packages("Matrix")
library(Matrix)
library(ggplot2)

# change default parameters
custom_defs <- endr_defs
# the number of weak learners
custom_defs$n_wl <- 1000
# the number of genes selected to construct weak learners
custom_defs$n_dm <- 10
# the number of nearest neighbors
custom_defs$n_neigs <- 15
# low-dimensional embedding optimization
custom_defs$opt <- TRUE
# random seed
custom_defs$seed <- 111
# the hash table size
custom_defs$H <- 101107
```

```
rna5c <- read.table("RNA-5c_log2CPM.csv", sep=",", header=T,
row.names=1)
rna5c.labels <- sapply(strsplit(names(rna5c), '\cr.'), func-
tion(x)x[2])
cell_line <- as.factor(rna5c.labels)

#wl 1k
rna5c.df <- t(rna5c)
rna5c.endr <- endr(rna5c.df,custom_defs)
f=data.frame(x = as.numeric(rna5c.endr[,1]), y = rna5c.endr
[,2])
#plotting tSNE
sp &lt;- ggplot(f, aes(x=x,y=y)) + geom_point(pch=19, cex=0.3,
aes(color = cell_line))
pdf("rna5c_edge_wl1k.pdf")
print(sp)
dev.off()
write.table(rna5c.endr, "rna5c.endr1k.csv", sep=",", col.
names=NA)

#wl 5k
custom_defs$n_wl <- 5000
rna5c.endr <- endr(rna5c.df,custom_defs)
f=data.frame(x = as.numeric(rna5c.endr[,1]), y = rna5c.endr
[,2])
#plotting tSNE
sp &lt;- ggplot(f, aes(x=x,y=y)) + geom_point(pch=19, cex=0.3,
aes(color = cell_line))
pdf("rna5c_edge_wl5k.pdf")
print(sp)
dev.off()
write.table(rna5c.endr, "rna5c.endr5k.csv", sep=",", col.
names=NA)

#wl 10k
custom_defs$n_wl <- 10000
rna5c.endr <- endr(rna5c.df,custom_defs)
f=data.frame(x = as.numeric(rna5c.endr[,1]), y = rna5c.endr
[,2])
#plotting tSNE
sp &lt;- ggplot(f, aes(x=x,y=y)) + geom_point(pch=19, cex=0.3,
aes(color = cell_line))
pdf("rna5c_edge_wl10k.pdf")
print(sp)
dev.off()
write.table(rna5c.endr, "rna5c.endr10k.csv", sep=",", col.
names=NA)
```

The output is a matrix defining the positions of each cell in the low dimensional space (*see* **Notes 9** and **10**).

3.5 General Considerations on Dimensionality Reduction

Dimensionality reduction is an important preliminary step in scRNAseq data analysis, since it provides denoising and simplification of high dimensional data at the same time. It is worthwhile mentioning the paper published by Nguyen and Holmes [13], which summarizes in ten points the elements to take into account when performing dimensionality reduction. Nguyen and Holmes paper focuses on critical points to be considered for the correct interpretation dimensionality reduction procedures. The points highlighted by Nguyen and Holmes, in our opinion, represent a useful checklist to be considered when applying a dimensionality reduction method.

4 Notes

1. RNA-5c-lowH1975 resembles a condition in which a rare cell type, i.e., H1975, is present.
2. Imputation aims to address the sparsity, i.e., the large number of zero observed in single-cell data. SAVER (single-cell analysis via expression recovery) is an expression recovery method for unique molecule index (UMI)-based scRNA-seq data borrowing information across genes and cells to provide accurate expression estimates for all genes [9].
3. A numerically stable method to homogenize RNAseq data is to set the size factor for each cell as the total counts divided by 10^6 (counts per million, CPM). CPM does not alter zeros, which dominate scRNA-Seq data. Log transformation is not possible for exact zeros, so a small pseudo-count, such as 1, is added to all normalized counts before taking the logarithm. Since the choice of pseudo-count is arbitrary and can introduce subtle biases in the transformed data [14], we transformed the counts in $\log_2$CPM after SAVER imputation.
4. The parameters configuration table, that can be modified in umap, is as follows:

```
## umap configuration parameters
## n_neighbors: 15
## n_components: 2
## metric: euclidean
## n_epochs: 200
## input: data
## init: spectral
## min_dist: 0.1
## set_op_mix_ratio: 1
## local_connectivity: 1
## bandwidth: 1
## alpha: 1
```

```
## gamma: 1
## negative_sample_rate: 5
## a: NA
## b: NA
## spread: 1
## random_state: NA
## transform_state: NA
## knn: NA
## knn_repeats: 1
## verbose: FALSE
## umap_learn_args: NA
```

We suggest increasing the number of epochs (n_epoch), e.g., 1000. Since UMAP finds a low dimensional embedding of the data that approximates an underlying manifold, increasing the number of epochs optimizes the above embedding. Furthermore, we suggest setting a defined random seed, e.g., 111, to provide reproducibility among replicated runs of umap. The n_components parameter indicates the dimension of the embedding space. In case of deciding to slice the low dimensional embedding in more than two dimensions, this number can be increased, e.g., 3, to generate a 3D plot. Another useful parameter is the min_dist ($0 \div 1$), which controls the packing among points. Notably, increasing this parameter helps to identify local structures persisting independently by the increment of this parameter. Furthermore, changing the metric used to evaluate the low dimension representation can help to identify local structures, which persist independently by the used metric. Cosine is an interesting metric to test instead of Euclidean. Cosine metric provides a measuring distance where the magnitude of the vectors is not considered.

Looking at the results shown in Fig. 1, it can be noted that, using the Euclidean, Fig. 1A, or cosine, Fig. 1B, distance, H2228, and HCC827 cells populations seem to be more homogeneous than A549, H1975, and H838, which seems to be constituted by few sub-populations. The presence of such intra-cell line heterogeneity is still clear for A549 and H838, using 0.5 and 0.8 minimal distance constrains, Fig. 1C, D, as instead intra-cell line heterogeneity becomes less evident for H1975.

5. The UMAP plots for RNA-5c and RNA-5c-lowH1975 are shown in Fig. 2. Notably, H1975 can be discriminated concerning the other cells, Fig. 2B black arrow, although its lower representation in cell number, i.e., 50 cells instead of the 436 cells present in RNA-5c dataset.

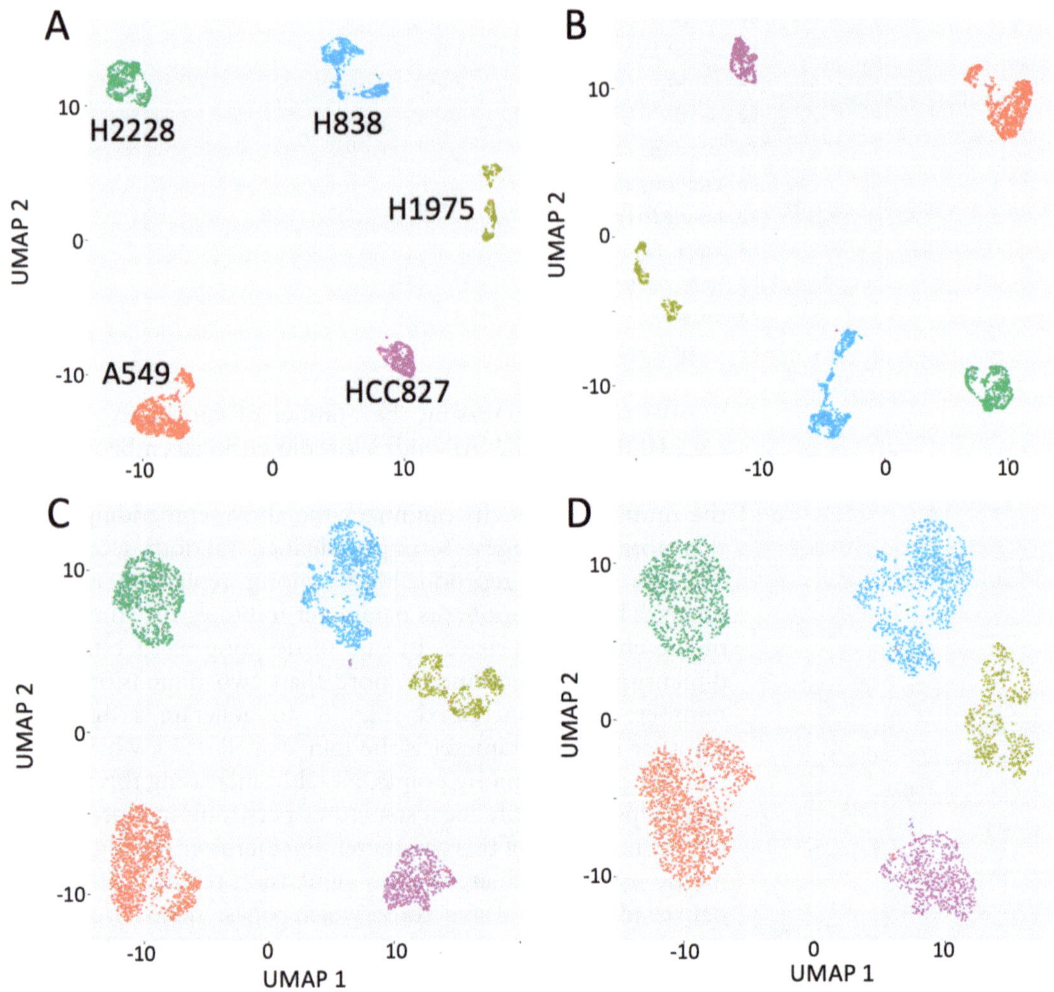

Fig. 1 UMAP plots for RNA-5c. (**a**) Euclidean distance. (**b**) Cosine distance. (**c**) Minimal distance among points 0.5. (**d**) Minimal distance among points 0.8

6. t-SNE is very flexible and can find structures where other dimensionality-reduction algorithms cannot. However, its flexibility can produce misleading results. The user should be careful in t-SNE data interpretation [15]. It is important to set seed before the analysis to provide reproducibility in repeating the t-SNE analysis. Furthermore, it is important to select a max_iter threshold (the default value is 1000) where the overall organization of the data becomes stable independently by a further increment of the max_iter parameter. The perplexity value provides the tread-off between local and global data structures. Thus, it is important to evaluate the stability of the low dimensional representation, i.e., defining the optimal value for max_iter, any time the perplexity value is changed. Figure 3B shows the t-SNE plot for RNA-5c with increasing values for the max_iter. After max_iter is set to 5000 or 1000

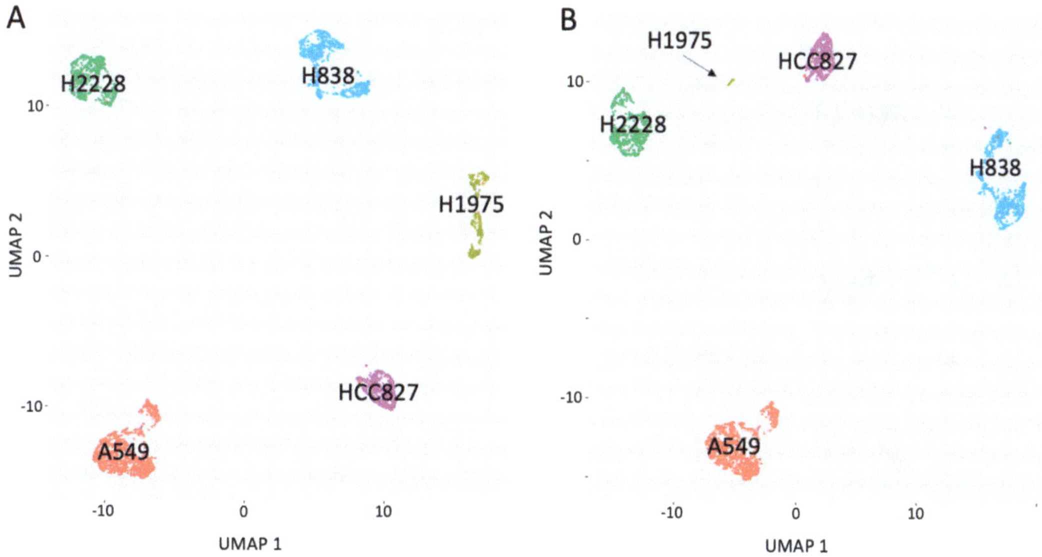

Fig. 2 UMAP plots. (**a**) RNA-5c. (**b**) RNA-5c-lowH1975

(Fig. 3C, D), the overall data structure is stable. It is important to note that the distance among elements in the t-SNE plot does not necessarily represent a real distance among the plotted elements.

Comparison of Fig. 1 with Fig. 3 shows that A549, H838, and H1975 are characterized by the presence of more than one cell sub-population.

The perplexity value is another important parameter that provides the tread-off between the local and global representation of the data. Unfortunately, there are no specific roles that help in setting the optimal perplexity value for a dataset. We suggest having a look at the tutorial developed by Wattenberg [15]. The perplexity value should be set in a way that Eq. 1 is satisfied.

$$3 \times \text{perplexity} < n \text{ of cells} - 1. \quad (1)$$

7. The structure of the t-SNE output is available at https://www.rdocumentation.org/packages/Rtsne/versions/0.15/topics/Rtsne.

8. The visualization of a rare population, i.e., H1975 in RNA-5c-lowH1975 (Fig. 4B), is more evident concerning what can be observed using UMAP (Fig. 2B), highlighting the limits of UMAP in representing rare cells populations and the ability of t-SNE in preserving local structure.

9. EDGE package [7] is based on an ensemble learning technique, which uses "weak learners." Weak learners are high bias models, i.e., in EDGE they are based on a few genes (default 10), which underfit but have low variance and perform a little better than a random model, i.e. error rate of 50%.

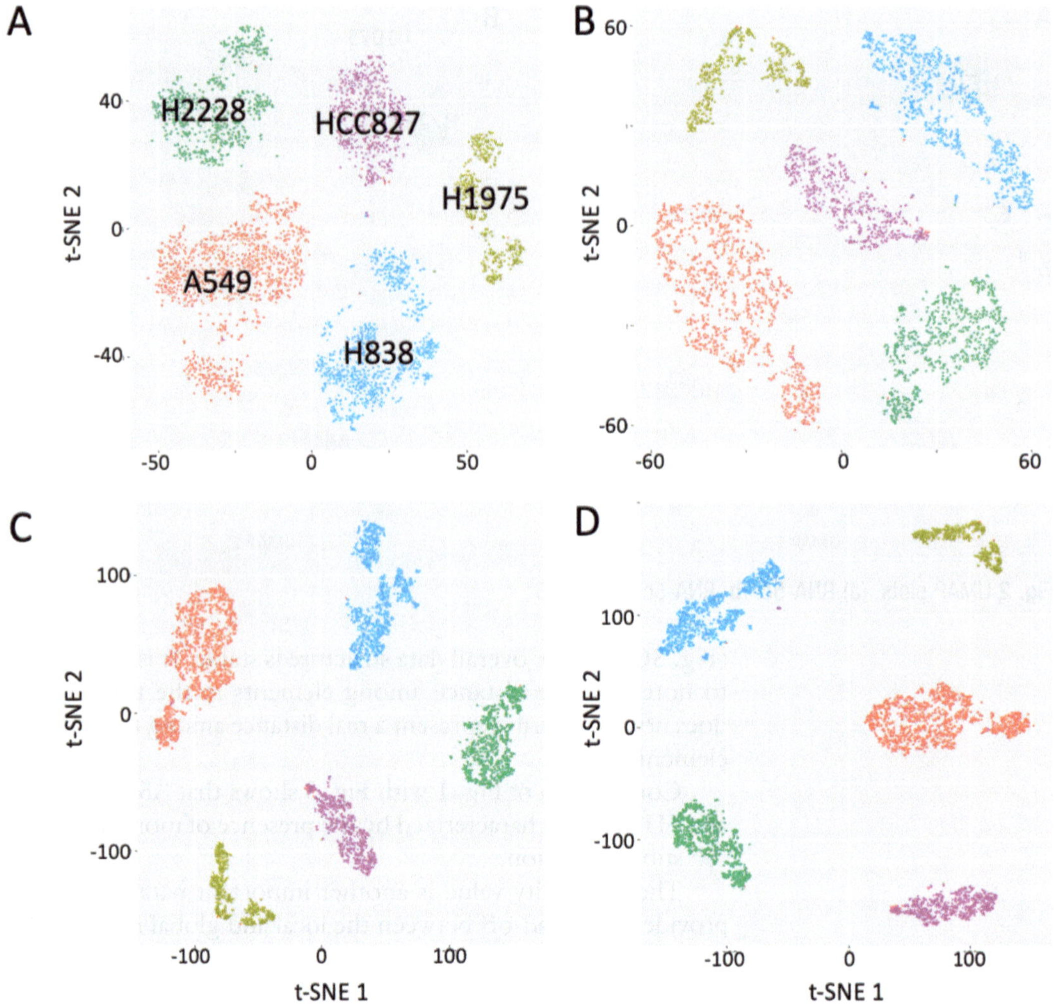

Fig. 3 t-SNE plots. (**a**) RNA-5c max_iter: 1000, no PCA selection, perplexity: 30. (**b**) RNA-5c max_iter: 1000, PCA: 5 dimensions, perplexity: 30. (**c**) RNA-5c max_iter: 5000, PCA: 5 dimensions, perplexity: 30. (**d**) RNA-5c max_iter 10,000, PCA 5 dimensions, perplexity: 30

EDGE combines weak learners to get a strong learner, which performs much better than all the individual weak learners. In Fig. 5, the effect of the progressive increase of the number of similarity weak learners can be observed. Specifically, in Fig. 5C, where the weak learners are set to 10,000, cell line H1975 seems to be composed of two different subsets of cells, where one seems consistently more similar to the HCC827 cell line.

10. Increasing the number of weak learners increase RAM consumption. For 10,000 weak learners, the analysis requires less than 16 GB RAM. For 50,000 weak learners, the analysis requires 124 GB RAM.

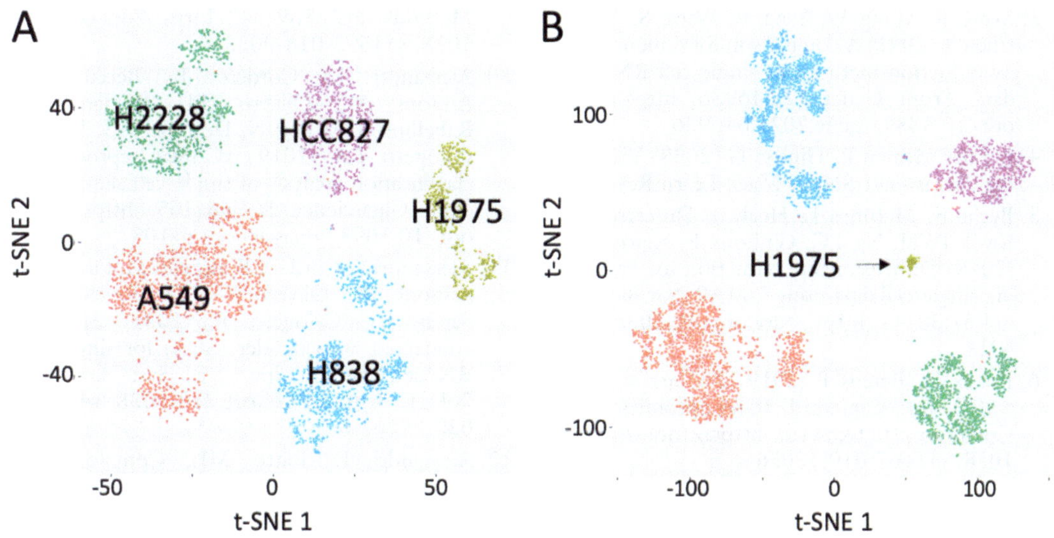

Fig. 4 t-SNE plots for RNA-5c and RNA-5c-lowH1975. (**a**) RNA-5c max_iter: 5000, PCA: 5 dimensions, perplexity: 30. (**b**) RNA-5c-lowH1975 max_iter: 5000, PCA: 5 dimensions, perplexity: 30

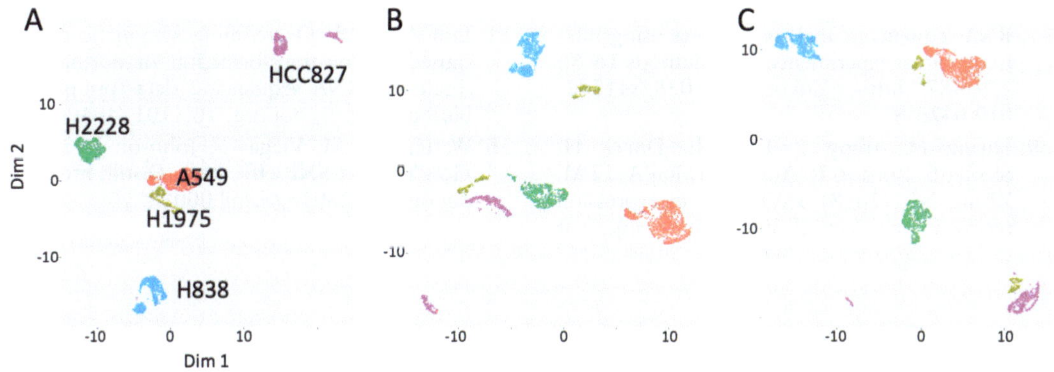

Fig. 5 EDGE plots for RNA-5c. (**a**) RNA-5c weak learners: 1000. (**b**) RNA-5c weak learners: 5000. (**c**) RNA-5c weak learners: 10,000

References

1. Lahnemann D, Koster J, Szczurek E, McCarthy DJ, Hicks SC, Robinson MD, Vallejos CA, Campbell KR, Beerenwinkel N, Mahfouz A, Pinello L, Skums P, Stamatakis A, Attolini CS, Aparicio S, Baaijens J, Balvert M, Barbanson B, Cappuccio A, Corleone G, Dutilh BE, Florescu M, Guryev V, Holmer R, Jahn K, Lobo TJ, Keizer EM, Khatri I, Kielbasa SM, Korbel JO, Kozlov AM, Kuo TH, Lelieveldt BPF, Mandoiu II, Marioni JC, Marschall T, Molder F, Niknejad A, Raczkowski L, Reinders M, Ridder J, Saliba AE, Somarakis A, Stegle O, Theis FJ, Yang H, Zelikovsky A, McHardy AC, Raphael BJ, Shah SP, Schonhuth A (2020) Eleven grand challenges in single-cell data science. Genome Biol 21(1):31. https://doi.org/10.1186/s13059-020-1926-6
2. Kim TH, Zhou X, Chen M (2020) Demystifying "drop-outs" in single-cell UMI data. Genome Biol 21(1):196. https://doi.org/10.1186/s13059-020-02096-y

3. Xiang R, Wang W, Yang L, Wang S, Xu C, Chen X (2021) A comparison for dimensionality reduction methods of single-cell RNA-seq data. Front Genet 12:646936. https://doi.org/10.3389/fgene.2021.646936
4. van der Maaten L, Hinton G (2008) Visualizing data using t-SNE. J Mach Learn Res 9:26
5. Becht E, McInnes L, Healy J, Dutertre CA, Kwok IWH, Ng LG, Ginhoux F, Newell EW (2018) Dimensionality reduction for visualizing single-cell data using UMAP. Nat Biotechnol 37:38–44. https://doi.org/10.1038/nbt.4314
6. Kobak D, Berens P (2019) The art of using t-SNE for single-cell transcriptomics. Nat Commun 10(1):5416. https://doi.org/10.1038/s41467-019-13056-x
7. Sun X, Liu Y, An L (2020) Ensemble dimensionality reduction and feature gene extraction for single-cell RNA-seq data. Nat Commun 11(1):5853. https://doi.org/10.1038/s41467-020-19465-7
8. Tian L, Dong X, Freytag S, Le Cao KA, Su S, JalalAbadi A, Amann-Zalcenstein D, Weber TS, Seidi A, Jabbari JS, Naik SH, Ritchie ME (2019) Benchmarking single cell RNA-sequencing analysis pipelines using mixture control experiments. Nat Methods 16(6): 479–487. https://doi.org/10.1038/s41592-019-0425-8
9. Huang M, Wang J, Torre E, Dueck H, Shaffer S, Bonasio R, Murray JI, Raj A, Li M, Zhang NR (2018) SAVER: gene expression recovery for single-cell RNA sequencing. Nat Methods 15(7):539–542. https://doi.org/10.1038/s41592-018-0033-z
10. Alessandri L, Cordero F, Beccuti M, Arigoni M, Olivero M, Romano G, Rabellino S, Licheri N, De Libero G, Pace L, Calogero RA (2019) rCASC: reproducible classification analysis of single-cell sequencing data. Gigascience 8(9):giz105. https://doi.org/10.1093/gigascience/giz105
11. Alessandri L, Cordero F, Beccuti M, Licheri N, Arigoni M, Olivero M, Di Renzo MF, Sapino A, Calogero R (2021) Sparsely-connected autoencoder (SCA) for single cell RNAseq data mining. NPJ Syst Biol Appl 7(1):1. https://doi.org/10.1038/s41540-020-00162-6
12. Alessandri L, Ratto ML, Contaldo SG, Beccuti M, Cordero F, Arigoni M, Calogero RA (2021) Sparsely connected autoencoders: a multi-purpose tool for single cell omics analysis. Int J Mol Sci 22(23):12755. https://doi.org/10.3390/ijms222312755
13. Nguyen LH, Holmes S (2019) Ten quick tips for effective dimensionality reduction. PLoS Comput Biol 15(6):e1006907. https://doi.org/10.1371/journal.pcbi.1006907
14. Lun A (2018) Overcoming systematic errors caused by log-transformation of normalized single-cell RNA sequencing data (pre-print). bioRxiv. https://doi.org/10.1101/404962
15. Wattenberg M, Viégas F, Johnson I (2016) How to use t-SNE effectively. Distill. https://doi.org/10.23915/distill.00002

Chapter 11

Functional-Feature-Based Data Reduction Using Sparsely Connected Autoencoders

Luca Alessandri and Raffaele A. Calogero

Abstract

Single-cell RNA sequencing (scRNA-seq) allows for the creation of large collections of individual cells transcriptome. Unsupervised clustering is an essential element for the analysis of these data, and it represents the initial step for the identification of different cell types to investigate the cell subpopulation structure of a biological sample. However, it is possible that the clustering aggregation features do not perfectly match the underlying biology since scRNA-seq data are characterized by high noise. In this chapter, we describe a functional feature-driven data reduction approach, which could provide a better link among cell clusters and their underlying cell biology.

Key words Data reduction, Transcription factor, Kinase, miRNA, Sparsely connected autoencoder

1 Introduction

In single-cell transcriptomics data analysis (scRNA-seq), the definition of how cells can be organized in cell sub-populations is probably the most important step. However, because of the intrinsic noise of the scRNA-seq data, it is possible that cell-specific biological features do not represent the main driving force for cell clustering. This is particularly true for immunological cells, which can be characterized by subtle functionalities, that some time cannot be effectively depicted by single-cell transcriptomics. Multimodal analysis, i.e., scRNA-seq combined with CITE-seq [1], is a way to increase the specificity of cell sub-population clustering. However, CITE-seq is not broadly applicable, especially in case of disease cells due to the lack of well-characterized cell-specific antibodies.

The biology of a cell type is a consequence of the gene expression modulation, which to a certain extent, is under the control of transcription factors (TFs), miRNAs, and kinases (Ks), which are respectively associated with gene expression modulation and

Raffaele A. Calogero and Vladimir Benes (eds.), *Single Cell Transcriptomics: Methods and Protocols*,
Methods in Molecular Biology, vol. 2584, https://doi.org/10.1007/978-1-0716-2756-3_11,

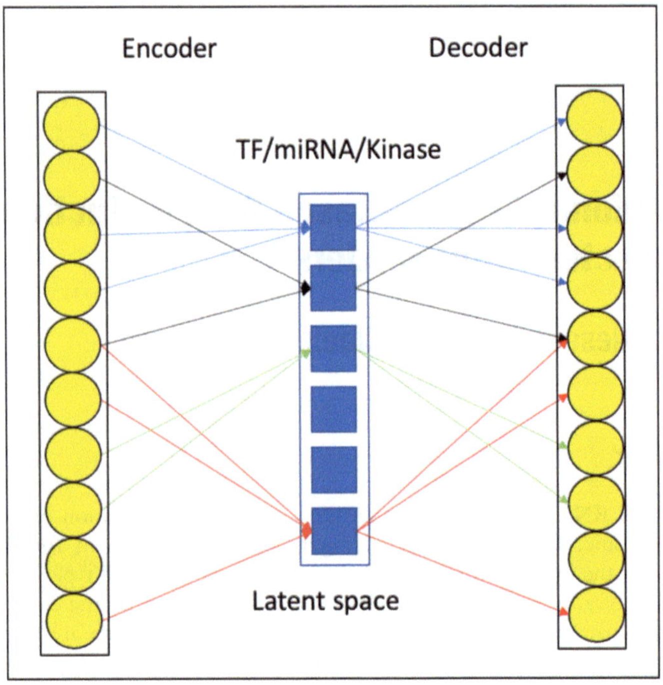

Fig. 1 Structure of a sparsely connected autoencoder (SCA). Input and output layers are connected, based on a specific biological feature, to a subset of nodes of the latent space. In this chapter, the latent space is based on the functional relationship existing among TFs, miRNAs, and kinases, but it can be modified on any custom relationship

protein activation. Nowadays, a still incomplete list of relationships among the above-mentioned regulators and their targets is available in open-source databases: (i) TRRUST [2], storing experimentally validated TFs to target genes relationships, (ii) miRTarBase [3], proving predicted and experimentally validated miRNAs to target genes relationships, and (iii) KinaseMD [4], storing experimentally validated Ks to target genes relationships. We only use, from the three databases, the experimentally validated relationships to guarantee high biological robustness of the analyses done using these features. As data reduction model, we decided to focus our attention to a particular type of deep-learning tool: autoencoder. Autoencoders have been used for data reduction, imputing, data denoising, and batch correction of single-cell RNAseq data. Recently, we have designed a new type of autoencoder model: sparsely connected autoencoder (SCA) [5–7]. SCA encoding/decoding functions consist of a single sparse hidden layer with connections based on known biological relationships [2–4] (Fig. 1), which only receives inputs from gene nodes associated with a specific relationship.

This reduction approach models data on the basis of the selected set of features, i.e., TFs, miRNAs, or Ks, which offers the possibility to represent scRNA-seq data using different facets of the same die. In this chapter, we describe how to use SCA to transform single-cell RNAseq data and to see how this transformation affects clustering results.

2 Materials

2.1 Minimal Hardware/Software Requirements

The analyses described in this chapter require a workstation with at least 64 GB RAM, 1 TB SSD, and 8 CPUs. All analytical steps require a UNIX-like environment (*see* **Notes 1** and **2**).

2.2 Exemplary Datasets

The exemplary dataset 390c_wctype [8] is available at the Github repository https://github.com/kendomaniac/single_cell_transcriptomics.git. This dataset refers to the sample 390c, extracted from the immune-gut atlas transcriptome (*see* **Note 3**). In 390c counts table, cell names include cell type assignment, i.e., TGACTAGGTTCCACAA.Treg (*see* **Note 3**). This dataset is part of 41,650 cells isolated from the cecum, transverse colon, and sigmoid colon of five individuals [6], which is one of the datasets constituting the immune-gut atlas transcriptome (*see* **Note 3**). To be analyzed by SCA, the count table needs to be imputed using SAVER tool [9]. The 390c SAVER imputed dataset, saver_390c_wctype.csv, is available at https://github.com/kendomaniac/single_cell_transcriptomics.

The exemplary dataset (RNA-5c) is available at the GitHub repository https://github.com/kendomaniac/single_cell_transcriptomics.git. This dataset includes the human lung adenocarcinoma cell lines A549, H1975, H2228, H838, and HCC827. RNA-5c is derived from GEO repository ID: GSM3618014 [10]. In this dataset count table, cell names and belonging cell lines are combined, e.g., Lib90_00000.HCC827 and Lib90_00002.H838. RNA-5c includes 1242 cells of A549, 436 cells of H1975, 749 cells of H2228, 879 cells of H838, and 598 cells of HCC827 cells.

All data are available as figshare repository: https://doi.org/10.6084/m9.figshare.19391150.

3 Methods

3.1 rCASC Installation

rCASC [5] is a single-cell analysis suite providing a computation reproducible infrastructure based on docker containers. The installation of rCASC is very simple. In an R session, type the code indicated below (*see* **Note 4**).

```
# Open an R session and type the following commands
library(devtools)
install_github("kendomaniac/rCASC", ref="master")
library(rCASC)
downloadContainers()# This step is not mandatory, since any
time a new docker container is required it is automatically
downloaded from docker hub repository.
```

rCASC provides an infrastructure to evaluate the overall quality of the clustering result via cell stability score (CCS, *see* **Note 5**).

3.2 Whole Transcriptome Clustering

The Louvain modularity clustering implemented in Seurat [11] was used to aggregate cells.

```
library(rCASC)
WD= "/data/single_cell_transcriptomics/single_cell_transcrip-
tomics/gut_atlas/"
SCRATCH= "/scratch/"
#Defining the PCA components
seuratPCAEval(group = "docker", scratch.folder=SCRATCH,
 file=paste(getwd(),"saver_390c_wctype.csv", sep="/"),
 separator=",", logTen = 0, seed = 111)

#Running clustering
seuratBootstrap(
 group = "docker",
 scratch.folder="/scratch",
 file=paste(getwd(),"saver_390c_wctype.csv", sep="/"),
 nPerm=40,
 permAtTime=10,
 percent=10,
 separator=",",
 logTen = 0,
 pcaDimensions=20,
 seed = 111,
 sparse = FALSE,
 format = "NULL",
 resolution = 0.6
)
```

The above function generates the following path: ~/Results/sample_names/number_clusters. In the folder defined by the number of clusters, there is a pdf ending with _Stability_Plot.pdf (Fig. 2a).

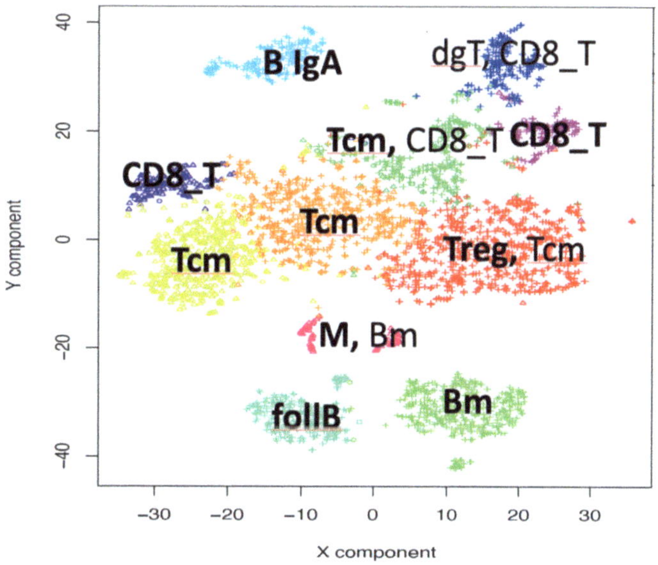

Fig. 2 Seurat clustering. Sample 390c at 0.6 resolution. The figure contains the main type of immunological cells constituting each cluster. In case there are two or more cell types, associated with a cluster, bold indicates the most represented cell type. Cell types: B cells producing IgA (B IgA), delta/gamma T cell (dgT), CD8 T cells (CD8_T), central memory T cells (Tcm), T regulatory cells (Treg), monocytes (M), B memory cells (Bm), follicular B cells (follB)

The clustering in Fig. 2 shows the presence of some clusters mainly constituted by a single cell type, whereas other clusters are heterogeneous. For example, Fig. 2 red cluster contains both T regulatory cells (Treg) and central memory cells (Tcm) and magenta cluster contains memory B cells and monocytes. Notably, memory T cells are split into different clusters.

3.3 Biology-Driven SCA Clustering

The function *autoencoder4clustering* generates a new dataset representing cells based on a specific type of biological features, e.g., TFs, which expressions are modeled based on their targets gene expression. The *permutation* parameter indicates the number of times the model is repeated. Each model generates a representation of the autoencoder hidden layer, which is a matrix having biological features as rows and cells as columns. The matrix at each permutation represents the modeling of each biological feature expression based on its gene targets expression.

```
library(rCASC)
autoencoder4clustering(group="docker",
 scratch.folder="/scratch",
 file=paste(getwd(),"saver_390c_wctype.csv", sep="/"),
 separator=",",
 bias="TF",
```

```
 permutation=40,
 nEpochs=1000,
 lr=0.01
 projectName="TF")
#IMPORTANT: rownames of the matrix should only be made by gene
symbol not by ensemblID:symbol format.
```

The *bias* parameter is the data reduction model to be applied: "TF", "mirna", "kinasi", "cytobands." *nEpochs* parameter indicates the number of times each model is trained (*see* **Note 6**). If the loss curve (*see* **Note 6**) is too steep or too shallow, the behavior of such curve can be modified by tuning the learning rate parameter (*lr*). After the execution of the *autoencoder4clustering*, the data reduction models are transformed in a cumulative sum matrix, which is used to perform Seurat clustering. In the present example, we have used a *resolution* of 0.6. *Resolution* is parameter, affecting the granularity of the clusters; as the resolution value increases, the number of detected clusters also increase. Resolution should be adapted in each experiment based on the cluster stability (*see* **Note 7**).

```
files=paste("./Results/TF/permutation/",
 list.files("./Results/TF/permutation/"),sep="")

temp=read.table(files[1],sep=",",header=TRUE,row.names=1)
for(i in files[-1]){
 temp=temp+read.table(i,header=TRUE,row.names=1,sep=",")
}
write.table(temp,"total.csv",col.names=NA,sep=",")
file=paste(getwd(),"total.csv",sep="/")
path=getwd()

scratch=paste(getwd(),"scratch",sep="/")
dir.create(scratch)
seuratBootstrap(group="docker",
 scratch.folder=scratch,
 file=file,
 nPerm=40,
 permAtTime=5,
 percent=10,
 separator=",",
 logTen=0,
 pcaDimensions=15,
 seed=111,
 sparse=FALSE,
 format="NULL",
 resolution=0.6)
```

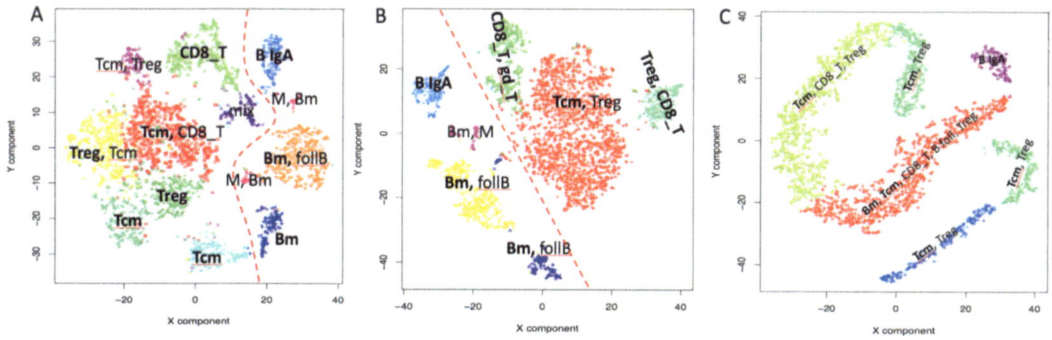

Fig. 3 Seurat clustering of SCA reduced data, resolution 0.6. (**a**) TFs-based data reduction of sample 390c. (**b**) miRNAs-based data reduction of sample 390c. (**c**) Kinases-based data reduction of sample 390c

3.3.1 TFs-Driven SCA Clustering

The results observable for the clustering performed using the SCA data reduction, modeled using TFs, miRNAs, and kinases, are summarized in Fig. 3.

With respect to what is observable using the full set of genes, Fig. 2, using the TFs-based data reduction, Fig. 3a, CD8 T cells are mainly associated to the green cluster, but the other clusters are not fully homogeneous. In the case of miRNA-based data reduction, we get more homogeneous clusters for gamma/delta T cells and T regulatory cells. Furthermore, both miRNAs- and TFs-based data reduction show a clear separation among T-cell and B-cell clusters. The clustering produced using Kinases is more intricated and different cell types are not very well separated (Fig. 3c).

3.3.2 Cytobands-Driven SCA Clustering

Another interesting way to transform data using SCA is the representation of the expression data using the genomics cytobands, which are the genomic regions associated with the Giemsa staining of metaphase chromosomes (heterochromatic regions stain darker than euchromatic regions). Although cytoband is an old-fashioned chromosome structure representation, it can be used to depict differences among cells clusters, looking at changes in block of genes associated with cytobands. As an example, we run SCA analysis on RNA-5c dataset. The overall analysis is identical to the one shown in the previous paragraph, but with the *bias* parameter of *autoencoder4clustering* function set to "cytoBands." Results of the clustering of the SCA output are summarized in Fig. 4.

It is notable that cell line H838 can be split into two highly stable clusters, indicating that the genomic transcription profile is different between the cells in clusters 4 and 5 (Fig. 4). Similar situation applies to A549 cell line, which can be split into three clusters (Fig. 4), clusters 1, 7, and 9, and cell line H1975, which is split into two clusters (Fig. 4), clusters 6 and 8. Especially in the oncological setting, this chromosome-driven clustering could represent an interesting approach to highlight transcription-linked structural chromosome reorganization.

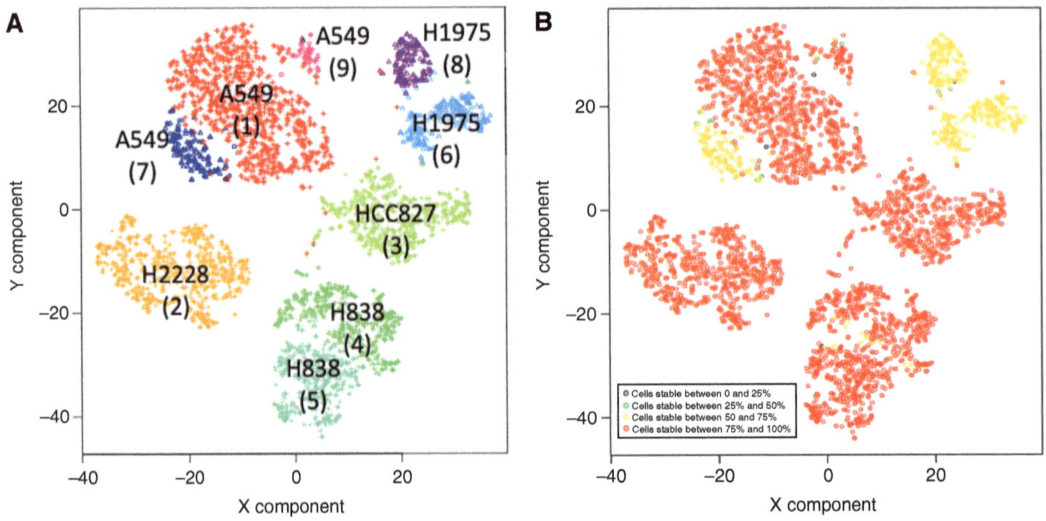

Fig. 4 Seurat clustering of SCA reduced data, resolution 0.6. (**a**) Clustering results of cytoband-based data reduction applied to RNA5c dataset. (**b**) Cell stability score plot. Cell stability score is a measure of the robustness of a cluster, since it indicates how many times a cell stays in a cluster independently by the presence of dataset perturbation, i.e., cells bootstrapping

4 Notes

1. The installation of docker is required. Docker is available for Windows, Mac, and UNIX infrastructures at https://docs.docker.com/get-docker/.
2. R installation is required (https://cran.r-project.org/). Specifically, the installation of devtools library is required.
3. The gut atlas datasets are available at https://www.gutcellatlas.org/.
4. The description of all rCASC functions is available at https://kendomaniac.github.io/rCASC/reference/index.html. The execution of the "downloadContainers()" command is not mandatory, since the first time a new docker container is required, it is downloaded from the repbioinfo docker.io repository.
5. CSS (cell stability score) describes cluster stability by estimating the persistence of a cell in a specific cluster upon jackknife resampling. A detailed description of the CSS metric is available in the supplementary data of rCASC paper [7].
6. It is important to check the behavior of the loss curve over training (Fig. 5). A progressive reduction of the loss is expected as the number of epochs progress.

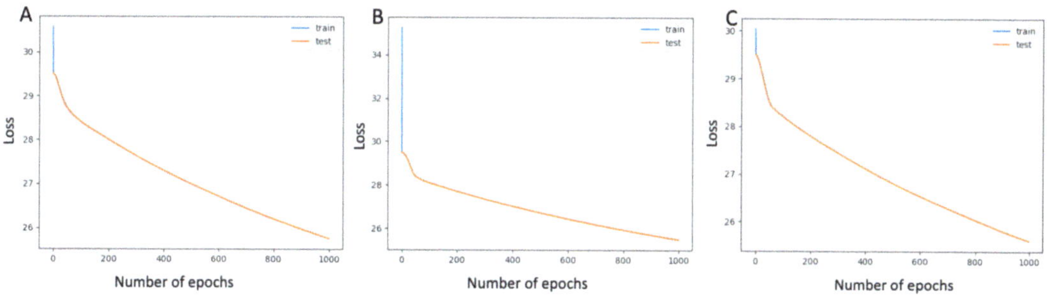

Fig. 5 Loss curve. (**a**) TFs-based modeling. (**b**) miRNAs-based modeling. (**c**) Kinases-based modeling. Training and test sets are superimposed because they are generated using the same input data, since the task of the approach is learning as much as possible about the datasets under analysis and not using the trained SCA to analyze a different dataset

7. CSS (cell stability score) describes cluster stability by estimating the persistence of a cell in a specific cluster upon jackknife resampling. A detailed description of the CSS metric is available in the supplementary data of rCASC paper [7].

References

1. Stoeckius M, Hafemeister C, Stephenson W, Houck-Loomis B, Chattopadhyay PK, Swerdlow H, Satija R, Smibert P (2017) Simultaneous epitope and transcriptome measurement in single cells. Nat Methods 14(9): 865–868. https://doi.org/10.1038/nmeth.4380
2. Han H, Cho JW, Lee S, Yun A, Kim H, Bae D, Yang S, Kim CY, Lee M, Kim E, Lee S, Kang B, Jeong D, Kim Y, Jeon HN, Jung H, Nam S, Chung M, Kim JH, Lee I (2018) TRRUST v2: an expanded reference database of human and mouse transcriptional regulatory interactions. Nucleic Acids Res 46(D1):D380–D386. https://doi.org/10.1093/nar/gkx1013
3. Huang HY, Lin YC, Li J, Huang KY, Shrestha S, Hong HC, Tang Y, Chen YG, Jin CN, Yu Y, Xu JT, Li YM, Cai XX, Zhou ZY, Chen XH, Pei YY, Hu L, Su JJ, Cui SD, Wang F, Xie YY, Ding SY, Luo MF, Chou CH, Chang NW, Chen KW, Cheng YH, Wan XH, Hsu WL, Lee TY, Wei FX, Huang HD (2020) miRTarBase 2020: updates to the experimentally validated microRNA-target interaction database. Nucleic Acids Res 48 (D1):D148–D154. https://doi.org/10.1093/nar/gkz896
4. Hu R, Xu H, Jia P, Zhao Z (2021) KinaseMD: kinase mutations and drug response database. Nucleic Acids Res 49(D1):D552–D561. https://doi.org/10.1093/nar/gkaa945
5. Alessandri L, Cordero F, Beccuti M, Licheri N, Arigoni M, Olivero M, Di Renzo MF, Sapino A, Calogero R (2021) Sparsely-connected autoencoder (SCA) for single cell RNAseq data mining. NPJ Syst Biol Appl 7(1):1. https://doi.org/10.1038/s41540-020-00162-6
6. Alessandri L, Ratto ML, Contaldo SG, Beccuti M, Cordero F, Arigoni M, Calogero RA (2021) Sparsely connected autoencoders: a multi-purpose tool for single cell omics analysis. Int J Mol Sci 22(23):12755. https://doi.org/10.3390/ijms222312755
7. Alessandri L, Cordero F, Beccuti M, Arigoni M, Olivero M, Romano G, Rabellino S, Licheri N, De Libero G, Pace L, Calogero RA (2019) rCASC: reproducible classification analysis of single-cell sequencing data. Gigascience 8(9):giz105. https://doi.org/10.1093/gigascience/giz105
8. James KR, Gomes T, Elmentaite R, Kumar N, Gulliver EL, King HW, Stares MD, Bareham BR, Ferdinand JR, Petrova VN, Polanski K, Forster SC, Jarvis LB, Suchanek O, Howlett S, James LK, Jones JL, Meyer KB, Clatworthy MR, Saeb-Parsy K, Lawley TD, Teichmann SA (2020) Distinct microbial and immune niches of the human colon. Nat Immunol 21(3):343–353. https://doi.org/10.1038/s41590-020-0602-z
9. Huang M, Wang J, Torre E, Dueck H, Shaffer S, Bonasio R, Murray JI, Raj A, Li M,

Zhang NR (2018) SAVER: gene expression recovery for single-cell RNA sequencing. Nat Methods 15(7):539–542. https://doi.org/10.1038/s41592-018-0033-z

10. Tian L, Dong X, Freytag S, Le Cao KA, Su S, JalalAbadi A, Amann-Zalcenstein D, Weber TS, Seidi A, Jabbari JS, Naik SH, Ritchie ME (2019) Benchmarking single cell RNA-sequencing analysis pipelines using mixture control experiments. Nat Methods 16(6):479–487. https://doi.org/10.1038/s41592-019-0425-8

11. Butler A, Hoffman P, Smibert P, Papalexi E, Satija R (2018) Integrating single-cell transcriptomic data across different conditions, technologies, and species. Nat Biotechnol 36(5):411–420. https://doi.org/10.1038/nbt.4096

Chapter 12

Single-Cell RNAseq Clustering

Marco Beccuti and Raffaele A. Calogero

Abstract

Single-cell RNA sequencing (scRNA-seq) allows the creation of large collections of individual cells transcriptome. Unsupervised clustering is an essential element for the analysis of these data, and it represents the initial step for the identification of different cell types to investigate the cell subpopulation organization of a sample. In this chapter, we describe how to approach the clustering of single-cell RNAseq transcriptomics data using various clustering tools, and we provide some information on the limitations affecting the clustering procedure.

Key words Unsupervised clustering, Single cell transcriptomics, Griph, SHARP, Seurat, Lovain modularity

1 Introduction

In single-cell transcriptomics data analysis, the definition of how cells can be organized in cell sub-populations is probably the most important step of an analysis. The identification of cell sub-populations is mainly based on the use of a computation method called clustering, aiming to segregate cells with similar transcription profiles and to assign them into homogeneous groups. Although unsupervised clustering is heavily used in single-cell analysis, there are many challenges involved [1] in its use. The course of dimensionality is one of the problems affecting clustering. The course of dimensionality describes the risk of overfitting due to the presence of more features, i.e., genes, than observations, i.e., cells. This issue can be moderated using approaches like feature selection, i.e., retaining only the genes being the most relevant to the data set structure, and/or dimensionality reduction, i.e., recombining the original variables into a new set of non-redundant variables. Dimensionality reduction helps a lot in visualizing clustering results. However, dimensionality reduction methods require the manual trimming of several parameters, which can strongly affect the visualization

Raffaele A. Calogero and Vladimir Benes (eds.), *Single Cell Transcriptomics: Methods and Protocols*,
Methods in Molecular Biology, vol. 2584, https://doi.org/10.1007/978-1-0716-2756-3_12,

of the results. A similar problem applies to clustering methods, which include parameters that must be chosen by the user to define the resolution of the clustering. The choice of such parameters can have a large effect on the clustering outcome. In this chapter, we describe how to use three clustering methods: griph [2], sharp [3], and Seurat [4]. We integrated the above-mentioned clustering tools in rCASC [5–7]. rCASC is a suite of programs devoted to single-cell transcriptomics data analysis and provides unique features like the cell stability score (CSS), which is a metric providing information on clustering stability, where stability refers to the ability of cells to remain associated with a specific cluster in presence of perturbation events, such as removal of a random subset of cell upon clustering.

2 Materials

2.1 Minimal Hardware/Software Requirements

The analyses described in this chapter require a workstation with at least 64 GB RAM, 2 TB SSD, 8 CPUs. All analytical steps require a UNIX-like environment (*see* **Notes 1** and **2**).

2.2 Exemplary Dataset

The exemplary dataset 390c_wctype [8] is available at the Github repository https://github.com/kendomaniac/single_cell_transcriptomics.git. This dataset refers to the sample 390c from the gut atlas transcriptome. In the counts table, cell names include the cell type assignment, i.e., TGACTAGGTTCCACAA.Treg (*see* **Note 3**). This dataset is part of 41,650 cells isolated from the cecum, transverse colon, and sigmoid colon of 5 individuals [6] (*see* **Note 3**).

All data are also available at figshare repository: https://doi.org/10.6084/m9.figshare.19390712.

3 Methods

3.1 rCASC Installation

rCASC [5] is a single-cell analysis suite providing a computation reproducible infrastructure based on docker containers. The installation of rCASC is very simple. In an R session, type the code indicated below (*see* **Note 4**).

```
# Open an R session and type the following commands
library(devtools)
install_github("kendomaniac/rCASC", ref="master")
library(rCASC)
downloadContainers()# This a non-mandatory step, since the software download, if needed, docker containers from docker.com repository.
```

rCASC provides an infrastructure to evaluate the overall quality of the clustering result via the cell stability score (CCS, *see* **Note 5**).

3.2 Griph Clustering

The griph algorithm [2] is based on Louvain modularity (*see* **Note 6**). The griph algorithm is closer to agglomerative clustering methods since every node is initially assigned to its own community and communities are subsequently built by iterative merging (*see* **Note 7**). The clustering is performed on SAVER [9] imputed 390c_wctype dataset (*see* **Note 8**).

```
Library(rCASC)
 griphBootstrap(
 group = "docker",
 scratch.folder="/scratch",
 file=paste(getwd(), "saver_390c_wctype.csv", sep="/"),
 nPerm=40,
 permAtTime=10,
 percent=10,
 separator=",",
 logTen = 0,
 seed = 111
)
```

The above function generates the following path: ~Results/sample_names/number_clusters. In the folder defined by the number of clusters, there is a pdf ending with _Stability_Plot.pdf. This pdf shows the cell organization in clusters and provides information about cluster stability. A file ending with _clustering.output.csv is also present, which provides the cell to cluster association and the coordinates of each cell in the above-mentioned pdf file (*see* **Note 9**).

3.3 SHARP Clustering

SHARP [3] is a clustering tool showing a very high speed and the ability to handle a large number of cells. It uses, as a data reduction approach, random projections. The clustering is based on the concept of divide and conquer since the dataset is spitted into various subsets, which are then clustered using weighted-based-metaclustering and all the clusters are subsequently aggregated using similarity-based-metaclustering. This approach makes the clustering tool extremely fast especially when a large number of cells are taken into account.

```
Library(rCASC)
sharpBootstrap(
 group = "docker",
 scratch.folder="/scratch",
 file=paste(getwd(), "saver_390c_wctype.csv", sep="/"),
 nPerm=40,
 permAtTime=10,
```

```
 percent=10,
 separator=",",
 logTen = 0,
 seed = 111
)
```

The output structure of sharp, embedded in rCASC, is identical to the one described for griph. A detailed description of the sharp output is available in **Note 10**.

3.4 Seurat Clustering

Seurat [4] is a suite of programs allowing multimodal analysis of single-cell omics data. Seurat clustering is based on Lovain modularity such as griph (*see* **Note 6**). The rCASC implementation of Seurat clustering requires, as a first step, that the user defines how much information he or she wishes to use in the clustering (*see* **Note 11**). To decide the amount of data variance that will be used for clustering, user must run *seuratPCAEval* function (*see* **Note 11**). Subsequently, the user applies the selected threshold of PC components within the pcadimension parameter of the *seuratBootstrap* function.

```
Library(rCASC)
WD="/data/single_cell_transcriptomics/single_cell_transcrip-
tomics/gut_atlas/"
SCRATCH="/scratch/"
#Defining the PCA components
seuratPCAEval(group = "docker", scratch.folder=SCRATCH, file=-
paste(getwd(), "saver_390c_wctype.csv", sep="/"), separa-
tor=",", logTen = 0, seed = 111)

#Running clustering
seuratBootstrap(
 group = "docker",
 scratch.folder="/scratch",
 file=paste(getwd(), "saver_390c_wctype.csv", sep="/"),
 nPerm=40,
 permAtTime=10,
 percent=10,
 separator=",",
 logTen = 0,
 pcaDimensions=20,
 seed = 111,
 sparse = FALSE,
 format = "NULL",
 resolution = 0.6
)
#resolution parameter was tested at 0.1, 0.3 and 0.6
```

The output structure of Seurat, embedded in rCASC, is identical to the one described for griph.

An important parameter for Seurat clustering is "resolution," which sets the granularity of the clustering; e.g., as the resolution value increases, the number of detected clusters also increase.The effect of three different resolutions on the 390c sample is shown in Fig. 1. Since cells' annotation is available for this dataset, this helps in understanding the effect of changing the resolution parameter values. As examples of the effect of the resolution parameter, we will follow the transformation of clusters observed at resolution 0.1: clusters 4 and 1. Cluster 4 at resolution 0.1 (Fig. 1a) is mainly constituted of CD8 T cells (193 cells) and gamma/delta T cells (95 cells). At resolution 0.3, cluster 3 (Fig. 1b) is made of the aggregation of cluster 4 (Fig. 1a) and a part of cluster 1 (Fig. 1a). In cluster 3, CD8 T cells increase (255 cells) as instead the gamma/delta T cells component is constant (95 cells). At resolution 0.6 (Fig. 1c), three clusters are generated from cluster 3 (Fig. 1b). Cluster 5 is mainly made of CD8 T cells (95 cells), cluster 10 mainly of CD8 T cells (130 cells), and cluster 8 of gamma/delta T cells (95 cells) and CD8 T cells (64 cells). Thus, considering the effect of different resolutions on CD8 T cells, resolution 0.3 seems to be the one that better represents the cell typology, although certain gamma/delta T cells are contaminated.

Applying the above-described analysis to cluster 1 (Fig. 1a), this is mainly made of central memory T cells (1209 cells), CD8 T cells (277 cells), T regulatory cells (521 cells). In Fig. 1b, the cluster 1 (Fig. 1a) is split into two parts: cluster 1, mainly made of CD8 T cells (199 cells) and central memory T cells (868 cells), and cluster 2, mainly made of T regulatory cells (489 cells) and central T

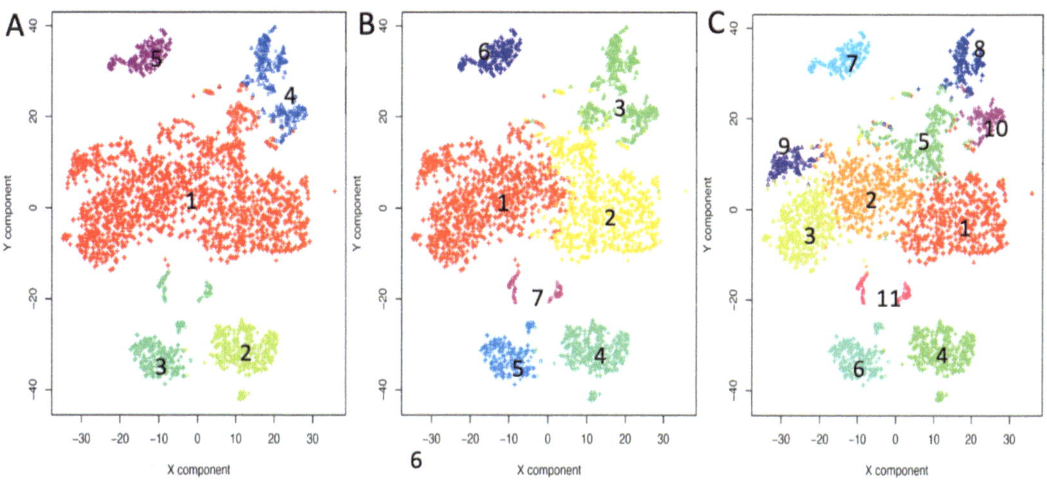

Fig. 1 Seurat clustering using different resolution thresholds. (**a**) Sample 390c at 0.1 resolution, (**b**) sample 390c at 0.3 resolution, and (**c**) sample 390c at 0.6 resolution

memory cells (321 cells). In Fig. 1c, the cluster 1 (Fig. 1a) is split into six parts: cluster 9 made of CD8 T cells (132 cells); cluster 3 made of central memory T cells (399 cells); cluster 2 made of central T memory cells (482 cells); cluster 5 made of central memory T cells, and cluster 1 made of T regulatory cells (478 cells) and central T memory T cells (137 cells). In this case, both resolutions 0.3 and 0.6 are not satisfying, because resolution 0.3 partially splits the different cell types, whereas resolution 0.6 divides the central T memory cells into three different clusters. These observations indicate that the clustering based only on transcription profiling does not guarantee the perfect separation among different cell types, especially if there are subtypes of the same cell type. The above examples take advantage of preexisting cell annotation that helps in describing the clustering limitation of some methods. However, in the real world, clustering is used to depict the cell characteristics present in a specific cluster. Unfortunately, no clearly defined paths are available to understand the best clustering tool for a specific dataset and the best set of parameters to be used. In this specific example, Griph seems to work best in discriminating among different cell types and it has the advantage of depicting by itself the optimal partition parameters. In the case of Seurat, the granularity can be defined by resolution parameter; however, understanding the optimal setting of this parameter is very challenging.

Both Griph and Seurat are however unable to aggregate together all the central T memory cells, which spread over different clusters, suggesting a high level of heterogeneity for this cell type.

4 Notes

1. The installation of docker is required. Docker for Windows, Mac, and UNIX infrastructures is available at https://docs.docker.com/get-docker/.
2. R installation is required (https://cran.r-project.org/). Specifically, the installation of devtools library is required.
3. The gut atlas datasets are available at https://www.gutcellatlas.org/.
4. The description of all rCASC functions is available at https://kendomaniac.github.io/rCASC/reference/index.html. The execution of the "downloadContainers()" command is not mandatory, since the first time a new docker container is required for an analysis it is downloaded from the repbioinfo docker.io repository.
5. CSS (cell stability score) describes cluster stability by estimating the persistence of a cell in a specific cluster upon jackknife resampling. A detailed description of the CSS metric is available in the supplementary data of rCASC paper [5].

6. The Louvain method for community detection is a method to extract communities from large networks. It detects the hierarchy of communities, which can be useful for understanding the global network organization. Communities are sets of nodes within a network, which show intra-nodes connectivity denser than other nodes. The modularity metric evaluates how much more densely connected are the nodes within a community with respect to a random network. The Louvain method of community detection detects communities in networks relying on a heuristic which maximizes the modularity.
7. The griph clustering tool is available at https://github.com/ppapasaikas/Griph. However, in this chapter, we use its implementation in rCASC, since it provides more information on clustering stability.
8. The 390c SAVER imputed dataset, saver_390c_wctype.csv, is available at https://github.com/kendomaniac/single_cell_transcriptomics.
9. Griph clustering generates 11 clusters (Fig. 2a). However, the cluster stability, measured with the CSS score (Fig. 2b), is of good quality, i.e., CSS greater than 50%, only for three clusters.

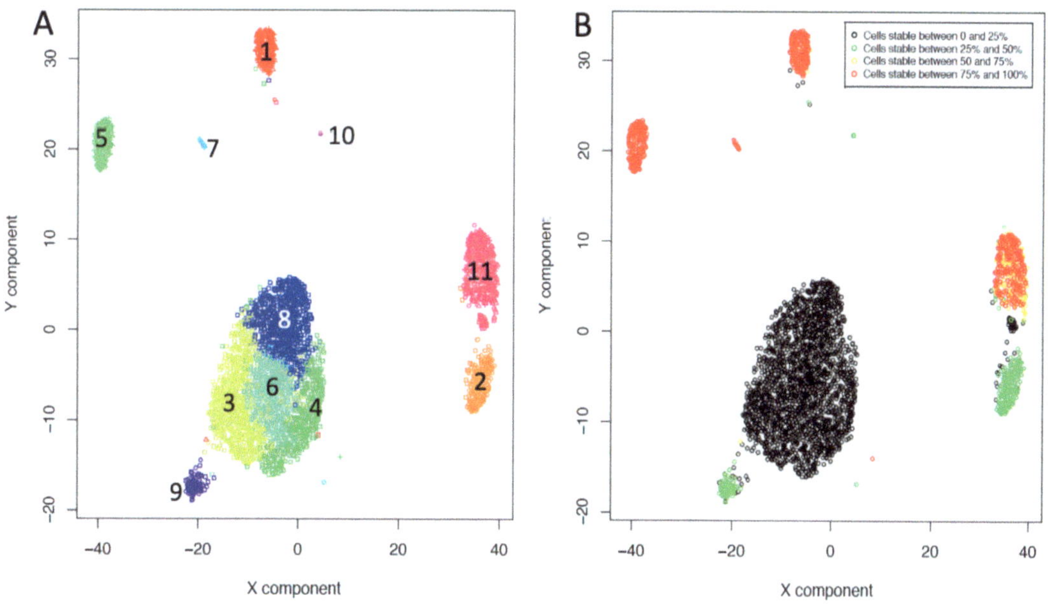

Fig. 2 Griph clustering results. (**a**) Cluster depicted by griph. (**b**) Cell cluster stability plot. The clusters showing good stability are 1, 5, and 11. Clusters 3, 4, 6, and 8 are, on the other end, extremely unstable. Cluster 1 is mainly made of CD8 T cells (77 cells) and gamma-delta T cells (95 cells). Cluster 2 is made mainly of follicular B cells (204 cells). Cluster 3 is mainly made of central memory T cells (379 cells) and T regulatory cells (185 cells). Cluster 4 is made of central memory T cells (202 cells) and T regulatory cells (100 cells). Cluster 5 is made of B cell IgA (207 cells). Cluster 6 is made of central memory T cells (224 cells) and T regulatory cells (169 cells). Cluster 7 is made of macrophage (35) cells. Cluster 8 is made of central memory T cells (404 cells), CD8 T cells (105 cells), and T regulatory cells (67 cells). Cluster 9 is made of CD8 T cells (115 cells). Cluster 10 is made of 11 mast cells and cluster 11 is made of memory B cells (417 cells)

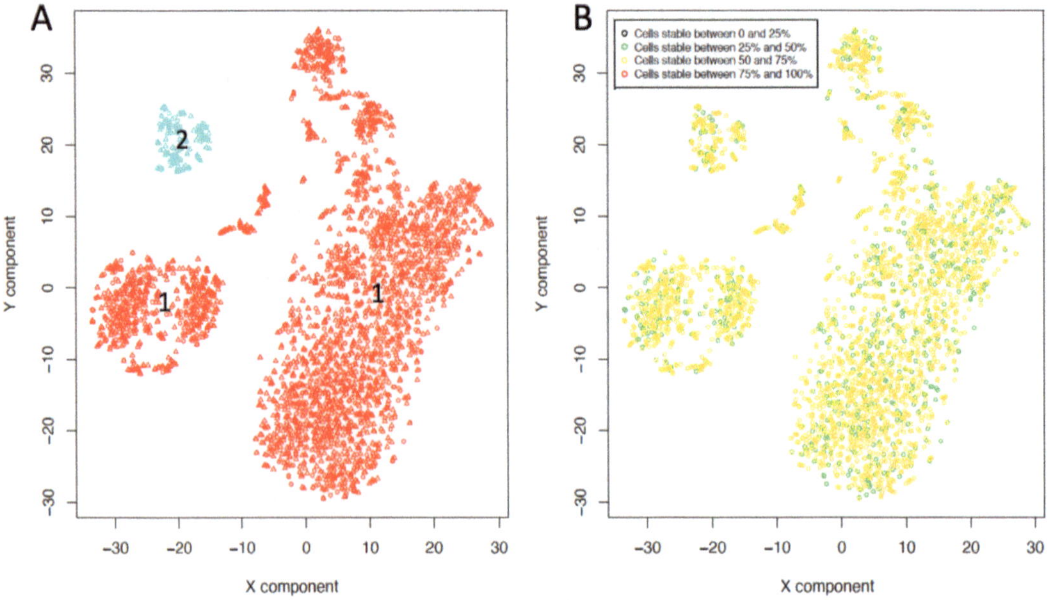

Fig. 3 Sharp clustering. (**a**) Cluster depicted by sharp. (**b**) Cell cluster stability plot

In particular, clusters 3, 4, 6, and 8 are extremely unstable suggesting that they belong to cells that are relatively similar. The dataset, we are using for this example, has the advantage of providing for each cell the cell type annotation. Thus, this information can help understand the overall characteristics of each cluster. Looking at the cell annotation for the most unstable clusters 3, 4, 6, and 8, (Fig. 2), the clusters are mainly composed of central memory T cells and T regulatory cells. This composition explains the low CSS stability of the clusters, since cells of the same type at each perturbation can shuffle among any of the nearby clusters. To better depict the diversity existing among cells of the same type, it would be useful to select the cells belonging to clusters 3, 4, 6, and 8, and then cluster them independently from the other cells. Taken together, the above observation suggests that the clustering provided by griph explains well the cell type diversity but fails in discriminating among sub-populations of the same cell type.

10. SHARP clustering generates two clusters (Fig. 3). The cell composition of the 390c gut sample is very complex, and two clusters are clearly an artifact, indicating that SHARP is unable to depict even the most representative cell populations in this dataset.

11. The PCA plot generated by the seuratPCAEval function provides a representation of the variance explained by the principal components, which helps the user to define the set of PC components to be used for clustering. The set of PCs refers

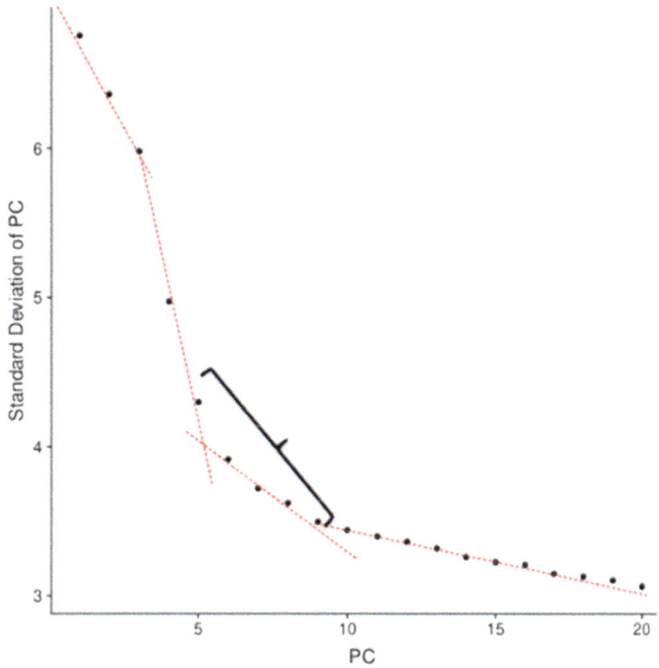

Fig. 4 PCA component selection

to the amount of variance that the user wishes to include in its analysis. In principle, the user should select as PC threshold the first point in which the explained variance becomes very little informative. In Fig. 4, the suggested threshold should be PC9. In this example, three slop changes are detectable: from PC1 to PC5, explaining most of the variance; from PC5 to PC9, where the slop is less steep, but still retains a certain amount of variance; and below PC9, where the explained variance become less and less important as moving to PC20.

References

1. Kiselev VY, Andrews TS, Hemberg M (2019) Challenges in unsupervised clustering of single-cell RNA-seq data. Nat Rev Genet 20(5): 273–282. https://doi.org/10.1038/s41576-018-0088-9
2. Serra D, Mayr U, Boni A, Lukonin I, Rempfler M, Challet Meylan L, Stadler MB, Strnad P, Papasaikas P, Vischi D, Waldt A, Roma G, Liberali P (2019) Self-organization and symmetry breaking in intestinal organoid development. Nature 569(7754):66–72. https://doi.org/10.1038/s41586-019-1146-y
3. Wan S, Kim J, Won KJ (2020) SHARP: hyper-fast and accurate processing of single-cell RNA--seq data via ensemble random projection. Genome Res 30(2):205–213. https://doi.org/10.1101/gr.254557.119
4. Butler A, Hoffman P, Smibert P, Papalexi E, Satija R (2018) Integrating single-cell transcriptomic data across different conditions, technologies, and species. Nat Biotechnol 36(5): 411–420. https://doi.org/10.1038/nbt.4096
5. Alessandri L, Cordero F, Beccuti M, Arigoni M, Olivero M, Romano G, Rabellino S, Licheri N, De Libero G, Pace L, Calogero RA (2019) rCASC: reproducible classification analysis of single-cell sequencing data. Gigascience 8(9): giz105. https://doi.org/10.1093/gigascience/giz105

6. Alessandri L, Cordero F, Beccuti M, Licheri N, Arigoni M, Olivero M, Di Renzo MF, Sapino A, Calogero R (2021) Sparsely-connected autoencoder (SCA) for single cell RNAseq data mining. NPJ Syst Biol Appl 7(1):1. https://doi.org/10.1038/s41540-020-00162-6
7. Alessandri L, Ratto ML, Contaldo SG, Beccuti M, Cordero F, Arigoni M, Calogero RA (2021) Sparsely connected autoencoders: a multi-purpose tool for single cell omics analysis. Int J Mol Sci 22(23):12755. https://doi.org/10.3390/ijms222312755
8. James KR, Gomes T, Elmentaite R, Kumar N, Gulliver EL, King HW, Stares MD, Bareham BR, Ferdinand JR, Petrova VN, Polanski K, Forster SC, Jarvis LB, Suchanek O, Howlett S, James LK, Jones JL, Meyer KB, Clatworthy MR, Saeb-Parsy K, Lawley TD, Teichmann SA (2020) Distinct microbial and immune niches of the human colon. Nat Immunol 21(3):343–353. https://doi.org/10.1038/s41590-020-0602-z
9. Huang M, Wang J, Torre E, Dueck H, Shaffer S, Bonasio R, Murray JI, Raj A, Li M, Zhang NR (2018) SAVER: gene expression recovery for single-cell RNA sequencing. Nat Methods 15(7):539–542. https://doi.org/10.1038/s41592-018-0033-z

Chapter 13

Identifying Gene Markers Associated with Cell Subpopulations

Maria Luisa Ratto and Luca Alessandri

Abstract

An important point of the analysis of a single-cell RNA experiment is the identification of the key elements, i.e., genes, characterizing each cell subpopulation cluster. In this chapter, we describe the use of sparsely connected autoencoder, as a tool to convert single-cell clusters in pseudo-RNAseq experiments to be used as input for differential expression analysis, and the use of COMET, as a tool to depict cluster-specific gene markers.

Key words Feature selection, Comet, Sparsely connected autoencoder

1 Introduction

Single cell RNA sequencing (scRNAseq) offers the possibility to identify cell sub-populations, based on the single-cell transcriptome. Clustering is the most used instrument to aggregate cells in groups, based on their transcription profile [1]. Notably, Yu and coworkers [1] highlight that good performance on cell clustering does not necessarily guarantee high accuracy in depicting different cell types. Furthermore, clustering algorithms can depict tiny differences among very similar cells, which are due to confounding factors, not necessarily linked to a functional difference among clusters, e.g., cell cycle and pre-apoptotic cells. On the other hand, cell type annotation, based on previously known cell-type specific markers [2], might define different cell types within the same cluster, raising the possibility that annotation was imprecise. For this reason, it is useful to integrate basic cell type annotation with cluster-specific gene markers, depicted by the analysis of cluster-specific transcriptomics data. In this chapter, we describe the use of two different approaches: SCA [3–5] (Sparsely Connected Autoencoder) and COMET [6] (COmbinatorial Marker dEtection from single-cell Transcriptomics).

Raffaele A. Calogero and Vladimir Benes (eds.), *Single Cell Transcriptomics: Methods and Protocols*, Methods in Molecular Biology, vol. 2584, https://doi.org/10.1007/978-1-0716-2756-3_13,

SCA can be used to generate cluster-specific pseudo-bulks resembling RNAseq data [3]. The overall idea of this approach is that genes characterizing a specific cluster will be the non-noise signal caught by the hidden layer of SCA.

COMET is a statistical method, based on XL-minimal Hyper-Geometric test (XL-mHG test), depicting cluster specific genes.

In this chapter, we show how to use the above-mentioned approaches for the analysis of scRNAseq clusters.

2 Materials

2.1 Minimal Hardware/Software Requirements

The analyses described in this chapter require a workstation with at least 64 GB RAM, 2 TB SSD, and 8 CPUs. All analytical steps require a UNIX-like environment (*see* **Notes 1** and **2**).

2.2 Exemplary Dataset

The exemplary dataset RNA-5c is available at the GitHub repository https://github.com/kendomaniac/single_cell_transcriptomics/tree/main/tian_et_al_2018. It includes the human lung adenocarcinoma cell lines A549, H1975, H2228, H838, and HCC827. It is derived from GEO repository, ID: GSM3618014 [7]. The count table is associated with cell names and cell lines, e.g., Lib90_00000.HCC827 and Lib90_00002.H838. RNA-5c includes 1242 cells of A549, 436 cells belong to H1975, 749 cells of H2228, 879 cells belong to H838, and 598 cells of HCC827 cells.

The exemplary dataset 390c_wctype [2] is available at the Github repository https://github.com/kendomaniac/single_cell_transcriptomics.git. This dataset refers to the sample 390c of the immune-gut atlas transcriptome and, in the counts table, cell names incorporate cell type assignment, i.e., TGACTAGGTTCCACAA.Treg. This dataset is part of 41,650 cells isolated from the cecum, transverse colon, and sigmoid colon of five individuals [2], which is one of the datasets constituting the immune-gut atlas transcriptome (*see* **Note 3**). All data are available as figshare repository: https://doi.org/10.6084/m9.figshare.19391234.

3 Methods

3.1 rCASC Installation

rCASC [5] is a single-cell analysis suite providing a computation reproducible infrastructure based on docker containers. The installation of rCASC is very simple. In an R session, type the code indicated below (*see* **Note 4**).

```
# Open an R session and type the following commands
library(devtools)
install_github("kendomaniac/rCASC", ref="master")
library(rCASC)
```

rCASC provides an infrastructure to evaluate the overall quality of the clustering result via the cell stability score (CCS, *see* **Note 5**).

3.2 Data Cleanup, Filtering, and Clustering

Before performing any analysis, we strongly suggest checking the overall quality of the data. Specifically, we suggest running the *mitoRibo* function implemented in rCASC. Furthermore, the dataset needs to be imputed using SAVER tool [8] and annotated with *scannoByGtf* rCASC function. To reduce the size of the count table under analysis, we suggest filtering the dataset, keeping only the subset of genes being the most expressed and characterized by the largest variance among cells. This filtering can be performed using the *topx* function available in rCASC.

```
# Folder where the counts table is located.
WD = "/mnt/cold1/calogero/5-lung_cell-lines/scRNAseq_5-lung_-
cell-lines"
INPUT="saver_GSM3618014_gene_count.csv"
SEPARATOR=","
GTF="Homo_sapiens.GRCh38.101.gtf"
SCRATCH="/scratch"

setwd(WD)
library(rCASC)
#filtering, selecting 5k most expressed genes
topx(group = "docker", file=paste(getwd(), paste("filtered_an-
notated_saver_ribomito", INPUT, sep="_"), sep="/"), thresh-
old=5000, separator=SEPARATOR, type="expression")
#filtering, selecting the 2.5k most variant genes
topx(group = "docker", file=paste(getwd(), paste("filtered_ex-
pression_filtered_annotated_saver_ribomito", INPUT, sep="_"),
sep="/"), threshold=2500, separator=SEPARATOR, type = "var-
iance")
system("mkdir VandE")
file.rename(from=paste("filtered_variance_filtered_expres-
sion_filtered_annotated_saver_ribomito", INPUT, sep="_"),
to="VandE.csv")
system("mv VandE.csv VandE")
```

In this exemplary analysis, the Louvain modularity clustering, implemented in Seurat [9], is used to aggregate cells in clusters for RNA-5c and 390c samples, but it is not mandatory to use this specific clustering tool.

```
library(rCASC)
seuratPCAEval(group = "docker", scratch.folder=SCRATCH, file=-
paste(getwd(), "VandE/VandE.csv", sep="/"), separator=",",
logTen = 0, seed = 111)

#Running clustering
seuratBootstrap(
  group = "docker",
  scratch.folder= "/scratch",
  file=paste(getwd(),"VandE.csv", sep="/"),
  nPerm=40,
  permAtTime=10,
  percent=10,
  separator=",",
  logTen = 0,
  pcaDimensions=10,
  seed = 111,
  sparse = FALSE,
  format = "NULL",
  resolution = 0.8
)
```

The above function generates the following path: ~/Results/sample_names/number_clusters. In the folder defined by the number of clusters, there is a pdf ending with _Stability_Plot.pdf (Fig. 1).

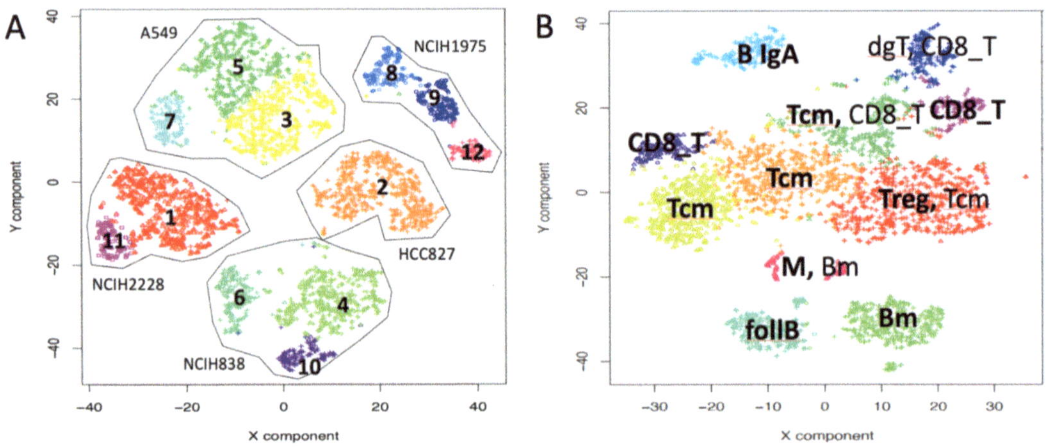

Fig. 1 Seurat clustering of scRNAseq data. (**a**) RNA-5c dataset at 0.8 resolution for RNA5c. Clusters belonging to a specific cell line are delimited by a black continuous line. (**b**) Sample 390c at 0.6 resolution. The figure contains the main type of immunological cells constituting each cluster. In case there are two or more cell types, associated to a cluster, bold indicates the most represented cell type. Cell types: B cells producing IgA (B IgA), delta/gamma T cell (dgT), CD8 T cells (CD8_T), central memory T cells (Tcm), T regulatory cells (Treg), monocytes (M), B memory cells (Bm), and follicular B cells (follB)

The clustering in Fig. 1a shows the presence of one cluster constituted by a single cell type, clusters 2 (HCC827 cell line), whereas all other cells lines are split into different clusters. Cell lines are a typical instrument for in vitro analysis of cell biological activity. In the field of oncology, cell lines are frequently used to study the responsivity to a specific drug. As drug response experiments are usually not done at single cell level, IC50 cannot be linked to a specific cell subpopulation response. scRNAseq offers the possibility to investigate if genes, specifically associated with cell subpopulations of a cell line, are different facet of the same die or represent a peculiar feature of the subpopulation, which might be related to a specific biological phenotype of the cell line, i.e., drug resistance.

The clustering in Fig. 1b shows the presence of a heterogeneous organization of immunological cell types in the detected clusters:

(i) Clusters sharing different cell types (Treg and Tcm in red cluster and Tcm and CD8_T in green cluster, Fig. 1b)

(ii) Same cell type in different clusters (Tcm in yellow, orange, green, and red clusters, Fig. 1b).

In the gut immune atlas [2], cell type annotation is done using the transcription expression, at single cell level, of previously known cell type gene markers and not by CITE-seq [10], which would have produced more robust results. Thus, it is necessary to investigate if erroneous assignment of the cell type is the reason why different cell types are associated with the same cluster (Fig. 1b, red and green clusters). It is also necessary to evaluate if Tcm cells present in different clusters (Fig. 1b yellow, orange, and green clusters) are transcriptionally different, thus representing sub-populations of the same cell type.

3.3 Sparsely Connected Autoencoders for Pseudo-Bulk Generation

The function *autoencoder4pseudoBulk* transforms each of the clusters in an RNAseq pseudo-bulk, which can be used to depict cluster specific genes. The *permutation* parameter indicates the number of pseudo-bulks that are generated.

```
library(rCASC)
WD = "/mnt/cold1/calogero/5-lung_cell-lines/scRNAseq_5-lung_-
cell-lines/VandE"
SEPARATOR=","
SCRATCH="/scratch"
CLS=12
PROJECTNAME="LUNG5CLS"

CLSOUT="/mnt/cold1/calogero/5-lung_cell-lines/scRNAseq_5-
lung_cell-lines/VandE/Results/VandE/12/VandE_clustering.out-
put.csv"
```

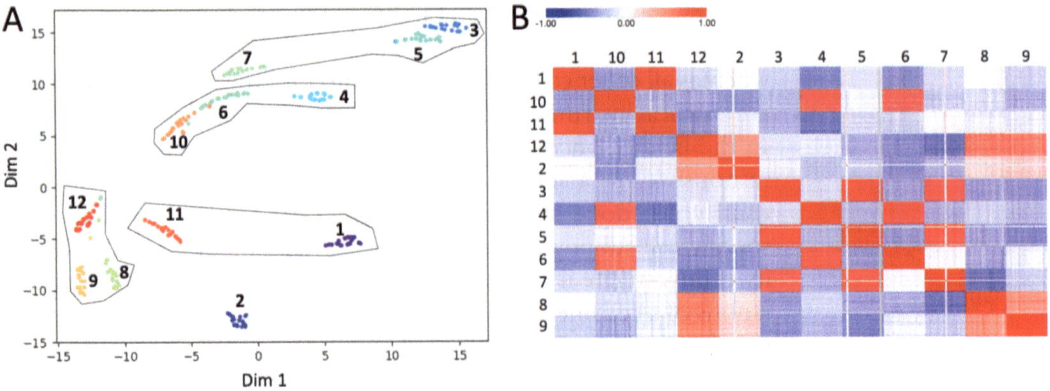

Fig. 2 Pseudo-bulks. (**a**) tSne plot of RNA-5c pseudo-bulks generated by sparsely connected-autoencoder. Pseudo-bulks belonging to a specific cell line are delimited by a black continuous line. (**b**) Pearson-based similarity score of RNA-5c Z-score log2CPM pseudo-bulks

```
autoencoder4pseudoBulk(group="docker",
    scratch.folder=SCRATCH,
    file=paste(getwd(),"VandE.csv", sep="/"),
    separator=SEPARATOR,
    permutation=20,
    nEpochs=1000,
    projectName=PROJECTNAME,
    bN=CLSOUT
)
```

The output of *autoencoder4pseudoBulk* function is a file called total.csv, located in the folder permutation. As part of the output, a tSne plot of the pseudo-bulks for the RNA-5c dataset is also generated (Fig. 2).

The pseudo-bulks tSne plot (Fig. 2a) provides a clear description of the cell line heterogeneity—i.e., HCC827 (cluster 2) is homogeneous, whereas all the other cell lines show a certain amount of heterogeneity. Similarity analysis among clusters, using Pearson correlation (Fig. 2b) indicates high similarity among clusters belonging to the same cell line, but in the case of cluster 12, a certain similarity with cluster 2 is observed, which belongs to a different cell line.

3.4 ANOVA-Like on RNA-5c Pseudo-Bulks Data

To depict genes specific to each cluster, we can perform differential expression analysis using edgeR ANOVA-like [11]. Pseudo-bulk data are converted in CPM (*see* **Note 6**) to provide a better similarity to RNAseq expression data, but retaining sample to sample variability.

```
counts2cpmAE <- function(fileIN, sep = ",", prefix){
       tmp <- read.table(fileIN, sep=sep, header=T)
       tmp <- t(tmp)
       tmp <- as.data.frame(tmp)
       col.names <- as.character(tmp[1,])
       tmp <- tmp[2:dim(tmp)[1],]
       names(tmp) <- col.names
       tmp <- t(tmp)
       tmp <- as.data.frame(tmp)
       col.sum <- apply(tmp, 1, sum)
       tmp1 <- tmp/col.sum
       tmp1 <- t(tmp1)
       tmp1 <- tmp1 * 1000000
         write.table(tmp1, paste(prefix, "cpm.csv", sep="_"),
sep=",", col.names=NA)
      write.table(log2(tmp1 + 1), paste(prefix, "log2cpm.csv",
sep="_"), sep=",", col.names=NA)
}
sample.folder="VandE"
pbAE.folder="LUNG5CLS"
dir(paste("./",sample.folder,"/Results/",pbAE.folder,"/permu-
tation/",sep=""))
whereis.total=paste("../",sample.folder,"/Results/",pbAE.
folder,"/permutation/total.csv",sep="")
prefix=pbAE.folder
counts2cpmAE(fileIN=whereis.total, sep=",", prefix=prefix)
```

To perform ANOVA-like, a reference sample is made by averaging all the clusters.

```
#Ordering columns, replicates of the same clusters are placed
near to each other
logcpm <- read.table("LUNG5CLS_log2cpm.csv", sep=",", head-
er=T, row.names=1)
tmp.n <- sapply(strsplit(names(logcpm), "\cr."), function(x)x
[1])
logcpm <- logcpm[,order(tmp.n)]
write.table(logcpm, "LUNG5CLS_log2cpm_ordered.csv", sep=",",
col.names=NA)

cpm <- read.table("LUNG5CLS_cpm.csv", sep=",", header=T, row.
names=1)
tmp.n <- sapply(strsplit(names(cpm), "\cr."), function(x)x
[1])
cpm <- cpm[,order(tmp.n)]
write.table(gcpm, "LUNG5CLS_cpm_ordered.csv", sep=",", col.
names=NA)
```

```
#building a reference groups for ANOVA-like
tmp.n <- sapply(strsplit(names(cpm), "\cr."), function(x)x
[1])
tmp.nu <- unique(tmp.n)
tmp.lst <- list()
for(i in tmp.nu){
  if(i == "X1"){
    tmp.lst[[i]] <- grep('^X1$', tmp.n)
  } else{
    tmp.lst[[i]] <- grep(i, tmp.n)
  }
}
tmp.lst <- as.data.frame(tmp.lst)

ref.cl <- list()
for(i in 1:dim(tmp.lst)[1]){
  clmns <- as.numeric(unlist(tmp.lst[i,]))
  ref.cl[[i]] <- trunc(apply(cpm[,clmns], 1, mean))
}
ref.cl <- as.data.frame(ref.cl)
names(ref.cl) <- paste("ref", seq(1,dim(tmp.lst)[1]),
sep=".")
names(ref.cl) <- paste(names(ref.cl), "Cov.0", sep="_")
ref.cl.na <- apply(ref.cl, 2, function(x)(sum(is.na(x))/
length(x)))
ref.cl <- ref.cl[,which(ref.cl.na < 0.2)]

for(i in 1:dim(tmp.lst)[2]){
   names(cpm)[tmp.lst[,i]] <- paste(names(cpm)[tmp.lst[,i]],
"_Cov", i, sep="")
}
#Adding covariates to column names
covar <- sapply(strsplit(names(cpm), "_"), function(x)x[2])
cpm <- cpm[,order(covar)]

#Removing columns with more than 20% of 0s, these columns refer
to low SCA quality models
cpm.cl.na <- apply(cpm, 2, function(x)(sum(is.na(x))/length
(x)))
cpm <- cpm[,which(cpm.cl.na < 0.2)]

#Formatting rownames to be compatible with docker4seq func-
tions
symb <- sapply(strsplit(rownames(cpm.new), "\cr."), function
(x)x[2])
ensembl <- sapply(strsplit(rownames(cpm.new), "\cr."), func-
tion(x)x[1])
rownames(cpm.new) <- paste(symb, ensembl, sep=":")
```

```
cpm.new <- data.frame(ref.cl, cpm)
write.table(cpm.new, "cpm_new.txt", sep="\t", col.names=NA)
#Running ANOVA-like
library(docker4seq)
anovaLike(
  group = "docker",
  file=paste(getwd(), "cpm_new.txt", sep="/"),
  logFC.threshold = 1,
  FDR.threshold=-0.1,
  logCPM.threshold = 4,
  plot = FALSE
)

anova.results <- read.table("ANOVAlike_cpm_new.txt",
sep="\t", header=T, row.names=1)
anova.results <- anova.results[which(anova.results$FDR <=
0.1),]
names(anova.results) <- gsub("groupsCov", "cl",names(anova.
results))

write.table(anova.results, "filtered_ANOVAlike_cpm_new.txt",
sep="\t", col.names=NA)
write.table(anova.results[,grep("cl", names(anova.results))],
"filtered_ANOVAlike_cpm_new_4_clustering.txt", sep="\t", col.
names=NA)
```

The output of this analysis are two files: filtered_ANOVAlike_cpm_new.txt, which contains the overall result of clusters compared with the reference set (*see* **Note 7**), and the file filtered_ANOVAlike_cpm_new_4_clustering.txt (*see* **Note 8**), which can be used to perform a hierarchical clustering using online tools, e.g., Morpheus from Broad Institute (https://software.broadinstitute.org/morpheus/).

The results of the hierarchical clustering performed on pseudo-bulks from the RNA-5c analysis are shown in Fig. 5. Notably, cluster-specific genes (Fig. 3b) show a different cluster-to-cluster similarity with respect to those depicted using the full transcriptome (Fig. 1a).

3.5 DESeq2 on Pseudo-Bulk Data from Gut Immune Atlas Sample 390c

In sample 390c from gut immunocell atlas (Fig. 1b), we observe that central-memory T cells (Tcm) are spread over four clusters: red, orange, yellow, and green. The task of this section is to understand how Tcm from different clusters are significantly transcriptionally different.

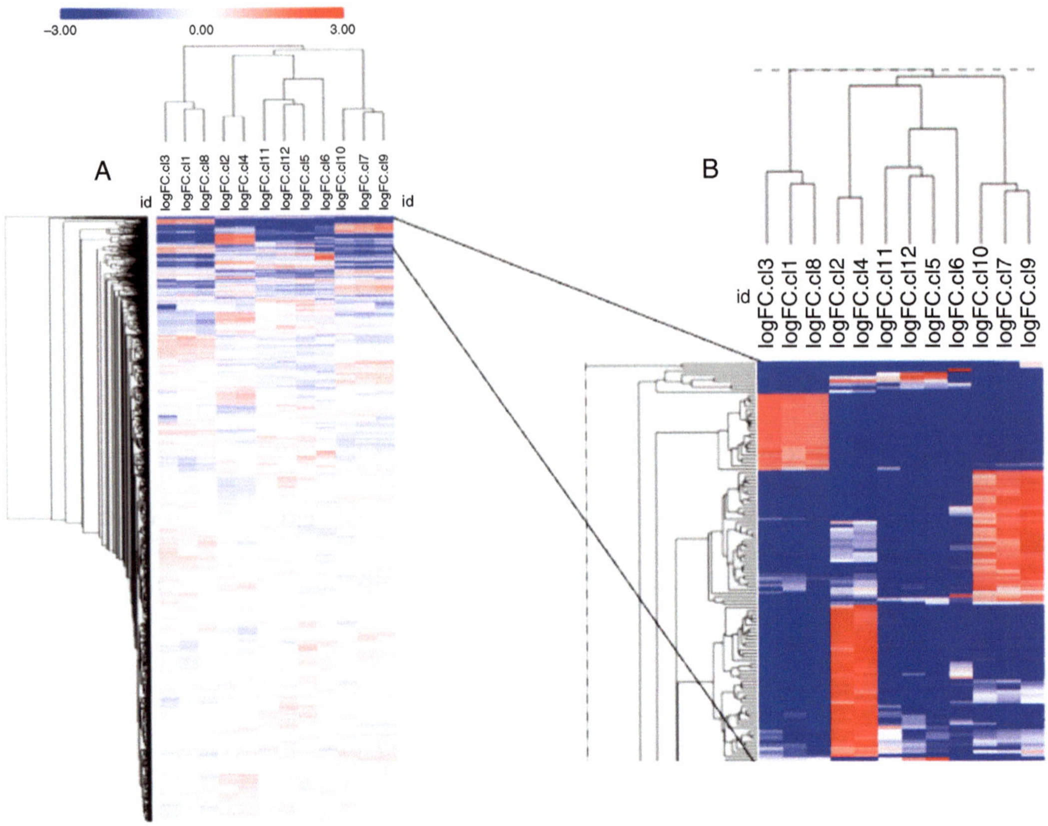

Fig. 3 Hierarchical clustering for ANOVA-like performed on RNA-5c pseudo-bulk. (**a**) Hierarchical clustering performed on the $\log_2$ fold change generated comparing each RNA-5c pseudo-bulks with respect to a reference dataset given by means of the overall pseudo-bulks, Clustering parameters: Euclidean distance and average linkage. (**b**) Zoom on the top of the clustering results

```
#Editing clustering.output file to create a set of new clusters
encompassing only Tcm derived respectively from red, orange,
yellow and green cluster.
cls.df <- read.table("/data/390c/Results/saver_390c_wctype/
11/saver_390c_wctype_clustering.output.csv", sep=",", head-
er=T, row.names=1)
#red cluster cl1
#orange cluster cl2
#yellow cluster cl3
#green cluster cl5

#Ordering cells based on the belonging cluster.
cls.df <- cls.df[order(cls.df$Belonging_Cluster),]

#stripping cell type annotation from the rowname
ctype <- sapply(strsplit(rownames(cls.df),"\cr."), function
(x)x[2])
cls.df$ctype <- ctype
```

```
#creating fake clusters for Tcm in cluster 1, 2, 3 and 5.
cls.df$Belonging_Cluster[which(cls.df$ctype == "Tcm" & cls.df
$Belonging_Cluster == 1)] <- 12
cls.df$Belonging_Cluster[which(cls.df$ctype == "Tcm" & cls.df
$Belonging_Cluster == 2)] <- 13
cls.df$Belonging_Cluster[which(cls.df$ctype == "Tcm" & cls.df
$Belonging_Cluster == 3)] <- 14
cls.df$Belonging_Cluster[which(cls.df$ctype == "Tcm" & cls.df
$Belonging_Cluster == 5)] <- 15
write.table(cls.df, "/data/390c/saver_390c_wctype_clustering.
output.csv", col.names=NA, sep=",")

library(rCASC)
autoencoder4pseudoBulk(
  group = "docker",
  scratch.folder="/scratch",
  file="/data/390c/saver_390c_wctype.csv",
  separator=",",
  permutation=20,
  nEpochs=1000,
  patiencePercentage = 5,
  seed = 1111,
  projectName="Tcm",
  bN="/data/390c/saver_390c_wctype_clustering.output.csv",
  lr = 0.01,
  beta_1 = 0.9,
  beta_2 = 0.999,
  epsilon = 1e-08,
  decay = 0,
  loss = "mean_squared_error",
  regularization = 10
)
```

The *autoencoder4pseudoBulk* function result table total.csv is in ~/Results folder/Tcm/permutation/ folder.

To detect the presence of cluster-specific genes, pseudo-bulk data are transformed in CPM. Then, pseudo-bulk data for red, orange, yellow, and green clusters are analyzed with DESeq2 [12], performing all possible permutations. The gene names of the differential expressed genes, depicted from each comparison, are then combined and used to filter the table encompassing all the pseudo-bulks for the red, orange, yellow, and green cell clusters.

```
# Transforming total.csv autoencoder4pseudoBulk result.
setwd("/data/390c/Results/Tcm/permutation")
counts2cpmAE <- function(fileIN, sep = ",", prefix){
       tmp <- read.table(fileIN, sep=sep, header=T)
       tmp <- t(tmp)
```

```
        tmp <- as.data.frame(tmp)
        col.names <- as.character(tmp[1,])
        tmp <- tmp[2:dim(tmp)[1],]
        names(tmp) <- col.names
        tmp <- t(tmp)
        tmp <- as.data.frame(tmp)
        col.sum <- apply(tmp, 1, sum)
        tmp1 <- tmp/col.sum
        tmp1 <- t(tmp1)
        tmp1 <- tmp1 * 1000000
          write.table(tmp1, paste(prefix, "cpm.csv", sep="_"),
sep=",", col.names=NA)
       write.table(log2(tmp1 + 1), paste(prefix, "log2cpm.csv",
sep="_"), sep=",", col.names=NA)
}
counts2cpmAE(fileIN="total.csv", sep=",", prefix="tcm")
cpm <- read.table("tcm_cpm.csv", sep=",", header=T, row.
names=1)
cpm <- trunc(cpm)
#setting rownames compatibile with docker4seq
cpm.embl <- sapply(strsplit(rownames(cpm), "\cr."), function
(x)x[1])
cpm.symb <- sapply(strsplit(rownames(cpm), "\cr."), function
(x)x[2])
rownames(cpm) <- paste(cpm.symb, cpm.embl, sep=":")

red <- cpm[,grep("X12\cr.", names(cpm))]
names(red) <- sub("^X", "red",names(red))
orange <- cpm[,grep("X13\cr.", names(cpm))]
names(orange) <- sub("^X", "orange",names(orange))
yellow <- cpm[,grep("X14\cr.", names(cpm))]
names(yellow) <- sub("^X", "yellow",names(yellow))
green <- cpm[,grep("X15\cr.", names(cpm))]
names(green) <- sub("^X", "green",names(green))

#Calculating cluster-specific differentially expressed genes,
|log2FC|≥2 and adjusted p-value ≤ 0.1. The thresholds used for
differential expression must be defined by the user. (see Note
9)
library(docker4seq)
de.pair <- function(dataset.ref, dataset.query, prefix){
   names(dataset.ref) <- paste(names(dataset.ref), "Cov.1",
sep="_")
  names(dataset.query) <- paste(names(dataset.query), "Cov.2",
sep="_")
  #DE
  tmp <- data.frame(dataset.ref, dataset.query)
  write.table(tmp, "comparison.txt", sep="\t", col.names=NA)
```

```
  wrapperDeseq2(
    output.folder=getwd(),
    group = "docker",
    experiment.table="comparison.txt",
    log2fc = 2,
    fdr = 0.1,
    ref.covar = "Cov.1",
    type = "gene",
    batch = FALSE
  )
   file.rename("log2normalized_counts.txt", paste(prefix,"-
log2normalized_counts.txt", sep="_"))
   file.rename("DEfiltered_gene_log2fc_2_fdr_0.1.txt", paste
(prefix,"DEfiltered_gene_log2fc_2_fdr_0.1.txt", sep="_"))
   file.rename("genes2GO.txt", paste(prefix,"genes2GO.txt",
sep="_"))
   file.rename("DEfull.txt", paste(prefix,"DEfull.txt",
sep="_"))
}
#detecting genes differentially expressed among clusters pairs
threshold |log2FC| >= 2, adj.pval <= 0.1
de.pair(dataset.ref=red, dataset.query=orange, prefix="re-
d_orange")
de.pair(dataset.ref=red, dataset.query=yellow, prefix="re-
d_yellow")
de.pair(dataset.ref=red, dataset.query=yellow, prefix="red_-
green")
de.pair(dataset.ref=orange, dataset.query=yellow, prefix="or-
ange_yellow")
de.pair(dataset.ref=orange, dataset.query=green, prefix="or-
ange_green")
de.pair(dataset.ref=yellow, dataset.query=green, prefix="yel-
low_green")

files <- dir()
de.set <- files[grep("DEfiltered_gene_log2fc_2_fdr_0.1.txt",
files)]
de.genes <- NULL
for(i in de.set){
  tmp.de <-read.table(i, sep="\t", header=T,row.names=1)
  de.genes <- c(de.genes, rownames(tmp.de))
}
de.genes <- unique(de.genes)
de.df <- data.frame(red, orange, yellow, green)
de.df <- de.df[which(rownames(de.df) %in% de.genes),]
write.table(de.df, "cluster_specific_genes.txt", sep="\t",
col.names=NA)
write.table(log2(de.df + 1), "log2_cluster_specific_genes.
txt", sep="\t", col.names=NA)
```

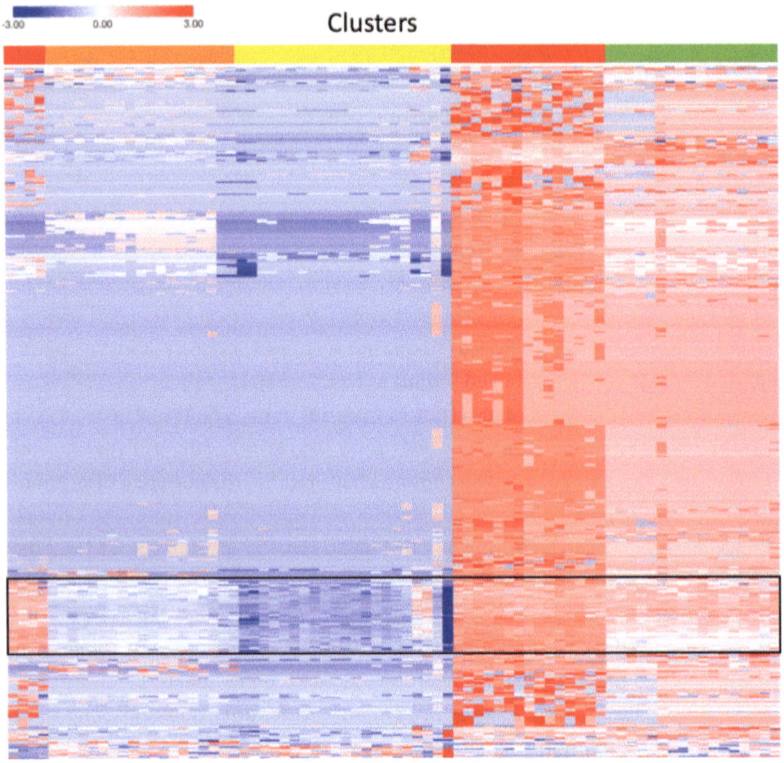

Fig. 4 Hierarchical clustering for genes detected a differentially expressed comparing to each other the pseudo-bulks for Tcm in red, orange, yellow and green cluster for sample 390c. Clustering parameters: Euclidean distance and average linkage

The hierarchical clustering of the gene subset of Tcm pseudo-bulks encompassing only differentially expressed genes is shown in Fig. 4. The analysis, using enrichR [13] and the Panglao database [14], of the gene signature in the black frame, Fig. 4, assign the red and green cluster to Treg cells. Thus, using the above-described approach, cell type assignment of red and green clusters could be refined.

3.6 COMET—A Tool to Identify Cluster-Specific Genes

COMET is a statistical method designed to select panels of gene markers from single-cell RNA-seq data [6]. COMET uses the minimal hypergeometric test to rank gene marker panels characterizing predefined cell populations (clusters) using single-cell RNA-seq data. COMET is available as standalone tool and it is implemented in rCASC [5], or as web application (http://www.cometsc.com/index). COMET is particularly effective in depicting few genes, from 1 to 4, which efficiently label a cell cluster. Thus, it results particularly interesting to identify set of genes to be used to further study and isolate cells belonging to a specific cluster.

```
WD = "/mnt/cold1/calogero/5-lung_cell-lines/scRNAseq_5-lung_-
cell-lines/VandE"
SEPARATOR=","
SCRATCH="/scratch"
CLS=12
CC="/mnt/cold1/calogero/5-lung_cell-lines/scRNAseq_5-lung_-
cell-line/annotated_GSM3618014_gene_count_cellCycle.csv"
CLSOUT="/mnt/cold1/calogero/5-lung_cell-lines/scRNAseq_5-
lung_cell-lines/VandE/Results/VandE/7/VandE_clustering.out-
put.csv"

setwd(WD)
library(rCASC)
cometsc(
        group="docker",
        file=paste(getwd(), "VandE.csv", sep="/"),
        scratch.folder=SCRATCH,
        threads=CLS, X=0.15, K=1, counts="True",
        skipvis="False", nCluster=CLS, separator=SEPARATOR
)
```

The main outputs of COMET are two folders: outputdata and outputvis. Outputdata contains comma separated files with the genes ranked as the most important for characterizing the cluster under analysis. Outputvis provides images for the 100 top most representative cluster-marker genes. In outputdata tables, the most interesting columns are TP and TN, which ranks from 0 to 1. The best markers are those showing values near 1 for both TP and TN. TP equal to 1 means that the gene can discriminate very efficiently the cells in the cluster under analysis from the other cells. TN equal to 1 means that the gene marker does not associate any cell outside the cluster to the cluster under analysis. Examples of COMET outputs are shown in Fig. 5 (*see* **Note 10**).

It is important to note that, in some clusters, many of the top ranked marked are defined as "NEGATION" indicating that the expression of that genes is higher in all clusters, but in the cluster under consideration. Usually, the presence of many expression negation genes is an indication that the cluster is not so different from the other clusters, since it is not possible to identify marker genes specifically expressed in the cluster under analysis. In these cases, it would be useful to perform autoencoder analysis (*see* Subheading 3.4) to identify which are the most similar clusters.

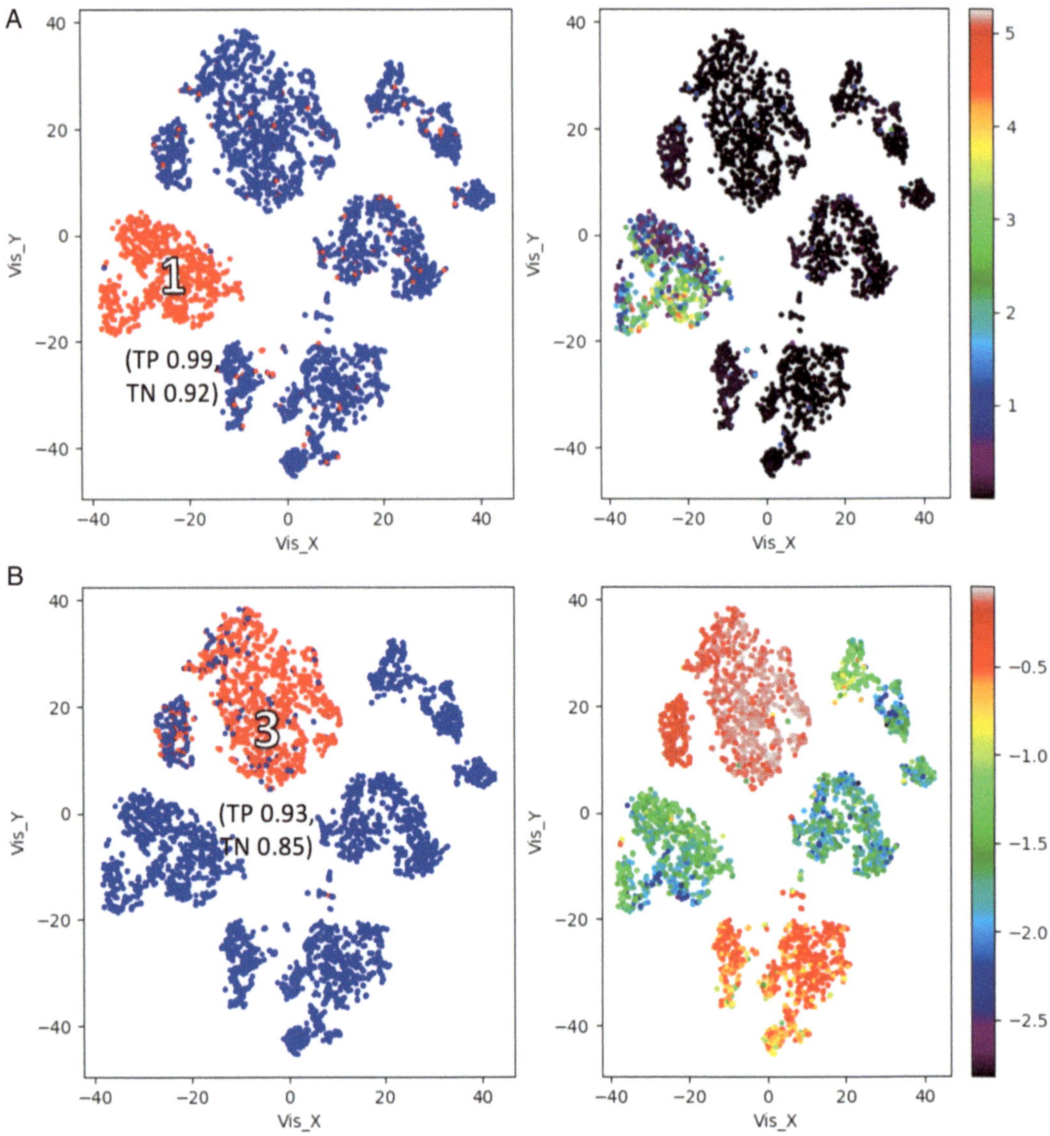

Fig. 5 Examples of COMET outputs. (**a**) Example of a top-ranked geneswith optimal TP and TN scores. (**b**) Example of expression negation; the gene is expressed at higher level in all sample, but the cluster under consideration

4 Notes

1. The installation of docker is required. Docker for Windows, Mac, and UNIX infrastructures is available at https://docs.docker.com/get-docker/.
2. R installation is required (https://cran.r-project.org/). Specifically, the installation of devtools library is required.
3. The gut atlas datasets are available at https://www.gutcellatlas.org/.

4. The description of all rCASC functions is available at https://kendomaniac.github.io/rCASC/reference/index.html. The execution of the "*downloadContainers*()" command is not mandatory. Since the first time a new docker container is required for an analysis, it is downloaded from the repbioinfo docker.io repository. The first time an rCASC function is used, it retrieves the docker image required for the analysis from docker.io/repbioinfo repository. Therefore, internet connection is required.
5. CSS (cell stability score) describes cluster stability, by estimating the persistence of a cell in a specific cluster upon jackknife resampling. A detailed description of the CSS metric is available in the supplementary data of rCASC paper [5].
6. Usually, CPM are not used as input to estimate differential expression, however, in this specific case, we are using CPM transformation to convert the output of the sparsely connected-autoencoder in a measure that is more like to what is expected to be a gene expression dataset.
7. The reference dataset is at the denominator of the fold change; thus, genes that have a positive fold change are the genes that are representative of a specific cluster with respect to the average reference cluster.
8. After having loaded the dataset to Morpheus, do not perform any transformation, since the data is already representing variation of each cluster with respect to the reference set.
9. We suggest testing combination of adjusted p-values (0.1, 0.05, 0.001) and absolute $\log_2$FC (1, 1.5, 2). Ideally the combination of adjusted p-values and $|\log_2FC|$ should produce less than 100 genes in total to depict the main players for each cluster.
10. Detailed description of COMET algorithm can be found at https://hgmd.readthedocs.io/en/latest/.

References

1. Yu L, Cao Y, Yang JYH, Yang P (2022) Benchmarking clustering algorithms on estimating the number of cell types from single-cell RNA-sequencing data. Genome Biol 23(1):49. https://doi.org/10.1186/s13059-022-02622-0
2. James KR, Gomes T, Elmentaite R, Kumar N, Gulliver EL, King HW, Stares MD, Bareham BR, Ferdinand JR, Petrova VN, Polanski K, Forster SC, Jarvis LB, Suchanek O, Howlett S, James LK, Jones JL, Meyer KB, Clatworthy MR, Saeb-Parsy K, Lawley TD, Teichmann SA (2020) Distinct microbial and immune niches of the human colon. Nat Immunol 21(3):343–353. https://doi.org/10.1038/s41590-020-0602-z
3. Alessandri L, Ratto ML, Contaldo SG, Beccuti M, Cordero F, Arigoni M, Calogero RA (2021) Sparsely connected autoencoders: a multi-purpose tool for single cell omics analysis. Int J Mol Sci 22(23):12755. https://doi.org/10.3390/ijms222312755
4. Alessandri L, Cordero F, Beccuti M, Licheri N, Arigoni M, Olivero M, Di Renzo MF, Sapino A, Calogero R (2021) Sparsely-connected autoencoder (SCA) for single cell

RNAseq data mining. NPJ Syst Biol Appl 7(1): 1. https://doi.org/10.1038/s41540-020-00162-6

5. Alessandri L, Cordero F, Beccuti M, Arigoni M, Olivero M, Romano G, Rabellino S, Licheri N, De Libero G, Pace L, Calogero RA (2019) rCASC: reproducible classification analysis of single-cell sequencing data. Gigascience 8(9):giz105. https://doi.org/10.1093/gigascience/giz105
6. Delaney C, Schnell A, Cammarata LV, Yao-Smith A, Regev A, Kuchroo VK, Singer M (2019) Combinatorial prediction of marker panels from single-cell transcriptomic data. Mol Syst Biol 15(10):e9005. https://doi.org/10.15252/msb.20199005
7. Tian L, Dong X, Freytag S, Le Cao KA, Su S, JalalAbadi A, Amann-Zalcenstein D, Weber TS, Seidi A, Jabbari JS, Naik SH, Ritchie ME (2019) Benchmarking single cell RNA-sequencing analysis pipelines using mixture control experiments. Nat Methods 16(6): 479–487. https://doi.org/10.1038/s41592-019-0425-8
8. Huang M, Wang J, Torre E, Dueck H, Shaffer S, Bonasio R, Murray JI, Raj A, Li M, Zhang NR (2018) SAVER: gene expression recovery for single-cell RNA sequencing. Nat Methods 15(7):539–542. https://doi.org/10.1038/s41592-018-0033-z
9. Butler A, Hoffman P, Smibert P, Papalexi E, Satija R (2018) Integrating single-cell transcriptomic data across different conditions, technologies, and species. Nat Biotechnol 36(5):411–420. https://doi.org/10.1038/nbt.4096
10. Stoeckius M, Hafemeister C, Stephenson W, Houck-Loomis B, Chattopadhyay PK, Swerdlow H, Satija R, Smibert P (2017) Simultaneous epitope and transcriptome measurement in single cells. Nat Methods 14(9): 865–868. https://doi.org/10.1038/nmeth.4380
11. Robinson MD, McCarthy DJ, Smyth GK (2010) edgeR: a Bioconductor package for differential expression analysis of digital gene expression data. Bioinformatics 26(1): 139–140. https://doi.org/10.1093/bioinformatics/btp616
12. Love MI, Huber W, Anders S (2014) Moderated estimation of fold change and dispersion for RNA-seq data with DESeq2. Genome Biol 15(12):550. https://doi.org/10.1186/s13059-014-0550-8
13. Chen EY, Tan CM, Kou Y, Duan Q, Wang Z, Meirelles GV, Clark NR, Ma'ayan A (2013) Enrichr: interactive and collaborative HTML5 gene list enrichment analysis tool. BMC Bioinformatics 14:128. https://doi.org/10.1186/1471-2105-14-128
14. Franzen O, Gan LM, Bjorkegren JLM (2019) PanglaoDB: a web server for exploration of mouse and human single-cell RNA sequencing data. Database (Oxford) 2019:baz046. https://doi.org/10.1093/database/baz046

Chapter 14

A Guide to Trajectory Inference and RNA Velocity

Philipp Weiler, Koen Van den Berge, Kelly Street, and Simone Tiberi

Abstract

Technological developments have led to an explosion of high-throughput single-cell data, which are revealing unprecedented perspectives on cell identity. Recently, significant attention has focused on investigating, from single-cell RNA-sequencing (scRNA-seq) data, cellular dynamic processes, such as cell differentiation, cell cycle and cell (de)activation. In particular, trajectory inference methods, by ordering cells along a trajectory, allow estimating a differentiation tree of cells. While trajectory inference tools typically work with gene expression levels, common scRNA-seq protocols allow the identification and quantification of unspliced pre-mRNAs and mature spliced mRNAs for each gene. By exploiting the abundance of unspliced and spliced mRNA, one can infer the RNA velocity of individual cells, i.e., the time derivative of the gene expression state of cells. Whereas traditional trajectory inference methods reconstruct cellular dynamics given a population of cells of varying maturity, RNA velocity relies on a dynamical model describing splicing dynamics. Here, we initially discuss conceptual and theoretical aspects of both approaches, then illustrate how they can be combined together, and finally present an example use case on real data.

Key words Trajectory inference, RNA velocity, Single-cell RNA sequencing, Transcription, Splicing, Gene regulation, Cell differentiation, Bioinformatics, Computational biology

1 Introduction

In biology, we have been successfully organizing cells into different cell types, representing tissues which in turn constitute organisms. However, oftentimes, the relationships between different cell states are not discrete, but continuous as in the case of cell differentiation or embryogenesis. Modeling these continuous spectra of cell states is the goal of trajectory inference (TI). Unlike clustering, which is the task of assigning categorical labels to cells, trajectory inference assigns continuous values called pseudo-time [1].

Pseudo-time values are designed to quantify a cell's position along a spectrum (Fig. 1). In the case of differentiation, or any

Kelly Street and Simone Tiberi contributed equally.

Raffaele A. Calogero and Vladimir Benes (eds.), *Single Cell Transcriptomics: Methods and Protocols*, Methods in Molecular Biology, vol. 2584, https://doi.org/10.1007/978-1-0716-2756-3_14,

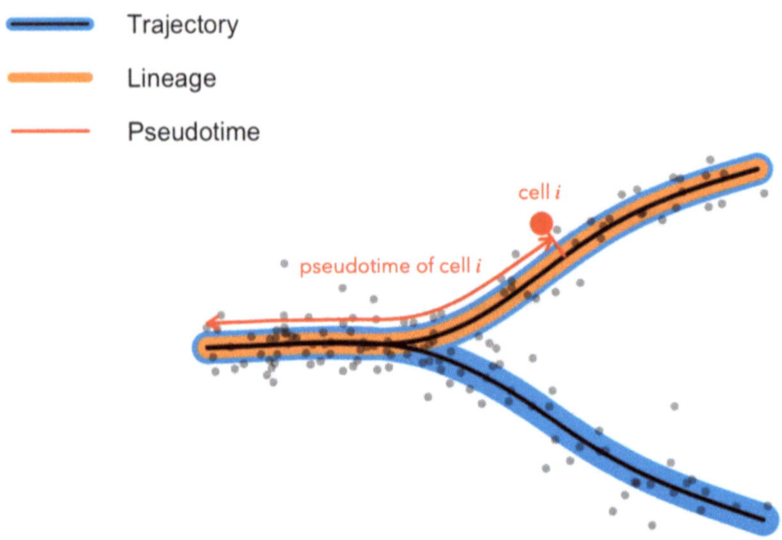

Fig. 1 Introduction to trajectory inference terminology. A trajectory is a collection of (one or more) lineages. A lineage is a single differentiation path. The pseudo-time of a cell is the length of the lineage from its inception up to the projection of the respective cell onto that lineage

process in which progression occurs in a single direction, a cell's pseudo-time value can be viewed as a proxy for how far along the spectrum a cell is relative to the progenitor state. However, pseudo-time is based solely on transcriptional information, so it cannot be interpreted as an estimator of the true time since initial differentiation. Rather, it is a high-resolution estimate of cell state, which is likely to be monotonically related to the true chronological time, but there is no guarantee that equivalent changes in transcriptional profiles follow a similar chronological time.

In Subheading 3.1, we will discuss the main conceptual ideas behind trajectory inference, including recommended upstream analyses. We will briefly introduce several types of trajectories, starting from a simple trajectory consisting of a single lineage (e.g., the differentiation of a unipotent stem cell to its mature cell state).

In the following, we will refer to a *trajectory* as a collection of lineages, which together describe a dynamic process (Fig. 1). A cell's pseudo-time, then, is defined as the length of the lineage from its inception up to the projection of a cell onto that lineage. If a trajectory consists of multiple lineages, cells may have multiple pseudo-time values, depending on which lineage the cell is projected upon.

Conversely to TI, RNA velocity describes the temporal change in mRNA over time by leveraging splicing dynamics [2]. These can be split into three main stages: transcription, splicing, and

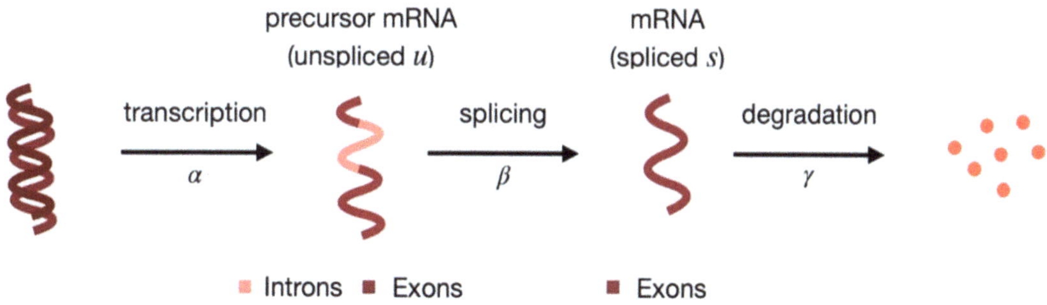

Fig. 2 The first step in splicing dynamics consists iof transcribing DNA to form unspliced, precursor mRNA, *u*. The included non-coding regions (introns) are removed through splicing. This process leaves only coding regions (exons) which form spliced mRNA, *s*, and are eventually degraded

degradation. A schematic overview of the process is given in Fig. 2. During transcription, DNA is synthesized into precursor messenger RNA (pre-mRNA). Pre-mRNA contains both expressed regions (exons) and non-coding regions (introns) which are redundant for translation. Consequently, introns are removed through splicing. This leaves only exons forming mature mRNA. Finally, the process is completed with the eventual degradation of mRNA. In the following, we will refer to pre-mRNA as unspliced mRNA u and to mRNA as spliced mRNA s. The rates of transcription, splicing, and degradation are denoted by α, β, and γ, respectively. Using this naming convention, the process of splicing has previously been modeled for each gene, independently of all others, by the following system of ordinary differential equations (ODEs) [3]:

$$\begin{aligned} \frac{du}{dt} &= \alpha - \beta u \\ \frac{ds}{dt} &= \beta u - \gamma s. \end{aligned} \tag{1}$$

By definition, the RNA velocity is the derivative of spliced counts with respect to time.

This chapter is organized as follows. Subheading 2 describes preprocessing steps for both TI and RNA velocity analyses. Subheading 3 illustrates the conceptual and theoretical aspects of both approaches, briefly explains how one can combine the two by inferring trajectories from the RNA velocity output, and concludes with an example usage, on a real dataset, of both TI and RNA velocity. Subheading 4 reports some notes and practical guidelines for actual implementation of both frameworks.

2 Materials

2.1 Trajectory Inference Data Preprocessing

Trajectory inference is most commonly applied to complex, large single-cell transcriptome sequencing datasets. Defining the trajectory in such high-dimensional datasets can be challenging and, therefore, often calls for upstream analyses that allow us to uncover the major biological variation. We assume that we are working with a high-quality dataset, where each measurement refers to a single cell, i.e., where empty droplets, doublets/multiplets, as well as low-quality cells have been removed. First, we should account for major sources of technical variation, such as differences in sequencing depth between individual cells, using normalization. Often, cell-specific size factors are calculated to account for this [4]. Since each cell has thousands of dimensions, even on normalized data, it is difficult to visualize the global biological structure of the dataset without using dimensionality reduction techniques. Indeed, owing to the so-called *curse of dimensionality*, all cells tend to become approximately equidistant in high-dimensional space [5]. To alleviate this issue, we try to reduce dimensionality using feature selection, where we select a limited subset of genes which we believe still explain a substantial amount of variation in our dataset. Indeed, one may view feature selection as the most basic dimensionality reduction technique. Finally, in order to infer the trajectory, we attempt to summarize and visualize the dynamic process in a low-dimensional space using dimensionality reduction techniques, such as principal component analysis (PCA), uniform manifold approximation and projection (UMAP) [6], or t-distributed stochastic neighbor embedding (t-SNE) [7]. Care must be taken to properly take into account the (count) properties of single-cell datasets, in order to ensure that this low-dimensional space is driven by biological signal and not by technical variation as shown in refs. 8, 9.

2.2 RNA Velocity Data Preprocessing

RNA velocity analyses are based on the counts of unspliced and spliced mRNA (*see* **Note 1**). Pipelines such as velocyto [2], alevin [10], alevin-fry [11], or kalisto/bustools [12] allow inferring them from scRNA-seq data. As in classical scRNA-seq analysis, measurements need to be preprocessed [13]. First, genes are filtered by expression level or by their occurrence in cells. Only genes with enough counts in both modalities should be retained. Following this first filter, each modality is normalized, and highly variable genes are identified and extracted, while all remaining genes are discarded.

Single-cell RNA sequencing data is notoriously noisy. To extract the relevant signal and infer RNA velocity, it needs to be smoothed first. This is done by calculating the first-order moment (mean) of an observation over its k nearest neighbors. Figure 3

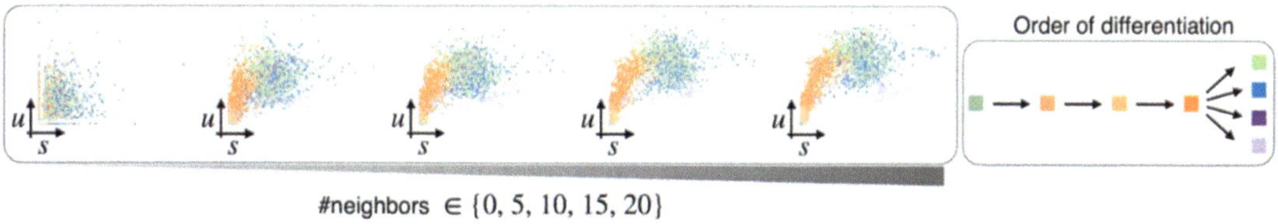

Fig. 3 Imputing measurements by first-order moments removes noise. With an increasing number of neighbors (from left to right), the phase portrait becomes less noisy and upregulation is visible. Starting from the origin, time increases when moving along the trajectory clockwise. Consequently, cell types should be ordered along the trajectory accordingly. Comparing this ordering of cell types with the biologically known order of differentiation, an upregulation is identified as the correct regulation type

shows how the phase portrait of gene Cpe changes, in a pancreatic endocrinogenesis dataset [14] (*see* Subheading 3.4), with an increasing number of neighbors. Whereas there is no curvature present in the raw data (leftmost plot), the upregulation in gene Cpe becomes visible when increasing the number of neighbors. This illustrates the importance of data imputation by moments for RNA velocity analysis. The phase portrait not changing dramatically after a certain number of neighbors k indicates robustness with respect to the choice of this parameter. Increasing k to large values is not recommended as it increases computational complexity and does not describe well the most similar observations. For large values of k, observations not similar to the reference would, thus, be included as well.

3 Methods

3.1 Trajectory Inference

3.1.1 Single Lineage

Many biological systems can be adequately described by a single continuum, such as one cell type transitioning into another or the spatial distribution of related cells along an organ. We can model these systems with a trajectory structure consisting of a single lineage. In this case, the trajectory inference problem can be seen as a dimensionality reduction to a single dimension: we want to find a single dimension that captures as much biologically relevant variation as possible.

In this framework, cells' loadings along the first principal component could be interpreted as pseudo-times. However, more specialized approaches generally provide better results. One reason for this is that the first principal component (PC1) may not always identify biological effects (e.g., sequencing depth or the fraction of zeros in each cell are often correlated with PC1) [9]. Similarly, PC1 captures distinct proportions of the overall variance in different datasets. Another issue is the so-called "horseshoe effect," where a continuous gradient appears as a U-shaped curve in the first two principal components [15].

Beyond PCA, many trajectory inference methods exist and are sufficiently adaptive to appropriately model a single lineage [16]. However, few are able to accurately infer directionality. With one lineage, this is a binary decision: which of the two ends should be the starting point? Without bringing in additional data or domain-specific knowledge, this can be a difficult question to answer; hence, most trajectory inference methods rely on users' supervision. A more data-driven approach may require annotated datasets and/or cell-type identification methods to select the cluster that most closely resembles stem cells; this cluster is then used as the starting point of the trajectory.

3.1.2 Diverging Lineages

Since trajectory inference is often applied to populations of differentiating cells, we are frequently interested in modeling multiple branching lineages. In this case, we not only need to determine each cell's pseudo-time but also its degree of commitment to a particular fate. Consider a bifurcating trajectory (i.e., consisting of two diverging lineages); cells in the starting population have not yet committed to a fate, whereas cells in the terminal populations have. In between these two extremes, there are likely to be different degrees of commitment to each of the two possible endpoints.

By their nature, these sorts of trajectories lend themselves well to tree-like structures. Hence, many TI methods construct a branching graph to summarize the layout of major cell types, with the root node representing the initial, undifferentiated population and (other) leaf nodes representing terminal populations [17–19]. For example, on the pancreas dataset [14] employed in Subheading 3.4, we used Slingshot [18] to construct a supervised graph on the set of cell clusters, resulting in a trajectory with four possible fates.

3.1.3 Converging Lineages

In some systems, cells from different progenitor populations converge into a single cell type. Any TI method that can model diverging lineages can also model simple converging lineages by selecting the terminal population as the "beginning" of the trajectory and then reversing the resulting pseudo-times. However, more complex systems that involve both divergence and convergence can be difficult to model and require methods that accommodate such complexity as PAGA [20] and Monocle3 [19].

3.1.4 Cycles

Another prominent feature of some trajectories is the cell cycle. On its own, a cyclic trajectory can be modeled in many ways, but as part of a larger trajectory, these can be difficult to accurately represent. Not only do cyclic components break the tree-like graph structure assumed by many TI methods but they also introduce some ambiguity regarding the interpretation of pseudo-time. Specifically, a cell which has undergone one full cycle may be transcriptionally indistinguishable from a cell which has not undergone any changes;

this may invalidate the interpretation of pseudo-time as being monotonically related to real time.

For instance, in the pancreas dataset (Subheading 3.4), there is a noticeable cycle in the first two clusters. However, since Slingshot assumes a tree-like structure, it is unable to accurately model this part of the trajectory.

3.1.5 Challenges and Limitations

All trajectory inference procedures attempt to recover a "best-fitting" trajectory on any dataset to which they are applied (*see* **Notes 2–4**), just like *k*-means clustering always provides *k* clusters. Care must be taken that the structure one attempts to model indeed represents a biological dynamic process. Poor pre-processing or improperly accounting for technical variation may give rise to spurious trajectories; for instance, doublet cells may be misinterpreted as "intermediate states" [21], or a hypothesized major source of biological variation may actually be reflecting technical effects, such as the fraction of zeros or the total number of counts for each cell [9].

For some datasets, there is a thin line between clustering the cells into discrete entities and modeling the structure as a trajectory. For example, depending on the parameters of one's preferred dimensionality reduction technique, the structure may seem discrete rather than continuous. Therefore, before estimating a trajectory on any dataset, domain-specific knowledge remains irreplaceable, e.g., through close examination of well-characterized marker genes. While computational algorithms may indeed be able to fit the observed structure, they will not be able to determine whether or not it is biologically plausible, and results may be misleading.

3.2 RNA Velocity

3.2.1 Splicing Dynamics

The solution of the simple, linear system (Eq. 1) can be found analytically and is given by:

$$u(t) = u_0 e^{-\beta(t-t_0)} + \frac{\alpha}{\beta}\left(1 - e^{-\beta(t-t_0)}\right)$$

$$s(t) = s_0 e^{-\gamma(t-t_0)} + \frac{\alpha}{\gamma}\left(1 - e^{-\gamma(t-t_0)}\right) + \frac{\alpha - \beta u_0}{\gamma - \beta}\left(e^{-\gamma(t-t_0)} - e^{-\beta(t-t_0)}\right), \tag{2}$$

when $\gamma \neq \beta$, and where *s_0* and *u_0* refer to the unspliced and spliced counts at time *t_0*, respectively, and *e^* indicates the exponential function.

Understanding how splicing dynamics evolve over time is fundamental for analyzing RNA velocity. During transcription, each modality increases until it reaches a steady state. As spliced mRNA is produced from unspliced mRNA, the abundance of the former increases prior to the latter. If transcription ends prematurely, steady states may not be reached.

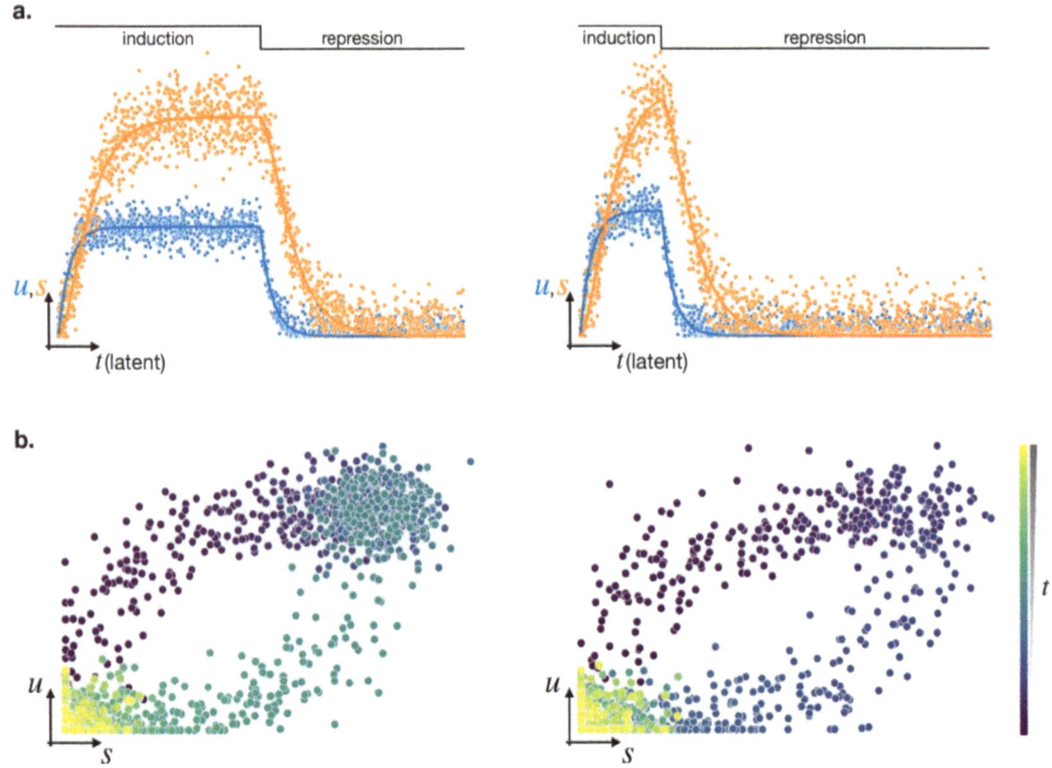

Fig. 4 Visualization of measured abundances. (**a**) Since spliced mRNA *s* is produced from unspliced mRNA *u*, the abundance of *s* increases (or decreases) with a temporal delay with respect to *u*. During DNA transcription (i.e., induction phase), both *s* and *u* increase. If this phase lasts long enough, a steady state is reached (left). Then, during repression, transcription is turned off and both modalities gradually decrease. (**b**) As single-cell measurements are snapshot data, time can neither be tracked over several measurements nor be associated with a single one. Unspliced and spliced counts are plotted against each other in a phase portrait. Its curvature increases as the temporal delay between *s* and *u* grows

Using the proposed system of ODEs (Eq. 2), this process can be simulated and abundances plotted against time (Fig. 4a). However, this cannot be done in real data measurements due to the destructive nature of scRNA-seq experiments. Consequently, phase portraits (Fig. 4b), where the abundance of unspliced mRNA is plotted against the one of spliced mRNA, are studied instead. In the ideal case, the full process of transcription, splicing, and degradation is observed leading to an almond-shaped phase portrait. The upper arc of the almond shape corresponds to the induction phase ending in a steady state in the upper right part of the portrait. Conversely, cells in the lower arc are in repression, i.e., transcription is turned off. If only part of the full cycle is observed during the experiment, the resulting portrait contains only the corresponding portion of the almond shape. Overall, this shape depends on the model parameters.

In the following, we will first discuss theoretical aspects of the RNA velocity analysis. This includes how scRNA-seq should be processed, two methods for inferring model parameters, and finally some RNA velocity-based downstream analyses. Then, in Subheading 3.4.2, these discussed facets will be applied to and showcased in a real-world dataset of endocrine development in the pancreas [14].

3.2.2 Parameter Inference

Rate parameters cannot be measured during a scRNA-seq experiment and need to be inferred instead. Since scRNA-seq data consist of snapshot measurements, time can neither be measured nor tracked. Consequently, time points assigned to each cell, as well as the time when the system switches from induction to repression, need to be estimated.

This section discusses how parameters are inferred and is divided into two parts: the first one discusses the estimation of the steady-state ratio using the so-called steady-state model and the second one outlines how the full set of observations and model parameters can be inferred using a different approach [2, 22] (*see* **Note 5**).

Steady-State Model

In a first attempt to estimate RNA velocity, splicing dynamics are assumed to have reached their steady states [2]. In these steady states, we consequently find that:

$$\frac{ds}{dt} = 0 \Leftrightarrow 0 = \beta u - \gamma s \Leftrightarrow u = \frac{\gamma}{\beta} s.$$

Note that this equation is only satisfied for cells in equilibrium, i.e., cells found in the lower and upper quantiles of the phase portrait. Assuming a common unit splicing rate β across all genes, we can estimate the steady-state ratio, $\gamma = \frac{\gamma}{\beta}$, through an extreme-quantile linear regression with closed form solution:

$$\tilde{\gamma} = \frac{u^{\mathrm{T}} s}{||s||^2},$$

where $||.||$ indicates the euclidean distance.The residual to the linear-regression fit of the unspliced mRNA in cell c (Fig. 5a), that we denote by v_c, is defined as follows:

$$v_c = u_c - \tilde{\gamma}\, s_c;$$

This residual is the estimated RNA velocity of cell c.

In other words, this simple estimation of RNA velocity requires both steady states of the system to be observed, as well as a common splicing rate β across all genes. While this approach proves fruitful in some cases, its assumptions represent serious limitations for its applicability. On the one hand, the estimated steady state ratio will deviate from the real (unobserved) ratio when not all steady states are observed; this occurs, for example, when

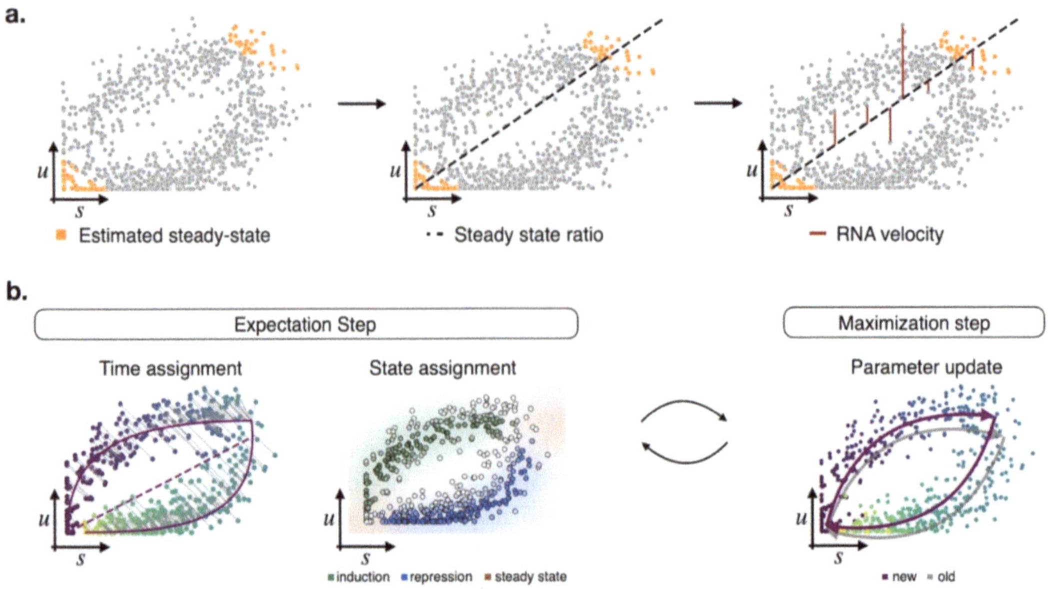

Fig. 5 Inferring RNA velocity and model parameters from scRNA-seq data. (**a**) The steady-state model assumes the system has reached its steady states and estimates the cells that have reached either steady state (left). A linear regression on these extreme quantiles is fitted (middle), and the RNA velocity is estimated as the residuals to this fit (right). (**b**) Solving the full splicing dynamics allows generalizing to transient genes. Parameters are inferred by expectation-maximization (EM). During the E-step (left), each observation is assigned both a time and a state (induction, repression, or steady state). The former assignment is found by minimizing the distance to the current estimate of the trajectory, the latter by the maximal likelihood of a state. Then, model parameters are updated by maximizing the overall likelihood during the M-step (right)

transcription ends prematurely. On the other hand, splicing is a gene-dependent property, and its rates can largely vary between genes. Furthermore, not inferring the full set of model parameters and only using a subset of the available data are additional drawbacks. Below, we discuss how these shortcomings can be addressed and resolved (*see* **Note 6**).

Dynamical Model

Having preprocessed the scRNA-seq measurements, we are interested in inferring model parameters α, β, γ of (Eq. 1). Since scRNA-seq experiments are destructive by nature (i.e., cells are destroyed when sequenced and, therefore, cannot be measured multiple times), observations cannot be associated with a specific time point. As such, the observations contain latent, i.e., hidden or unknown, variables; these are time point t and state k (induction, repression, or steady state) of an observation. Consequently, we cannot estimate model parameters via maximum likelihood, but turn to expectation-maximization (EM) [23], as proposed by [22] and implemented in the Python package *scVelo*, instead. The algorithm is divided into two steps: the E-step (i.e., expectation step) and the M-step (i.e., maximization step). Hidden variables are updated during the E-step, while the remaining parameters are updated during the M-step (Fig. 5b). To infer model parameters, we assume constant rates across time. This requires splitting (Eq. 2) into two parts: induction and repression. During induction the transcription rate α is positive, while during repression, it is equal to zero. Finding the optimal time assignment is computationally expensive. To boost performance, the optimal solution can be approximated, leading to a 30-fold speed-up [22]. For this, we rewrite the analytic solution (Eq. 2) for s as

$$s(t) = \tilde{\beta}\, u(t) + s_\infty - \tilde{\beta}\, u_\infty + \left(s_0 - s_\infty - \tilde{\beta}\,(u_0 - u_\infty) \right) e^{-\gamma\tau}, \quad (3)$$

with $\tilde{\beta} = \frac{\beta}{\gamma - \beta}$, and s_∞ and u_∞ being the values of the systems at equilibrium (i.e., as time goes to infinity). Defining $\tilde{s}(t) = s(t) - \tilde{\beta} u(t)$ and initial and steady states accordingly, (Eq. 3) can be solved for τ as

$$\tau = -\frac{1}{\gamma} \log \left(\frac{\tilde{s} - \tilde{s}_\infty}{\tilde{s}_0 - \tilde{s}_\infty} \right).$$

In case $\beta \leq \gamma$, the inverse of $u(t)$ is considered instead, i.e.,

$$\tau = \frac{1}{\beta} \log \left(\frac{u - u_\infty}{u_0 - u_\infty} \right).$$

The EM algorithm estimates gene-specific reaction rates and assigns gene-specific time points to each observation. These parameters are only identifiable up to a proportionality constant λ. It is easy to see that multiplying all rates by λ and dividing time by the same value do not change the solution. If the true time scale $[t_0, t_{\max}]$ was known, the scaling parameter λ could be factored out. Here, λ is estimated by aligning the inferred time assignment across all genes to a common scale.

3.2.3 From Velocities to Transition Probabilities

RNA velocities, representing the true underlying splicing dynamics, allow predicting the future state of cells. Given the gene profile of cell j at time t, c_j, and its future gene profile at time $t + \Delta t$ (with $\Delta t > 0$), c, the change in its gene profile is estimated by $c - c_j$. Since future gene profiles cannot be assigned to each cell, their neighbors are considered as candidates. The probability of transitioning to one neighbor is high if the heuristic displacement and velocity vector point into a similar direction (Fig. 6a). This similarity is captured by cosine similarity, i.e.,

$$\tilde{\pi}_{jk} = \frac{< \delta_{jk}, v_j >}{||\delta_{jk}||\ ||v_j||},$$

with displacement $\delta_{jk} = c_k - c_j$ and velocity vector v_j for cell j. As visualized in Fig. 6b, π_{jk} is maximal when displacement and velocity

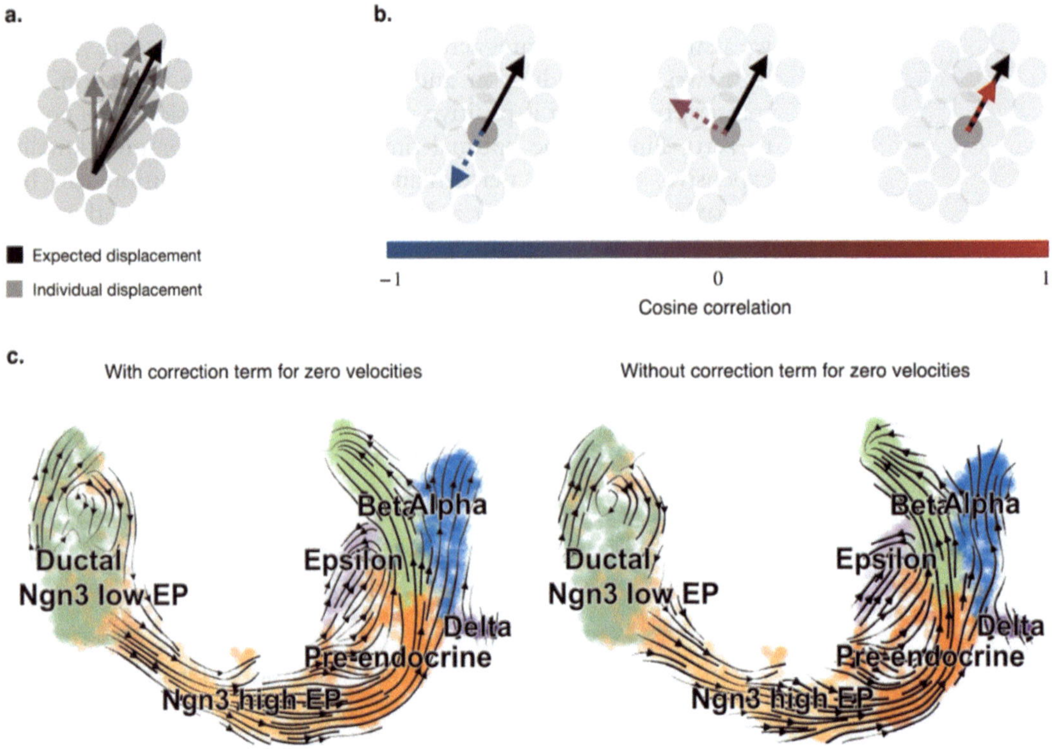

Fig. 6 Visualization of velocities (application example taken from ref. 22). (**a**) The expected displacement of a reference cell indicates the direction of its future state. Here, the expected displacement is given by a weighted average of the displacement between a reference and its neighbors. Transition probabilities are used as weights. (**b**) The cosine correlation measures the similarity between two vectors. Here, displacement and velocity vectors are compared to estimate transition probabilities. (**c**) To account for vanishing or orthogonal velocities, the expected displacement is corrected by the mean displacement (left). The correction term particularly affects, compared to omitting it (right), the border of the embedding

vector point into the same direction. In case of orthogonality or opposite directions, $\tilde{\pi}_{jk}$ vanishes or takes its minimum, respectively. The graph associated with this similarity matrix is called velocity graph (*see* **Note 7**).

Similarity measures cannot necessarily be interpreted as probabilities. The cosine similarity, for example, falls in the interval $[-1, 1]$. An exponential kernel is applied to transform cosine similarities to transition probabilities, i.e.,

$$\pi_{jk} = \frac{1}{z_j} \exp\left(\frac{\tilde{\pi}_{jk}}{\sigma_j}\right),$$

with normalization constant z_j and parameter σ_j adjusting the kernel width. These transition probabilities define the elements of the transition matrix Π, which allows determining descendants and ancestors of a given cellular distribution. Visualizing these in a low-dimensional embedding of a latent space can be a powerful tool to easily and intuitively study cellular differentiation. Common latent space representations are UMAP and t-SNE [6, 7]. However, two- or three-dimensional representations should be interpreted with caution as they contain limited information compared to higher dimensional representations [24]. As such, they should not form the sole basis of quantitative modeling.

Differentiation paths can be visualized by an adjusted expected lower-dimensional displacement. Given the embedding vectors s_j and s_k, the expected displacement is given by

$$E_{\Pi}\left[\frac{s_k - s_j}{||s_k - s_j||}\right] = \sum\nolimits_{j \neq k} \pi_{jk} \frac{s_k - s_j}{||s_k - s_j||}. \tag{4}$$

In case of vanishing or orthogonal (with respect to the displacement) velocities, transition probabilities are uniform and the heuristically expected displacement is zero; nonetheless, the value of (Eq. 4) is not null. For consistency, (Eq. 4) is therefore adjusted as:

$$\sum\nolimits_{j \neq k} \left(\pi_{jk} - \frac{1}{n}\right) \frac{s_k - s_j}{||s_k - s_j||},$$

with n being the number of possible transitions (Fig. 6c).

3.2.4 Gene-Shared Latent Time

The inferred model parameters and time assignments are gene specific. In reality, however, each cell is associated with a time. As such, gene-specific times need to be pooled together into a single, gene-shared latent time for every cell. To mitigate the effect of noise, only a subset of genes, for example well-fitted ones, are considered. In addition to the goodness of fit, the correlation between the inferred time and a velocity graph-based pseudo-time is a decisive factor for selecting genes.

To infer a velocity-based pseudo-time, as a first step, initial and terminal states are identified. They can heuristically be estimated by the stationary distribution μ^* satisfying

$$\mu^* = \mu^* \Pi^{\mathrm{T}}.$$

For a more sophisticated approach of identifying these states, the Python library *CellRank* can, for example, be used [25] (*see* Subheading 3.3) (*see* **Note 8**).

Given an initial state, the average number of steps (in a random walk) to reach a given cell from the starting cell defines, similar to diffusion pseudo-time [26], a pseudo-time. Compared to this approach, diffusion pseudo-time uses a similarity-based diffusion kernel and does not implicitly infer terminal states. Then, genes, for which the correlation between their time assignments and the constructed pseudo-time does not exceed a given threshold, are neglected from further analysis.

Finally, for each estimated terminal state, p-quantiles of time increments (with respect to a given initial state) across all genes are calculated. This is motivated by cell density being not uniformly distributed across time. Instead, the density usually increases in late time points. The gene-shared latent time is then given by the mean p-quantile across all starting cells.

3.2.5 Challenges and Limitations

While RNA velocity has had tremendous success and has been applied successfully to various datasets, the current framework fails in some scenarios. Even though extending the steady-state model to solving the full splicing kinetics generalized the concept to transient populations, the model still relies on constant parameters. Examples of violations of this assumption are transcriptional bursts, i.e., when the transcription rate increases abruptly after some time, or multiple kinetics. Transcriptional bursts can, for example, occur in hematopoiesis [27, 28] and multiple kinetics in dentate gyrus [2, 22, 29].

Another limitation of the current formulation is that it relies on phase portraits exhibiting at least a partial almond shape. In reality, gene measurements are noisy, i.e., do not always lead to such shapes in the phase portraits. In this case, model parameters may be inferred incorrectly leading to wrong conclusions. This can, for example, be shown on a dataset of peripheral blood mononuclear cells including only mature cells [29, 30]. In this case, albeit no differentiation is expected, the velocity stream has a clear, but arbitrary, direction.

See ref. 29 for a more in-depth discussion on existing pitfalls and possible extensions to the existing framework.

3.3 Combining Trajectory Inference and RNA Velocity

Whereas RNA velocity assigns directed, dynamic information to each cell, trajectory inference models the continuous differentiation of cell types on the phenotypic manifold. To combine the two approaches, *CellRank* has been proposed for RNA velocity informed TI [25]. For this, *CellRank* describes state transitions as a Markov chain to, first, identify initial, intermediate, and terminal states and, then, determine fate probabilities of each cell, as well as expression trends and putative decision drivers in different trajectories.

Markov chain models consist of states and transition probabilities between states. In *CellRank*, each state corresponds to a single cell. The possible transitions and initial estimates of their probabilities are given by a similarity-based k-nearest neighbor (kNN) graph. To not only rely on transcriptional similarity, *CellRank* considers the directional information provided by RNA velocity as well. For this, a transition matrix similar to the one defined in Subheading 3.2.3 is used and uncertainties are propagated. Instead of cosine correlation, *CellRank* centers gene expression and velocities to zero mean, resulting in Pearson's correlation. Both transition matrices (i.e., the similarity and velocity-based ones) are combined by a weighted mean for a more robust estimate of transition probabilities. While *CellRank* has originally been developed to combine RNA velocity with transcriptomic similarity, it has meanwhile been extended to include other estimates of directionality, such as pseudo-time, as well.

Since scRNA-seq is high dimensional and noisy, the transition matrix cannot be used as defined to solve the problem of trajectory inference. Instead, the system is first coarse grained into states, so-called macrostates, forming initial, intermediate, and terminal states, based on Generalised Perron Cluster Cluster Analysis (GPCCA). These states can be automatically classified on the basis of coarse-grained transition probabilities, and the probability to transition into each terminal state, the fate probability, can be quantified as well. Finally, *CellRank* uses these results to study fate commitment or identify putative driver genes specific to individual trajectories.

3.4 Hands-On Real Data Analyses

3.4.1 Trajectory Inference Analysis Pipeline

We demonstrate some of the ideas presented above with a hands-on analysis of a single-cell pancreatic endocrinogenesis dataset taken from ref. 14. We perform this analysis in *R*, making use of multiple Bioconductor packages, most notably *Slingshot* for the task of trajectory inference.

The data can be easily downloaded in a usable format here: https://github.com/theislab/scvelo_notebooks/tree/master/data/Pancreas/endocrinogenesis_day15.h5ad. After downloading the data, we read it into *R* as a SingleCellExperiment object via the *zellkonverter* package [31]. This creates an object that contains not only the count matrix but also cluster labels and dimensionality reductions from the original publication.

```
library(zellkonverter)
sce <- readH5AD('endocrinogenesis_day15.h5ad')
```

We are now ready to perform trajectory inference. We provide *Slingshot* with the UMAP dimensionality reduction and the cluster labels, as well as some biological supervision. In this case, the only supervision that impacts our results is the specification of the "Alpha" cluster as an endpoint; all other endpoints and the starting cluster are correctly inferred by *Slingshot*. We also increase the flexibility of the resulting curves by adjusting the "*spar*" parameter. This parameter is passed to the smoothing spline function which underlies *Slingshot*'s curves, and it allows us to account for the elongated nature of the UMAP, where most of the differentiation between lineages occurs toward the ends.

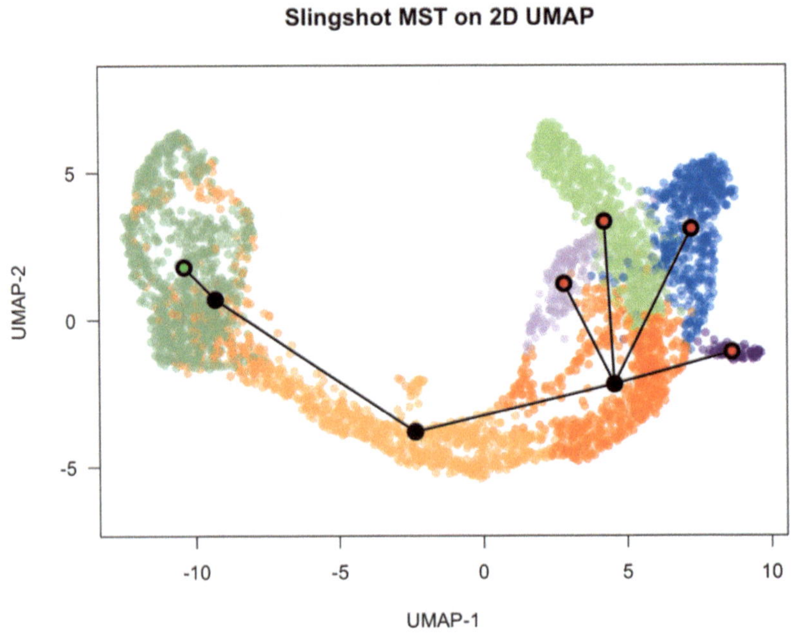

Fig. 7 The minimum spanning tree, computed by Slingshot, visualized on a two-dimensional UMAP embedding. Green (left), black (middle), and red (right) circles indicate initial, intermediate, and terminal clusters, respectively

```
library(slingshot)
pto <- slingshot(reducedDim(sce,'X_umap'),
    clusterLabels = sce$clusters, start.clus = 'Ductal',
    end.clus = c('Alpha', 'Beta', 'Delta', 'Epsilon'),
    spar = 1.1)
```

We can then plot *Slingshot*'s minimum spanning tree (MST) on the 2-dimensional UMAP (Fig. 7).

```
library(scales)
plot(reducedDim(sce,'X_umap'), asp=1,
     col = alpha(metadata(sce)$clusters_colors[sce$clusters],
alpha = .5),
    pch = 16,  xlab = 'UMAP-1', ylab = 'UMAP-2',
    main = 'Slingshot MST on 2D UMAP')
lines(SlingshotDataSet(pto), type = 'lineage', show.con-
straints = TRUE)
```

From the two-dimensional UMAP in Fig. 7, we can see the cluster labels (representing cell types) and four distinct endpoints. However, there is significant overlap between the terminal clusters on the right-hand side of the plot, indicating that two dimensions might not be enough to fully characterize this complex trajectory structure.

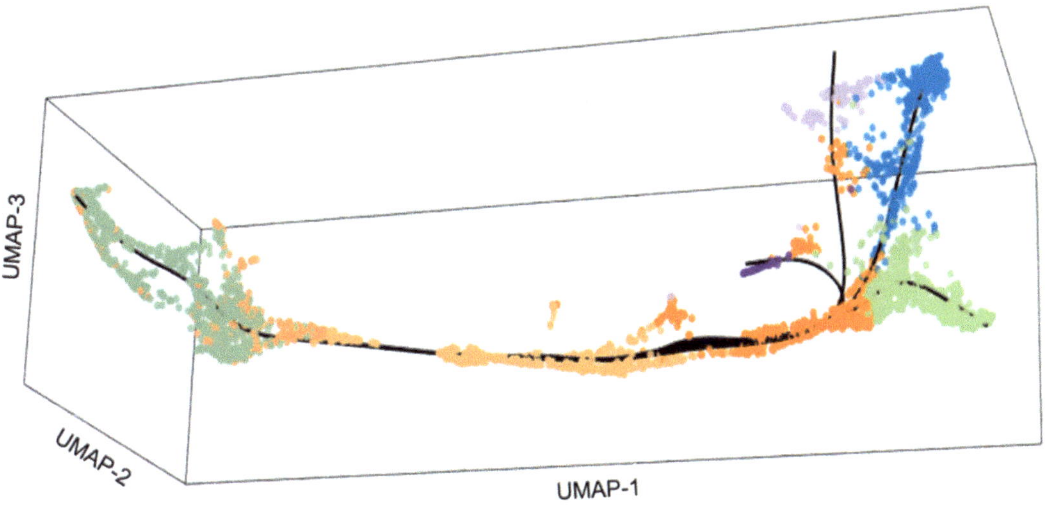

Fig. 8 Visualization of the trajectory computed by Slingshot, on a three-dimensional UMAP embedding

Therefore, we produced a three-dimensional UMAP (calculated from the precomputed top 50 PCA components) and re-run *Slingshot* on this embedding.

This provides us with a reduced-dimensional space in which the four terminal clusters are more clearly separated (Fig. 8).

```
# calculate 3D UMAP:
library(uwot)
umap3d <- umap(reducedDim(sce,'X_pca'), n_components = 3)

# Compute trajectories on 3D UMAP embedding:
library(slingshot)
pto_3D <- slingshot(umap3d, clusterLabels = sce$clusters,
    start.clus = 'Ductal',
    end.clus = c('Alpha', 'Beta', 'Delta', 'Epsilon'), spar =
1.1)
# Plot inferred trajectories on 3D-UMAP:
require(rgl)
plot3d(umap3d, aspect = 'iso', size=5,
    col = metadata(sce)$clusters_colors[sce$clusters])
plot3d.SlingshotDataSet(SlingshotDataSet(pto_3D), add = TRUE,
type='c', lwd=3)
```

For visualization purposes, we also produce an embedding of these curves in the original, two-dimensional space. Note that this process does not change the pseudo-time values, but it is merely a way to visualize our results in a different space. The sharp turn in the Delta lineage (Fig. 9) reinforces our earlier notion that the two-dimensional UMAP embedding was insufficient to adequately represent this trajectory.

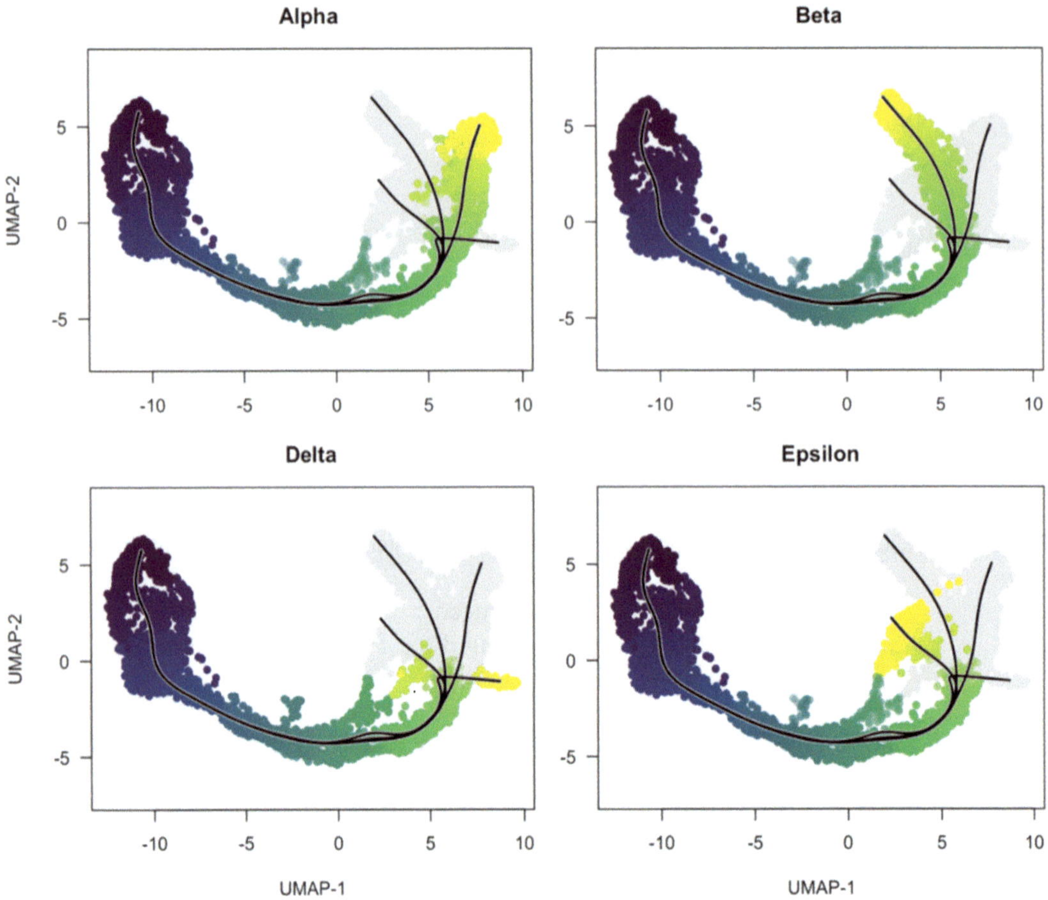

Fig. 9 The trajectory and pseudo-time computed by Slingshot, visualized on a two-dimensional UMAP embedding. Each image refers to an end cluster: Alpha (top left), Beta (top right), Delta (bottom left), and Epsilon (bottom right)

```
# Create embedding of trajectories in 2D-UMAP
emb <- embedCurves(pto_3D, reducedDim(sce,'X_umap'),
spar=1.1)

# Plot inferred lineages on 2D-UMAP
layout(matrix(1:4,2,2))
par(mar=c(4,4,3,1))
for(i in 1:4){
    title <- c('Alpha','Delta','Beta','Epsilon')[i]
    xl <- rep(c('','UMAP-1'), 2)[i]
    yl <- rep(c('UMAP-2',''), each = 2)[i]
    plot(reducedDim(sce,'X_umap'), asp=1, col='grey90',
    xlab=xl, ylab=yl,
    main = title, pch = 16)
    points(reducedDim(sce,'X_umap'), pch = 16,
           col = alpha(viridisLite::viridis(100)[cut(sling-
```

```
Pseudotime(pto_3D)[,i], breaks=100)],
        alpha = slingCurveWeights(pto_3D)[,i]))
    lines(SlingshotDataSet(emb), lwd=3, col='white')
    lines(SlingshotDataSet(emb), lwd=2)
}
```

3.4.2 RNA Velocity Analysis Pipeline

Here, we combine all RNA velocity aspects discussed into a single pipeline to run RNA velocity analysis on the dataset of endocrine development in the pancreas [14], which has previously been analyzed in ref. 22 as well. For this, we will use *scVelo* [22], the Python package implementation accompanying the dynamical model which also includes an efficient implementation of the steady-state model. For documentation and instructions on how to install *scVelo, see scvelo.org*.

The dataset to be analyzed can easily be loaded using *scVelo*:

```
import scvelo as scv
adata = scv.datasets.pancreas()
```

In the resulting AnnData object, estimated unspliced and spliced counts are stored as layers. The X slot, based on which we will perform principal component analysis and calculate the kNN graph, contains the spliced counts as well.

Then, we preprocess the data, as discussed before. Here, we filter genes which share less than 20 counts across modalities and normalize gene counts by their median. Finally, we only retain the 2000 most highly variable genes. To mitigate the effect of outliers for PCA and kNN graph calculation, spliced counts are log1p transformed as well, i.e., $\log(s + 1)$. The transformation is, however, only applied to the X slot, not the entry in layers. In *scVelo*, we jointly perform all these steps by simply calling the function *pp.filter_and_normalize.*

```
scv.pp.filter_and_normalize(adata,min_shared_counts=20,
n_top_genes=2000)
```

Next, counts need to be imputed by first-order moments to reduce the effect of noise. As outlined in Subheading 2.2, this requires a dimension reduction into PCA space and kNN graph construction. Both steps, including the moment calculation, can be applied using the *pp.moments* function.

```
scv.pp.moments(adata, n_pcs=30, n_neighbors=30)
```

The data is now in a format such that we can estimate RNA velocity. We first calculate velocities using the steady-state model and then with the dynamical one. For the former, we simply need to call the velocity function with *mode = 'deterministic'*.

```
scv.tl.velocity(adata, mode='deterministic')
```

For the dynamical model, we first infer the parameters using *tl. recover_dynamics*, then calculate the velocities by specifying *mode* = *'dynamical'*. To speed up parameter inference, users can also increase the number of cores used to parallelize calculations, via *n_jobs* parameter; below, we use 4 cores. It is worth noting that setting *n_jobs* to −1 would use all available cores in the machine.

```
scv.tl.recover_dynamics(adata, n_jobs=4)
scv.tl.velocity(adata, mode='dynamical')
```

Then, we compute a velocity graph and project the velocity stream onto the UMAP embedding. Note that we did not calculate the UMAP embedding as it is already provided in the AnnData object loaded via *scVelo*. If users were to calculate a UMAP embedding for their dataset, they could, for instance, use the *pl.umap* function from *Scanpy* Python package.

```
scv.tl.velocity_graph(adata)
scv.pl.velocity_embedding_stream(adata, basis='umap')
```

Now, we can compare the two approaches based on their velocity streams. As shown in Fig. 10a, b, overall, both approaches yield similar results that agree with known cellular differentiation. However, there are also differences between the two cases that are worth mentioning. In the results from the steady-state model, there is no flow going into the cluster of Epsilon cells and there is an inconsistent backflow in Alpha cells.

It is worth highlighting that even in the absence of dynamics, velocity stream embeddings can, and will, display directionality. As such, instead of studying the embeddings, we should focus on phase portraits. One important advantage of the dynamical model (over the steady-state model) is that it provides a likelihood score for each model. Genes with high likelihood contribute more to the overall dynamics compared to those with low scores. In our case, genes with the highest likelihood are informative, i.e., most of their phase portraits display the discussed almond shape curvature. These phase portraits can be plotted using *scVelo*'s *pl.scatter* function. Four exemplary genes with high likelihood are shown in Fig. 10c. This confirms that RNA velocity analysis can successfully be applied on this dataset.

```
top_genes=adata.var['fit_likelihood'].sort_values(ascending=-
False).index
scv.pl.scatter(adata, basis=top_genes[:10], ncols=5, fra-
meon=False)
```

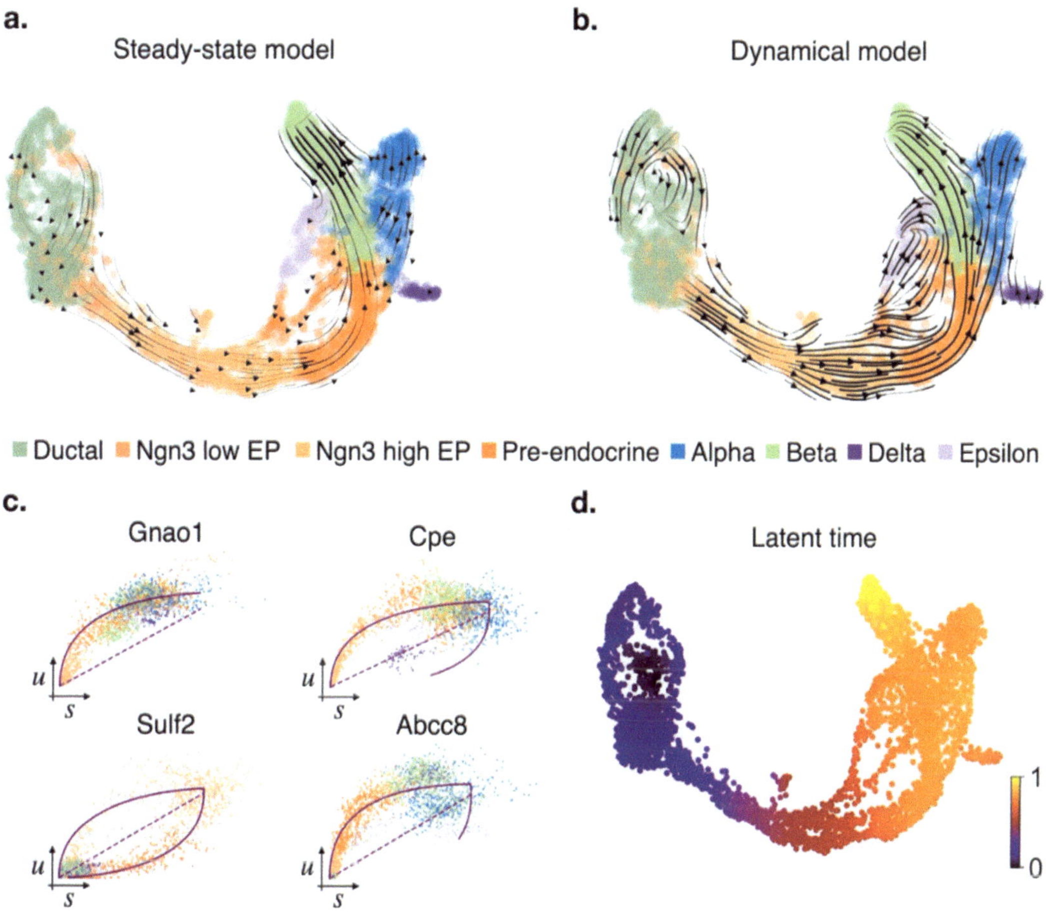

Fig. 10 Velocity stream in UMAP-based embedding. (**a**) The steady-state model does not recover a flow in Epsilon cells and includes an inconsistent backflow in Alpha cells. (**b**) Using the dynamical model, flows are more consistent and the dynamics in Epsilon and Delta cells are recovered. (**c**) Each gene is assigned a likelihood score indicating its importance in the overall dynamics. Phase portraits of exemplary genes with high likelihood include the inferred trajectory (solid purple line) as well as an estimated steady-state ratio (dashed purple). (**d**) Gene-specific time assignments can be pooled into a single gene-shared latent time, which recovers cellular differentiation in time and agrees with prior biological knowledge

Finally, latent time can be calculated using *scVelo*'s *latent_time* function. As displayed in Fig. 10d, the estimated latent time agrees with both prior biological knowledge on cellular differentiation in pancreas and the velocity stream embedding. Early cell types (ductal) are assigned early time points, where mature cells (Beta, Alpha, Delta, Epsilon) are associated with later ones.

```
scv.tl.latent_time(adata)
scv.pl.scatter(adata,color='latent_time', color_map='gnu-
plot', colorbar=True)
```

4 Notes

1. For RNA velocity analysis, unspliced and spliced reads need to be preprocessed similarly to standard scRNA-seq workflows (i.e., filtering lowly abundant genes, normalizing counts, and subsetting highly variable genes). Furthermore, missing data may be imputed to mitigate the effect of noise.
2. While we have introduced several types of trajectories, it may not always be clear how to choose among them. For example, the structure of a complex dataset may be overfitted, leading to many bifurcations and/or cycles, while the same dataset could also be described by a more parsimonious trajectory. The balance to strike may be considered analogous to the classical bias-variance tradeoff in statistics.
3. Many methods are not flexible enough to allow for different types of trajectories within the same dataset. For example, the pancreas dataset that we have analyzed (Subheading 3.4) may be better represented by a combination of a cyclic and a diverging trajectory; however, *Slingshot* is not able to fit such a trajectory.
4. Dynamic processes are subject to change depending on environmental or experimentally induced conditions; the comparison of trajectories between such conditions has recently gained increasing interest. One may compare global features, such as a trajectory's topology across conditions, as well as smaller changes, such as gene expression across the dynamic, between conditions. It may similarly be of interest to compare RNA velocity estimates across conditions. Such comparisons may hint at, for example, treatment-induced faster progression.
5. To infer RNA velocity, three models exist: steady-state, dynamical, and stochastic. The first two have been discussed here, and the latter is a generalization of the first (for more details *see* ref. 22).
6. The RNA velocity steady-state model assumes splicing dynamics to have reached it equilibria within the experimental time scale, as well as a constant splicing rate across all genes. As a consequence, only the steady-state ratio can be estimated. In contrast to this, the dynamical model solves the full splicing dynamics by considering all available data.
7. While velocity streams on low-dimensional embeddings are useful visualization means, they should not be over-interpreted. Instead, phase portraits should be studied first to ensure they exhibit (at least partially) the required almond shape such that RNA velocity can be inferred confidently.

8. To infer RNA velocity using any of the proposed models, the Python package *scVelo* can be used. Instead, *CellRank* should be used for further downstream analyses, such as estimating transition probabilities, automatically identifying initial and terminal states, and quantifying fate probabilities as well as gene expression and putative driver genes in different lineages.

Acknowledgments

The authors would like to acknowledge Marius Lange and Fabian Theis for their precious comments and suggestions. KVdB is a postdoctoral fellow of the Belgian American Educational Foundation (BAEF) and is supported by the Research Foundation Flanders (FWO) grants 1246220N, G062219N, and V411821N.

References

1. Trapnell C, Cacchiarelli D, Grimsby J, Pokharel P, Li S, Morse M et al (2014) The dynamics and regulators of cell fate decisions are revealed by pseudotemporal ordering of single cells. Nat Biotechnol 32:381–386
2. La Manno G, Soldatov R, Zeisel A, Braun E, Hochgerner H, Petukhov V et al (2018) RNA velocity of single cells. Nature 560:494–498
3. Zeisel A, Köstler WJ, Molotski N, Tsai JM, Krauthgamer R, Jacob-Hirsch J et al (2011) Coupled pre-mRNA and mRNA dynamics unveil operational strategies underlying transcriptional responses to stimuli. Mol Syst Biol 7:529
4. Vallejos CA, Risso D, Scialdone A, Dudoit S, Marioni JC (2017) Normalizing single-cell RNA sequencing data: challenges and opportunities. Nat Methods 14:565–571
5. Kiselev VY, Andrews TS, Hemberg M (2019) Challenges in unsupervised clustering of single-cell RNA-seq data. Nat Rev Genet 20: 273–282
6. McInnes L, Healy J, Melville J (2018) UMAP: uniform manifold approximation and projection for dimension reduction. arXiv
7. Van der Maaten L, Hinton G (2008) Visualizing data using t-SNE. J Mach Learn Res 9: 2579–2605
8. Hicks SC, Townes FW, Teng M, Irizarry RA (2018) Missing data and technical variability in single-cell RNA-sequencing experiments. Biostatistics 19:562–578
9. Townes FW, Hicks SC, Aryee MJ, Irizarry RA (2019) Feature selection and dimension reduction for single-cell RNA-Seq based on a multinomial model. Genome Biol 20:295
10. Srivastava A, Malik L, Smith T, Sudbery I, Patro R (2019) Alevin efficiently estimates accurate gene abundances from dscRNA-seq data. Genome Biol 20:65
11. He D, Zakeri M, Sarkar H, Soneson C, Srivastava A, Patro R (2021) Alevin-fry unlocks rapid, accurate, and memory-frugal quantification of single cell RNA-seq data. bioRxiv
12. Melsted P, Booeshaghi AS, Liu L, Gao F, Lu L, Min KHJ et al (2021) Modular, efficient and constant-memory single-cell RNA-seq preprocessing. Nat Biotechnol 39:813–818
13. Luecken MD, Theis FJ (2019) Current best practices in single-cell RNA-seq analysis: a tutorial. Mol Syst Biol 15:e8746
14. Bastidas-Ponce A, Tritschler S, Dony L, Scheibner K, Tarquis-Medina M, Salinno C et al (2019) Comprehensive single cell mRNA profiling reveals a detailed roadmap for pancreatic endocrinogenesis. Development 146: dev173849
15. Diaconis P, Goel S, Holmes S (2008) Horseshoes in multidimensional scaling and local kernel methods. Ann Appl Stat 2:777–807
16. Saelens W, Cannoodt R, Todorov H, Saeys Y (2019) A comparison of single-cell trajectory inference methods. Nat Biotechnol 37: 547–554
17. Ji Z, Ji H (2016) TSCAN: Pseudo-time reconstruction and evaluation in single-cell RNA-seq analysis. Nucleic Acids Res 44:e117

18. Street K, Risso D, Fletcher RB, Das D, Ngai J, Yosef N et al (2018) Slingshot: cell lineage and pseudotime inference for single-cell transcriptomics. BMC Genomics 19:477
19. Cao J, Spielmann M, Qiu X, Huang X, Ibrahim DM, Hill AJ et al (2019) The single-cell transcriptional landscape of mammalian organogenesis. Nature 566:496–502
20. Wolf FA, Hamey FK, Plass M, Solana J, Dahlin JS, Göttgens B et al (2019) PAGA: graph abstraction reconciles clustering with trajectory inference through a topology preserving map of single cells. Genome Biol 20:59
21. Amezquita RA, Lun ATL, Becht E, Carey VJ, Carpp LN, Geistlinger L et al (2020) Orchestrating single-cell analysis with Bioconductor. Nat Methods 17:137–145
22. Bergen V, Lange M, Peidli S, Wolf FA, Theis FJ (2020) Generalizing RNA velocity to transient cell states through dynamical modeling. Nat Biotechnol 38:1408–1414
23. Hastie T, Tibshirani R, Friedman J (2013) The elements of statistical learning: data mining, inference, and prediction. Springer Science & Business Media
24. Chari T, Banerjee J, Pachter L (2021) The specious art of single-cell genomics. bioRxiv
25. Lange M, Bergen V, Klein M, Setty M, Reuter B, Bakhti M et al (2020) CellRank for directed single-cell fate mapping. bioRxiv
26. Haghverdi L, Büttner M, Wolf FA, Buettner F, Theis FJ (2016) Diffusion pseudotime robustly reconstructs lineage branching. Nat Methods 13:845–848
27. Setty M, Kiseliovas V, Levine J, Gayoso A, Mazutis L, Pe'er D (2019) Characterization of cell fate probabilities in single-cell data with Palantir. Nat Biotechnol 37:451–460
28. Barile M, Imaz-Rosshandler I, Inzani I, Ghazanfar S, Nichols J, Marioni JC et al (2021) Coordinated changes in gene expression kinetics underlie both mouse and human erythroid maturation. Genome Biol 22:1–22
29. Bergen V, Soldatov RA, Kharchenko PV, Theis FJ (2021) RNA velocity-current challenges and future perspectives. Mol Syst Biol 17:e10282
30. Zheng GXY, Terry JM, Belgrader P, Ryvkin P, Bent ZW, Wilson R et al (2017) Massively parallel digital transcriptional profiling of single cells. Nat Commun 8:14049
31. Zappia L, Lun A (2021) zellkonverter: Conversion between scRNA-seq objects. R package version 1.4.0

Chapter 15

Integration of scATAC-Seq with scRNA-Seq Data

Ivan Berest and Andrea Tangherloni

Abstract

Single-cell studies are enabling our understanding of the molecular processes of normal cell development and the onset of several pathologies. For instance, single-cell RNA sequencing (scRNA-Seq) measures the transcriptome-wide gene expression at a single-cell resolution, allowing for studying the heterogeneity among the cells of the same population and revealing complex and rare cell populations. On the other hand, single-cell Assay for Transposase-Accessible Chromatin using sequencing (scATAC-Seq) can be used to define transcriptional and epigenetic changes by analyzing the chromatin accessibility at the single-cell level. However, the integration of multi-omics data still remains one of the most difficult tasks in bioinformatics. In this chapter, we focus on the combination of scRNA-Seq and scATACSeq data to perform an integrative analysis of the single-cell transcriptome and chromatin accessibility of human fetal progenitors.

Key words scRNA-seq, scATAC-seq, Multi-omics, Integrative analysis, Bioinformatics

1 Introduction

The increasing number of single-cell studies in the last decade has facilitated our understanding of the molecular processes of normal cell development, as well as the onset of several pathologies [1, 2]. Among the different approaches, single-cell RNA sequencing (scRNA-Seq) has been heavily applied to measure and study the transcriptome-wide gene expression at single-cell resolution, becoming the "Method of the Year" in 2013, enabling the possibility of investigating the heterogeneity among the cells of the same cell type [3]. In addition, novel cell types have been identified and known cell types have been redefined, thanks to the huge amount of scRNA-Seq data produced in the last years [4–6]. In order to better understand human health and diseases, studying the molecular intricacy and variations at multiple levels (i.e., genome, epigenome, transcriptome, proteome, and metabolome) is crucial [7], but it still represents one of the most difficult challenges [8]. This can be achieved at a single cell level, thanks to the rapidly increasing in throughput and quality of other single-cell omics, which can

Raffaele A. Calogero and Vladimir Benes (eds.), *Single Cell Transcriptomics: Methods and Protocols*,
Methods in Molecular Biology, vol. 2584, https://doi.org/10.1007/978-1-0716-2756-3_15,

provide novel and fundamental insights about the regulatory interactions of complex phenotype, by also linking the genotype to the phenotype [9]. In this chapter, we focus on the analysis and integration of transcriptomics and epigenetics data; in particular, on the integration of scRNA-Seq and single-cell Assay for Transposase-Accessible Chromatin using sequencing (scATAC-Seq) data. Before sequencing the cells, the cell populations of interest are generally isolated using specific marker panels, composed of a few cell surface proteins and clusters of differentiation molecules, previously identified by expert biologists manually or by automatic methods [10, 11]. The isolation of the desired cell populations using succinct marker panels still maintains the heterogeneity among cells of the same cell type, a characteristic that makes difficult the integration of multiple scRNA-Seq datasets [8]. The main problems that can arise during the integration of scRNA-Seq datasets, also containing the same cell types, are the technical non-negligible batch effects that may exist among the datasets [8, 12–15]. It is worth reminding that there exist mainly three different types of batches, which might be present during the integration of multiple experiments [14]. The first type of batch effect can arise when the same cell types are sequenced using different sequencing technologies. The second type of batch effect might appear when non-identical cell types are sequenced with both the same sequencing technology and different sequencing technologies, while the last type of batch effect is generally obtained when more than half a million cells are sequenced in the same experiment. Nowadays, a plethora of tools have been proposed to tackle this problem, each one with its own advantages and drawbacks, and most of them can be used for the integration of experiments maintaining cell-type separation. Note that a good batch-removal algorithm should map the cells of the same cell type into a single point, by maintaining the separability among the different cell types so that every cell type can be detected in a single cluster [13]. The interested reader is referred to [13, 14] for further details and a comparison among the existing approaches. Finally, a few approaches have been specifically developed to integrate different scRNA-Seq. Among them, it is worth mentioning Ingest [16], which was designed to integrate a new dataset into an annotated reference dataset capturing the biological variability of interest. Ingest projects the query dataset into the reference dataset by using a model fitted on the reference dataset (i.e., Principal Component Analysis (PCA) [17] or Uniform Manifold Approximation and Projection (UMAP) [18]). The authors of the Seurat package [19] proposed an approach that allows for mapping and annotating a query dataset [20] by finding similar cells (i.e., anchors) across the datasets. Starting from each anchor cell, the nearest anchors within the reference dataset and its pairs on the query dataset are determined, which are then used to integrate the query into the reference dataset. Finally, Scanorama [21] was proposed for data integration by creating a combined representation of multiple datasets. In order to merge similar cells from different datasets, Scanorama automatically finds and stitches shared cell types among the datasets. Specifically, Scanorama uses an approximated mutual nearest neighbor-matching strategy to find similar elements among the datasets while generating (or correcting) a cell-by-gene low-dimensional space expression matrix. The methods described above work well with horizontal integration of the scRNA-Seq (i.e., when one modality is profiled for batches of cells), but they are not necessarily well suited for the integration of multiple modalities. The development of new single-cell assays allows us to measure various cellular molecular phenotypes from all levels, from genotype to phenotype; however, a biological relevant integration of such multimodal data still remains a bottleneck in the field. Several recently published reviews summarized the current challenges and computational strategies for single-cell data integration [22–24]. To perform the integration of paired unmatched scRNA-Seq and scATAC-Seq, where the two modalities are generated independently from different cells, certain assumptions are required. As there is not a well-annotated feature space of regulatory elements, associated with the genes of the different cell types, one strategy is mapping the gene body or gene promoter chromatin accessibility to the respective gene expression, so that it is possible to work with the same set of features between the modalities. For example, using this assumption, we can use the previously described Seurat v3 method to pair unmatched datasets from different modalities based on anchors [20]. LIGER uses a similar gene-centric assumption to assign "metagenes" and integrate data applying a non-negative matrix factorization (iNMF) [25]. Similarly to these approaches, MAESTRO calculates the regulatory potential for each gene using scATAC-seq data by taking into account, for each chromatin peak, the exponential decay of the sequencing reads from the gene TSS (Transcription Start Site) and aligns the gene expression and regulatory potential in the Canonical Correlation Analysis (CCA) space [26]. Both Seurat v3 and MAESTRO first perform a CCA dimensionality reduction step and then identify mutual nearest neighbors, following the Mutual Nearest Neighbor (MNN) algorithm [27] between the modalities, treating them as different batches of one modality. Such approaches work better when there is high level of biological similarity between the unmatched datasets. However, these assumptions are not always correct because of the complex relationships between gene expression and gene regulation. Detailed information about cell-type specific networks of the gene expression regulation, modulated by enhancers and transcription factors, is needed to improve the accuracy of the unpaired scRNA-Seq and scATAC-Seq data integration. A comprehensive analysis and accumulation of the joint profiling of chromatin accessibility and gene expression from different cells (i.e., SHARE-seq [28], 10× Multiome) is essential for understanding the relationships between these two modalities.

2 Materials

A recently published single-cell multiomics dataset [29] composed of 18 fetal samples is used throughout this chapter to show how scRNA-Seq and scATAC-Seq can be analyzed and combined.

2.1 Exemplary Datasets

Ranzoni et al. recently published a study where scRNA-Seq and scATACSeq data were used to study the regulation of human fetal hematopoiesis [29]. The authors sequenced blood cells from liver and bone marrow of 18 different fetuses, which allowed them to define the transcriptional and epigenetic changes during blood differentiation. Fifteen fetuses were used to collect the scRNA-Seq datasets, while the remaining three fetuses were used to generate the scATAC-Seq datasets.

2.2 Computational Hardware

The following computational analyses were performed on a computer equipped with an Intel i7 CPU (2.7 GHz, 8 threads) and 32 GB of RAM. It is worth noting that such computational resources are well suited for our analysis (up to 5000 cells in both modalities); however, considering much more cells, more RAM is required, especially for the clustering and integration steps.

2.3 Computational Software

The downstream analysis of the scRNA-Seq datasets was mainly performed in Python (v.3.6.9) using the SCANPY package (v.1.4.5.1) [16] and scAEspy [30]. For the analysis of scATAC-seq data we used R (v.3.6.1) and, specifically, Seurat (v.3.1.4) [20], Signac (v.0.2.4) [31], and Harmony (v.0.1.0) [32] R packages.

2.4 Data Availability

The fastq files and the cell assignment of the scRNA-Seq data can be found in ArrayExpress with accession code E-MTAB-9067, while the fastq files and cell assignment of the scATAC-Seq data can be downloaded from ArrayExpress with accession code E-MTAB-9068. The GitLab repository https://gitlab.com/cvejic-group/integrative-scrna-scatac-human-foetal contains the count matrices of the 15 scRNA-Seq experiments, as well as the Scanpy and Seurat objects.

3 Methods

The library preparation of the scRNA-Seq experiments was obtained following the smart-Seq2 method [33], with some of the modifications proposed in [34]. Bioanalyzer (Agilent) was used to assess the quality of libraries, and the good-quality ones were then quantified with KAPA Library Quantification Kit (Roche), and finally submitted for sequencing. These libraries were multiplexed with Nextera Index sets A, B, C, and D (v.2, Illumina) and sequenced on a HiSeq4000 and a NovaSeq6000 (Illumina), in pair-end mode. The quality of the demultiplex fastq files was checked, and then the fastq files were aligned and quantified using a scRNA-Seq upstream analysis pipeline based on STAR (v.2.5.4a) [35] with default para meters. The Ensembl release 91 of the GRCh38 human reference genome was used for the index and annotation. STAR was also used with the option quantMode GeneCounts to quantify both the transcript and gene counts. Regarding the scATAC-Seq experiments, the library preparation was done following the method described by Chen et al. [36], the library traces were evaluated with Bioanalyzer, and then these libraries were sequenced using a HiSeq4000 in paired-end mode. samtools (v1.9) [37], bedtools (v2.27.1) [38], Picard tools [39], and BWA (v0.7.17) [40] were used for the upstream analysis of the scATAC-Seq data. The fastq files were aligned to the Ensembl release 91 of the GRCh38 human reference genome. For each cell, the duplicates were marked using the MarkDuplicates function of the Picard tools, and then removed using samtools view with -F 1804 parameter. The bam files were transformed into bed files using bamtobed bedtools function in bedpe mode, and only the fragments that are no bigger than 1000 bp were kept. Following the SnapATAC strategy [41], only the clusters with more than 50 cells were considered. For each cell, macs2 [42] with parameters --nomodel --shift 100 --ext 200 --qval 5e-2 -B was called to obtain the peaks.

3.1 Integration of Multiple scRNA-Seq Datasets

We used the following workflow to integrate the 15 scRNA-Seq experiments.

1. A quality control (QC) step, based on the number of identified counts, number of expressed genes, and the fraction of mitochondrial genes, for each experiment independently. The following listing shows the steps required to perform the QC of a scRNA-Seq dataset. We assume here that the counts are saved on a tab-separated value (TSV) file, where each row represents a gene and each column a cell.

```
import pandas as pd
import scanpy as sc

rawData = pd.read_csv("Counts.tsv", delimiter="Ät")
adata = sc.AnnData(rawData.T)

adata.var['mt'] = adata.var_names.str.startswith('MT-')
sc.pp.calculate_qc_metrics(adata, qc_vars=['mt'],
                           percent_top=None,
                           log1p=False,
                           inplace=True)

min_counts = 10000
max_counts = 1750000
min_genes = 200
max_genes = 9000
max_pctMito = 0.4

sc.pp.filter_cells(adata, min_counts = min_counts)
sc.pp.filter_cells(adata, max_counts = max_counts)
sc.pp.filter_cells(adata, min_genes = min_genes)
sc.pp.filter_cells(adata, max_genes = max_genes)
adata = adata[adata.obs['percent_mito'] < max_pctMito]
sc.pp.filter_genes(adata, min_cells=1)
```

2. The filtered count gene expression matrices were combined to obtain a single count gene expression matrix. The following listing concatenates the AnnData objects of the scanpy package to merge all the scRNA-Seq experiments.

```
merged = adata1.concatenate(adata2, join='outer')
merged = merged.concatenate(adata3, join='outer')
merged = merged.concatenate(adata4, join='outer')
merged = merged.concatenate(adata5, join='outer')
merged = merged.concatenate(adata6, join='outer')
merged = merged.concatenate(adata7, join='outer')
merged = merged.concatenate(adata8, join='outer')
merged = merged.concatenate(adata9, join='outer')
merged = merged.concatenate(adata10, join='outer')
merged = merged.concatenate(adata11, join='outer')
merged = merged.concatenate(adata12, join='outer')
merged = merged.concatenate(adata13, join='outer')
merged = merged.concatenate(adata14, join='outer')
merged = merged.concatenate(adata15, join='outer')
```

3. The genes expressed in less than 10 cells were removed, as shown in the following listing.

```
sc.pp.filter_genes(merged, min_cells=10)
```

4. A data normalization was applied to normalize the cells. Specifically, each cell was normalized by the total counts over all genes so that all cells had the same total counts after the normalization. Then, the normalized expression was log-transformed. As depicted in the following listing, firstly the counts are added to a new layer named counts; then, the expression of cells is normalized and log-transformed. Afterward, the log-transformed values are saved in the .raw attribute of the AnnData object. Finally, these values are also added to a new layer named log.

```
merged.layers["counts"] = merged.X
sc.pp.normalize_per_cell(merged,
 counts_per_cell_after=1e4)
sc.pp.log1p(merged)
merged.raw = merged
merged.layers["log"] = merged.raw.X
```

5. Detection of the top Highly Variable Genes (HVGs), which were selected separately within each batch and then merged. Note that each batch corresponded to a specific experiment. As shown in the listing, we calculate the top 1000 HVGs by selecting them separately within each batch. We assume that, for each cell, there is a variable named batch indicating the experiment.

```
sc.pp.highly_variable_genes(merged,
 n_top_genes=1000,
 batch_key="batch")
```

6. A dimensionality reduction step was performed using autoencoders (AEs), provided by scAEspy [30]. Specifically, an AE was used on the HVG space by considering the counts of the selected HVGs. The listing below shows how to select the counts of the HVGs. Then, the parameter settings of scAEspy are set to use the Gaussian-Mixture Maximum Mean Discrepancy AE (GMMMDAE). In such a case, we trained the GMMMDVAE for 100 epochs with a batch size equal to 100, a single hidden layer composed of 64 neurons, a latent space of 16 neurons, 15 different Gaussian distributions, learnable prior distribution, constrained Poisson as loss function, and sigmoid activation function. In addition, the Adam optimizer [43] was used for the optimization step because it was

specifically designed for first-order gradient-based optimization of stochastic objective functions. Finally, we generated the latent space representation of the cells using the function latentRepresentation and saved it into the field X_latent of the AnnData object.

```
HVGs = merged.var['highly_variable']
data = merged[:, HVGs].layers["counts"]

original_dim = data.shape[1]
hidden_layers = [64]
latent_space = 16
num_gaussians = 15
learnable_prior = True
rec_loss = "Poisson"
constrained = True
activation ='sigmoid'

alpha = 0
lambd = 2

GMMMDVAE = scAEspy(original_dim,
 hidden_layers=hidden_layers,
 latent_layer=latent_space,
 learnable_prior=learnable_prior,
 constrained=constrained,
 num_gaussians=num_gaussians,
 activation=activation,
 rec_loss=rec_loss,
 alpha=alpha,
 lambd=lambd,
 seed=42,
 verbose=True)

GMMMDVAE.build()
GMMMDVAE.train(data, data, epochs=100, batch_size=100)

cellsLatent = GMMMDVAE.latentRepresentation(data)
merged.obsm['X_latent'] = cellsLatent
```

7. In order to remove the possible batch effects among the samples, the batch-balanced *k* nearest neighbors (BBKNN) approach [44] was applied on the latent space generated by GMMMDVAE. It is worth mentioning that each batch corresponded to a specific experiment. The following listing explains how BBKNN can be applied to the latent space generated by an

AE to correct the batch effects. As mentioned before, we assume that, for each cell, there is a variable named batch indicating the experiment.

```
n_pcs = merged.obsm["X_latent"].shape[1]

bbknn.bbknn(merged,
 n_pcs = n_pcs,
 batch_key = "batch",
 use_rep = "X_latent",
 metric = "euclidean",
 approx = False,
 use_faiss = False,
 trim = None)
```

After the integration of the 15 experiments, the cells were clustered using the Leiden algorithm, applied to the neighborhood graph generated by BBKNN. The obtained clusters were manually annotated by considering well-known cell-type specific genes, as well as Differentially Expressed Genes (DEGs), which were calculated using the Wilcoxon rank-sum (i.e., Mann–Whitney U test) test with the Bonferroni correction [45–47] to adjust the *p*-values for multiple testing. In such a case, the Wilcoxon rank-sum test is used to compare each cluster to the union of all the other clusters. Different progenitor populations were identified, including hematopoietic stem cell/multipotent progenitors (HSC/MPPs), cycling HSC/MPPs (HSC/MPPs-cycle), lympho-myeloid progenitors (LMPs), megakaryocyte erythroid-mast progenitors (MEMPs), cycling MEMPs (MEMPs-cycle), and granulocytic progenitors (GPs). Figure 1 shows the UMAP [18] spaces calculated from the latent space generated by GMMMDVAE (left panels) and calculated from the corrected neighborhood graph generated by BBKNN (right panels). These plots clearly show that BBKNN allowed for removing the batch effects, correctly intermixing the samples containing the same phenotypically defined cell types. In this figure, we also reported the CD34+ CD38- non-committed cells that were used for the integration of scRNA-Seq and scATAC-Seq data.

3.2 Integration of scRNA-Seq and scATAC-Seq Datasets

We used the following code to integrate the 3 scATAC-seq fetal samples and to annotate them based on the reference CD34+ CD38- non-committed cells from the 15 scRNA-Seq experiments.

3.2.1 Quality Control Analysis and scATAC-Seq Datasets Merging

For the downstream analysis of scATAC-seq data, we used a consensus peakset of 152,282 peaks generated across all the three samples. We combined all samples into a single BAM file and transformed it to the corresponding BED format, which contained the fragments of the open chromatin linked to each individual cell.

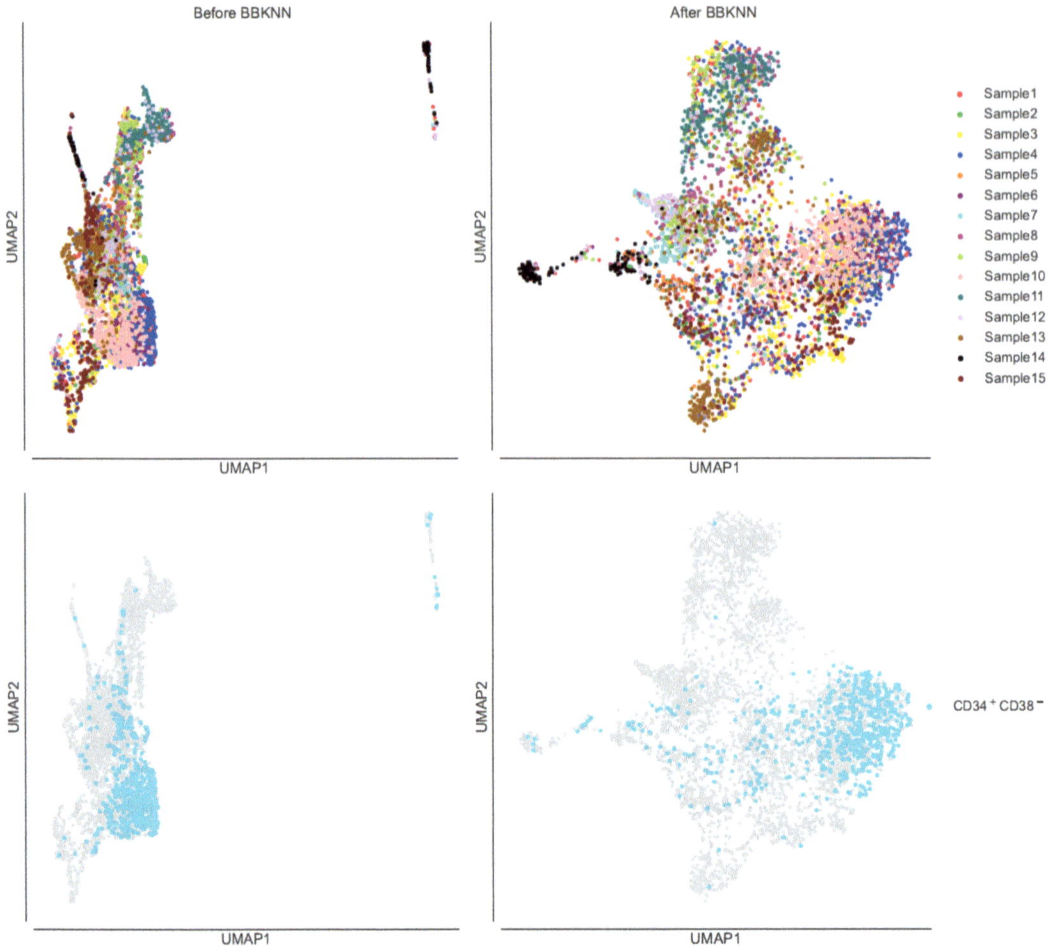

Fig. 1 UMAP plots showing the 15 scRNA-Seq samples and the CD34+ CD38-fetal non-committed cells before (left panels) and after (right panels) the application of BBKNN

1. We started with importing our scATAC data in R as a Seurat object using the CreateSeuratObject() function. It is important to note that starting from Signac (v.1.0.0), the CreateChromatinAssay() function should be used before the CreateSeuratObject() function for the scATAC data. To perform the following analyses, we binarized the counts per cell using Signac BinarizeCounts() function.

```
library(tidyverse)
library(GenomicRanges)
library(Seurat)
library(Signac)
library(BSgenome.Hsapiens.UCSC.hg38)
# read in corresponding metadata
metadata <- read_tsv("ATAC.metadata.tsv", col_names = T)
```

```
# create peak-by-cell matrix
peaks.df <- read_tsv("consenusPeaks.bed",
col_names <- c("chr","start","end","score","strand"))
peaks.GR <- makeGRangesFromDataFrame(peaks.df)
fragment.path <- "merged.fragments.sort.bed.gz"
peak_matrix <- FeatureMatrix(fragments = fragment.path,
 features = peaks.GR)
# create genomic bin-by-cell matrix
genome <- BSgenome.Hsapiens.UCSC.hg38
seq.gen <- seqlengths(genome)
bin_matrix <- FeatureMatrix(fragments = fragment.path,
genome = seq.gen, binsize = 5000, chunk = 100)
# create Seurat object
merged <- CreateSeuratObject(counts = peak_matrix,
 assay = 'peaks',
 project = 'merged',
 min.cells = 1,
 meta.data = metadata)
merged[["bins"]] <- CreateAssayObject(counts = bin_matrix)
merged <- SetFragments(object = merged, file = fragment.path)
 merged <- BinarizeCounts(object = merged, assay = c
("peaks","bins"))
```

2. To filter out low-quality cells from our data, we computed the ATAC-specific quality measures, such as FRiP (fraction of fragments in peaks), percentage of reads in ENCODE blacklisted regions [48], nucleosome signal, and TSS enrichment. The nucleosome signal highlights how strong is the nucleosomal pattern (fragments with a periodicity of 200 bp) based on the fragment size of each cell. In essence, this is the ratio of mono-nucleosomal to nucleosome-free fragments. TSS enrichment is based on the number of fragments located at the TSS sites of the genes of interest (in our case protein coding) compared to the reads at the TSS flanking sites.

```
# compute fraction of fragments in the peaks
merged <- FRiP(object = merged,
 peak.assay = "peaks", bin.assay = "bins")
# compute ratio of reads in hg38 blacklisted regions
merged$blacklist_ratio <- FractionCountsInRegion(object =
merged,
 assay = 'peaks', regions = blacklist_hg38)
# compute nucleosome signal
merged <- NucleosomeSignal(object = merged)
merged$nucleosome_group <- ifelse(merged$nucleosome_signal >
10,
 'NS > 10', 'NS < 10')
```

```
# add TSS annotation of protein coding genes
library(EnsDb.Hsapiens.v86)
gene.ranges <- genes(EnsDb.Hsapiens.v86,
 filter = GeneBiotypeFilter("protein_coding"))
tss.ranges <- GRanges(seqnames = seqnames(gene.ranges),
 ranges = IRanges(start = start(gene.ranges), width = 2),
 strand = strand(gene.ranges))
seqlevelsStyle(tss.ranges) <- 'UCSC'
tss.ranges <- keepStandardChromosomes(tss.ranges,
 pruning.mode = 'coarse')
# compute TSS enrichment score
merged <- TSSEnrichment(object = merged, tss.positions = tss.
ranges)
merged$high.tss <- ifelse(merged$TSS.enrichment > 2, 'High',
'Low')
```

We filtered out cells with less than 1000 detected peaks and an overall FRiP lower than 50%. Additionally, we expected the final set of cells to have less than 5% of fragments in the hg38 blacklisted regions, as well as the nucleosome signal, for each cell, to be higher than 10. Finally, we proceeded with the cells having a TSS enrichment score higher than 2 (i.e., containing at least twice the number of fragments overlapping the TSS in comparison to the flanking regions 2000 bp away), ending up with 3, 611 cells.

```
merged.filt = subset(merged,
 subset = nFeature_peaks > 1000 &
 FRiP > 0.5 &
 blacklist_ratio < 0.05 &
 nucleosome_signal < 10 &
 TSS.enrichment > 2)
```

3. After the preprocessing of the scATAC data, we observed strong batch effects in the data (see the left panel of Fig. 2). Thus, we decided to apply the Harmony algorithm [32] to remove the batch effects associated with the sample preparation, sequencing batch, and biological origin of data (i.e., femur or liver). To run Harmony, we firstly normalized the peak data using TF-IDF [49], and then we used all the peaks and ran the Latent Semantic Indexing (LSI) dimensionality reduction approach. We excluded the first LSI component as it was highly correlated with the sequencing depth.

```
library(harmony)
merged.filt <- merged.filt %>%
 RunTFIDF( method = 2) %>%
```

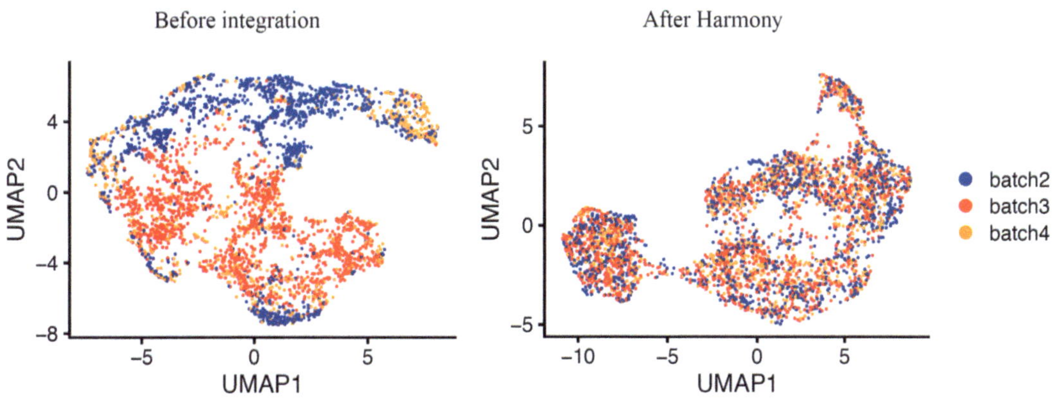

Fig. 2 UMAP plots highlighting the effect of the batch removal from the scATAC data

```
FindTopFeatures( min.cutoff = 'q0') %>%
RunSVD(assay = 'peaks', reduction.name = 'lsi')
# remove 1st LSI dimension
lsi.dim = 2:50
merged.int <- harmony::RunHarmony(object = merged.filt,
 group.by.vars = c("SAMPLE","batch","origin"),
 reduction = 'lsi', assay.use = 'peaks',
 project.dim = FALSE, dims.use = lsi.dim,
 max.iter.harmony = 20, max.iter.cluster = 200,
 sigma = 0.25,theta = c(2,4,4))
```

Then, we applied the RunHarmony() function to our Seurat object, using the 49 LSI dimensions as input. We expected to have around 20 Harmony iterations setting the σ parameter to 0.25. We increased σ (default: 0.1) to expand the distance between the cells and centroids for a softer k-means clustering. We also increased the value of the θ (default: 2) parameter to 4 for the sequencing batches and biological origin of the data, as we observed stronger batch effects for them. Ultimately, we were able to successfully remove unwanted variations from the data (see the right panel of Fig. 2).

3.2.2 Annotating scATAC Data Based on the Shared scRNA-Seq

1. For the integration of the two modalities, we followed the methodology described by Stuart et al. [20], finding the anchors between the gene expression and chromatin-based gene activity, assigning the annotations to the chromatin data based on the clustering of scRNA-Seq. We started by generating the gene activity matrix, counting fragments overlapping

the gene bodies (Ensembl 86 release) extended upstream and downstream by 3000 bp. We tested several extension sizes but we decided to use a 3 kilobase extension as it showed the highest prediction score per cell on average (*see* Fig. 3b) after the integration. We normalized the gene activity data using the LogNormalize strategy from the Seurat NormalizeData() function, taking the median of the gene activity counts as scale. factor.

```
library(EnsDb.Hsapiens.v86)
# assign gene body coordinates and extend them
gene.coords <- genes(EnsDb.Hsapiens.v86,
 filter = ~gene_biotype == "protein_coding")
seqlevelsStyle(gene.coords) <- 'UCSC'
ext <- 3000
genebodyExt.coords <- Extend(x = gene.coords,
 upstream = ext, downstream = ext)
genebodyExt.coords <- keepStandardChromosomes(genebodyExt.co-
ords,
 pruning.mode = 'coarse')
# compute gene activities based on ATAC data
fragment.path <- "merged.fragments.sort.bed.gz"
gene.activities <- FeatureMatrix(fragments = fragment.path,
 features = genebodyExt.coords, chunk = 20,
 cells = rownames(merged.int@meta.data))
# change rownames from regions to gene names
gene.key <- genebodyExt.coords$gene_name
names(gene.key) <- GRangesToString(grange = genebodyExt.co-
ords)
rownames(gene.activities) <- gene.key[rownames(gene.activ-
ities)]
gene.activities <- gene.activities[unique(rownames(gene.ac-
tivities)),]
```

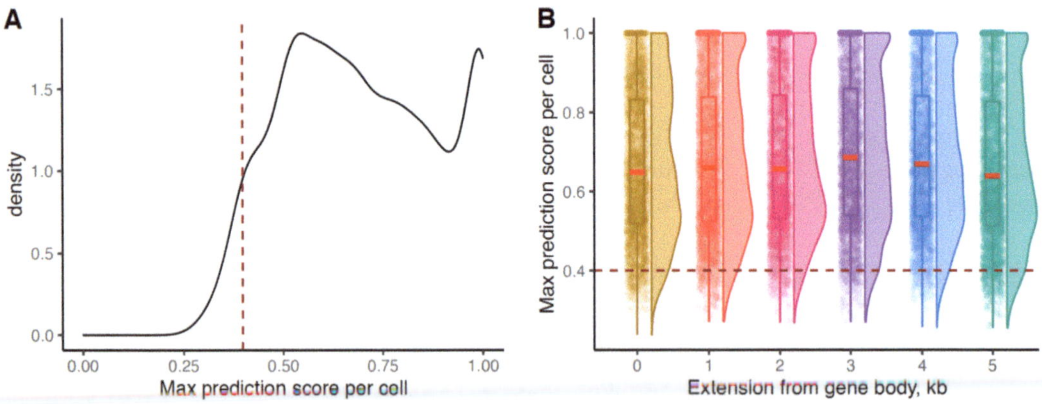

Fig. 3 The prediction scores vary depending on the gene body extension: (**a**) distribution of predictions scores based on the 3 kb upstream and downstream extended gene body integration; (**b**) distributions of the prediction scores for tested gene body extension sizes

```
# create new assay in Seurat object
assayName <- paste0("GeneAct_",ext,"bp")
merged.int[[assayName]] <- CreateAssayObject(counts = gene.
activities)
# normalize gene activity data
merged.int <- NormalizeData(object = merged.int, assay =
assayName,
 normalization.method = 'LogNormalize',
scale.factor = median(merged.int@meta.data[,paste0("nFea-
ture_",assayName)]))
```

2. As reference data, we used the previously scRNA-Seq dataset, described in Subheading 3.1. Importantly, we took for annotation only the CD38- fetal cells and cell types that have at least 20 cells in the defined subset of cells, ending up with 511 cells and 6 distinct cell types (i.e., HSC/MPPs, HSC/MPPs-cycle, MEMPs, MEMPs-cycle, GPs, and LMPs). Afterwards, we used the FindTransferAnchors() function for both the scRNA-Seq and the gene activity scATAC-Seq data to find anchor points (features with a shared variation) between the modalities in a shared latent space using CCA. To predict the cell-type labels for the gene activity scATAC data, we used the TransferData() function, taking as input the cell-type annotation from the scRNA-Seq data and the specified Harmony dimensionality reduction for the calculation of anchor weights. We then assigned the nearest cell anchors to the number of detected cell types in the reference scRNA-Seq data (k.weight = 6), as we observed that increasing this parameter does not improve the prediction accuracy, but rather it led to a decreased variability of the predicted cell types (i.e., most of the cells were annotated as HSC/MPPs). We were able to annotate 95% of cells that have prediction score higher than 0.4 (*see* Fig. 3a). Overall, as shown in Fig. 4b, we observed similar proportions of cell types in scATAC-Seq data compared to reference scRNA-Seq data, with a clear separation between MEMPs, LMPs, and GPs on the UMAP space (*see* Fig. 4a).

```
# load RNA dataset
merged.RNA = readRDS("RNA_all_samples.Rds")
# subset CD38- cell types with more than 20 cells
cellsRNA = merged@meta.data[which(merged$gate == "38-"),"an-
notation_merged"]
celltypes = names(table(cellsRNA))[table(cellsRNA) > 20]
mergedRNA.filt = subset(merged.RNA, subset = gate == "38-" &
 annotation_merged %in% celltypes)
# find anchors between RNA and ATAC gene activities
transfer.anchors <- FindTransferAnchors(reference = mergedRNA,
 query = merged.int,reduction = 'cca',
```

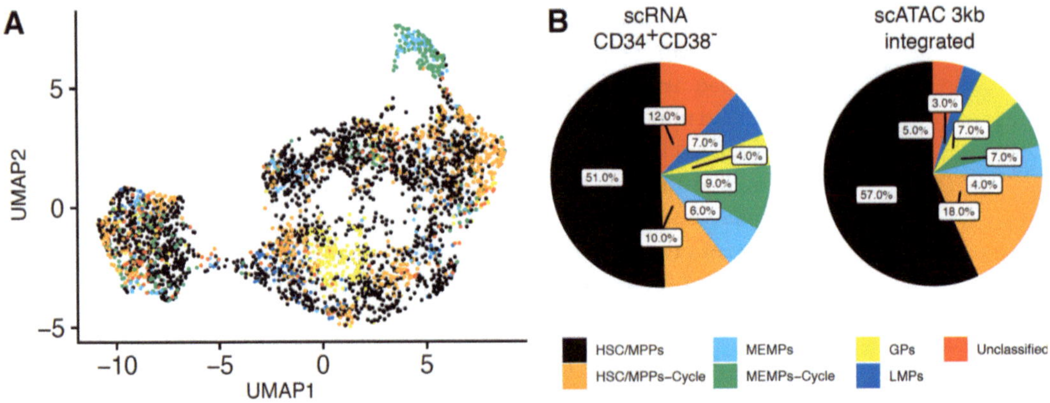

Fig. 4 Annotated scATAC-Seq data: (**a**) UMAP plot of the 3611 annotated cells from scATAC-Seq; (**b**) pie charts visualizing proportions of the cell types in the reference scRNA-Seq dataset (left) and the integrated scATAC-Seq data (right)

```
 query.assay = assayName,verbose = T,
 features = rownames(test[[assayName]]))
# predict cell types annotation of the ATAC data
predicted.labels <- TransferData(anchorset = transfer.anchors,
 refdata = mergedRNA.filt$annotation_merged,
 weight.reduction = merged.int[['harmony']],
 dims = 1:50, k.weight = 6)
# assign cells not passing prediction score threshold
predictions.df = as.data.frame(predicted.labels, keep.row-
names = T)
predictions.df$id = ifelse(predictions.df$prediction.score.
max < 0.4,
 "notsign", predictions.df$predicted.id)
rownames(predictions.df) = predictions.df$rn
# add prediction results to the ATAC Seurat object
merged.int <- AddMetaData(object = merged.int, metadata =
predictions.df)
```

References

1. Gladka MM, Molenaar B, De Ruiter H, Van Der Elst S, Tsui H, Versteeg D, Lacraz GPA, Huibers MMH, Van Oudenaarden A, Van Rooij E (2018) Single-cell sequencing of the healthy and diseased heart reveals cytoskeleton-associated protein 4 as a new modulator of fibroblasts activation. Circulation 138(2): 166–180
2. Keren-Shaul H, Spinrad A, Weiner A, Matcovitch-Natan O, Dvir-Szternfeld R, Ulland TK, David E, Baruch K, Lara-Astaiso D, Toth B et al (2017) A unique microglia type associated with restricting development of Alzheimer's disease. Cell 169(7):1276–1290
3. Paul F, Arkin Y'a, Giladi A, Jaitin DA, Kenigsberg E, Keren-Shaul H, Winter D, Lara-Astiaso D, Gury M, Weiner A et al (2015) Transcriptional heterogeneity and lineage commitment in myeloid progenitors. Cell 163(7):1663–1677

4. Baron M, Veres A, Wolock SL, Faust AL, Gaujoux R, Vetere A, Ryu JH, Wagner BK, Shen-Orr SS, Klein AM et al (2016) A single-cell transcriptomic map of the human and mouse pancreas reveals inter-and intra-cell population structure. Cell Syst 3(4):346–360
5. Villani A-C, Satija R, Reynolds G, Sarkizova S, Shekhar K, Fletcher J, Griesbeck M, Butler A, Zheng S, Lazo S et al (2017) Single-cell RNA--seq reveals new types of human blood dendritic cells, monocytes, and progenitors. Science 356(6335):eaah4573
6. Kernfeld EM, Genga RMJ, Neherin K, Magaletta ME, Xu P, Maehr R (2018) A single-cell transcriptomic atlas of thymus organogenesis resolves cell types and developmental maturation. Immunity 48(6):1258–1270
7. Subramanian I, Verma S, Kumar S, Jere A, Anamika K (2020) Multi-omics data integration, interpretation, and its application. Bioinform Biol Insights 14:1177932219899051
8. Lahnemann D, Koster J, Szczurek E, McCarthy DJ, Hicks SC, Robinson MD, Vallejos CA, Campbell KR, Beerenwinkel N, Mahfouz A et al (2020) Eleven grand challenges in single-cell data science. Genome Biol 21(1): 1–35
9. Forcato M, Romano O, Bicciato S (2021) Computational methods for the integrative analysis of single-cell data. Brief Bioinform 22(1):20–29
10. Janeway CA, Donald Capra J, Travers P, Walport M (1999) Immunobiology: the immune system in health and disease. Taylor & Francis Group, Abingdon-on-Thames
11. Tangherloni A, Riva SG, Spolaor S, Besozzi D, Nobile MS, Cazzaniga P (2021) The impact of representation on the optimization of marker panels for single-cell RNA data. In: Proceedings of the IEEE congress on evolutionary computation. IEEE, pp 1423–1430
12. Luecken MD, Theis FJ (2019) Current best practices in single-cell RNAseq analysis: a tutorial. Mol Syst Biol 15(6):e8746
13. Luecken MD, Buttner M, Chaichoompu K, Danese A, Interlandi M, Mueller MF, Strobl DC, Zappia L, Dugas M, Colome-Tatche M et al (2020) Benchmarking atlas-level data integration in single-cell genomics. BioRxiv
14. Tran HTN, Ang KS, Chevrier M, Zhang X, Lee NYS, Goh M, Chen J (2020) A benchmark of batch-effect correction methods for single-cell RNA sequencing data. Genome Biol 21(1): 1–32
15. Leek JT, Scharpf RB, Bravo HC, Simcha D, Langmead B, Johnson WE, Geman D, Baggerly K, Irizarry RA (2010) Tackling the widespread and critical impact of batch effects in highthroughput data. Nat Rev Genet 11(10):733–739
16. Alexander Wolf F, Angerer P, Theis FJ (2018) SCANPY: large-scale single-cell gene expression data analysis. Genome Biol 19(1):15
17. Wold S, Esbensen K, Geladi P (1987) Principal component analysis. Chemom Intell Lab Syst 2(1–3):37–52
18. Becht E, McInnes L, Healy J, Dutertre C-A, Kwok IWH, Ng LG, Ginhoux F, Newell EW (2019) Dimensionality reduction for visualizing single-cell data using UMAP. Nat Biotechnol 37(1):38
19. Satija R, Farrell JA, Gennert D, Schier AF, Regev A (2015) Spatial reconstruction of single-cell gene expression data. Nat Biotechnol 33(5):495
20. Stuart T, Butler A, Hoffman P, Hafemeister C, Papalexi E, Mauck WM, Hao Y, Stoeckius M, Smibert P, Satija R (2019) Comprehensive integration of single-cell data. Cell 177(7): 1888–1902.e21
21. Hie B, Bryson B, Berger B (2019) Efficient integration of heterogeneous single-cell transcriptomes using scanorama. Nat Biotechnol 37(6):685–691
22. Forcato M, Romano O, Bicciato S (2020) Computational methods for the integrative analysis of single-cell data. Brief Bioinform 22(3):bbaa042
23. Argelaguet R, Cuomo ASE, Stegle O, Marioni JC (2021) Computational principles and challenges in single-cell data integration. Nat Biotechnol 39(10):1202–1215
24. Miao Z, Humphreys BD, McMahon AP, Kim J (2021) Multi-omics integration in the age of million single-cell data. Nat Rev Nephrol 17(11):710–724
25. Liu J, Gao C, Sodicoff J, Kozareva V, Macosko EZ, Welch JD (2020) Jointly defining cell types from multiple single-cell datasets using LIGER. Nat Protoc 15(11):3632–3662
26. Wang C, Sun D, Huang X, Wan C, Li Z, Han Y, Qin Q, Fan J, Qiu X, Xie Y, Meyer CA, Brown M, Tang M, Long H, Liu T, Liu XS (2020) Integrative analyses of single-cell transcriptome and regulome using MAESTRO. Genome Biol 21(1):198
27. Haghverdi L, Lun ATL, Morgan MD, Marioni JC (2018) Batch effects in single-cell RNA-sequencing data are corrected by matching mutual nearest neighbors. Nat Biotechnol 36(5):421–427
28. Ma S, Zhang B, LaFave LM, Earl AS, Chiang Z, Yan H, Ding J, Brack A, Kartha VK, Tay T, Law T, Lareau C, Hsu Y-C, Regev A, Buenrostro JD (2020) Chromatin potential identified by shared single-cell

profiling of RNA and chromatin. Cell 183(4): 1103–1116.e20

29. Ranzoni AM, Tangherloni A, Berest I, Riva SG, Myers B, Strzelecka PM, Xu J, Panada E, Mohorianu I, Zaugg JB, Cvejic A (2021) Integrative single-cell RNA-seq and ATACseq analysis of human developmental hematopoiesis. Cell Stem Cell 28(3):472–487.e7
30. Tangherloni A, Ricciuti F, Besozzi D, Liò P, Cvejic A (2021) Analysis of single-cell RNA sequencing data based on autoencoders. BMC Bioinform 22(1):1–27
31. Stuart T, Srivastava A, Madad S, Lareau CA, Satija R (2021) Single-cell chromatin state analysis with signac. Nat Methods 18(11): 1333–1341
32. Korsunsky I, Millard N, Fan J, Slowikowski K, Zhang F, Wei K, Baglaenko Y, Brenner M, Loh PR, Raychaudhuri S (2019) Fast, sensitive and accurate integration of single-cell data with harmony. Nat Methods 16(12):1289–1296
33. Picelli S, Faridani OR, Bjorklund AK, Winberg G, Sagasser S, Sandberg R (2014) Full-length rna-seq from single cells using smart-seq2. Nat Protoc 9(1):171–181
34. Macaulay IC, Svensson V, Labalette C, Ferreira L, Hamey F, Voet T, Teichmann SA, Cvejic A (2016) Single-cell RNA-sequencing reveals a continuous spectrum of differentiation in hematopoietic cells. Cell Rep 14(4): 966–977
35. Dobin A, Davis CA, Schlesinger F, Drenkow J, Zaleski C, Jha S, Batut P, Chaisson M, Gingeras TR (2013) Star: ultrafast universal rna-seq aligner. Bioinformatics 29(1):15–21
36. Chen X, Miragaia RJ, Natarajan KN, Teichmann SA (2018) A rapid and robust method for single cell chromatin accessibility profiling. Nat Commun 9(1):1–9
37. Li H, Handsaker B, Wysoker A, Fennell T, Ruan J, Homer N, Marth G, Abecasis G, Durbin R (2009) The sequence alignment/map format and samtools. Bioinformatics 25(16): 2078–2079
38. Quinlan AR (2014) Bedtools: the swiss-army tool for genome feature analysis. Curr Protoc Bioinform 47(1):11–12
39. Broad Institute (2018) Picard tools. http://broadinstitute.github.io/picard/
40. Li H, Durbin R (2009) Fast and accurate short read alignment with Burrows–Wheeler transform. Bioinformatics 25(14):1754–1760
41. Fang R, Preissl S, Yang L, Hou X, Lucero J, Wang X, Motamedi A, Shiau AK, Zhou X, Xie F et al (2021) Comprehensive analysis of single cell ATAC-seq data with snapatac. Nat Commun 12(1):1–15
42. Zhang Y, Liu T, Meyer CA, Eeckhoute J, Johnson DS, Bernstein BE, Nusbaum C, Myers RM, Brown M, Li W et al (2008) Model-based analysis of ChiP-seq (MACS). Genome Biol 9(9):1–9
43. Kingma DP, Ba J (2014) Adam: a method for stochastic optimization. arXiv preprint arXiv:1412.6980
44. Polanski K, Young MD, Miao Z, Meyer KB, Teichmann SA, Park J-E (2020) BBKNN: fast batch alignment of single cell transcriptomes. Bioinformatics 36(3):964–965
45. Mann HB, Whitney DR (1947) On a test of whether one of two random variables is stochastically larger than the other. Ann Math Stat 18:50–60
46. Wilcoxon F (1992) Individual comparisons by ranking methods. In: Breakthroughs in statistics. Springer, pp 196–202
47. Dunn OJ (1961) Multiple comparisons among means. J Am Stat Assoc 56(293):52–64
48. Amemiya HM, Kundaje A, Boyle AP (2019) The ENCODE blacklist: identification of problematic regions of the genome. Sci Rep 9(1): 9354
49. Cusanovich DA, Hill AJ, Aghamirzaie D, Daza RM, Pliner HA, Berletch JB, Filippova GN, Huang X, Christiansen L, DeWitt WS, Lee C, Regalado SG, Read DF, Steemers FJ, Disteche CM, Trapnell C, Shendure J (2018) A single-cell atlas of in vivo mammalian chromatin accessibility. Cell 174(5):1309–1324.e18

Chapter 16

Using "Galaxy-rCASC": A Public Galaxy Instance for Single-Cell RNA-Seq Data Analysis

Pietro Mandreoli, Luca Alessandri, Raffaele A. Calogero, Marco Antonio Tangaro, and Federico Zambelli

Abstract

rCASC is a modular workflow providing an integrated environment for single-cell RNA-seq (scRNA-Seq) data analysis exploiting Docker containers to achieve functional and computational reproducibility. It was initially developed as an R package usable also through a Java GUI. However, the Java frontend cannot be employed when running rCASC on a remote server, a typical setup due to the significant computational resources commonly needed to analyze scRNA-Seq data.

To allow the use of rCASC through a graphical user interface on the client side and to harness the many advantages provided by the Galaxy platform, we have made rCASC available as a Galaxy set of tools, also providing a dedicated public instance of Galaxy named "Galaxy-rCASC." To integrate rCASC into Galaxy, all its functions, originally implemented as a set of Docker containers to maximize reproducibility, have been extensively reworked to become independent from the R package functions that launch them in the original implementation. Furthermore, suitable Galaxy wrappers have been developed for most functions of rCASC. We provide a detailed reference document to the use of Galaxy-rCASC with insights and explanations on the platform functionalities, parameters, and output while guiding the reader through the typical rCASC analysis workflow of a scRNA-Seq dataset

Key words Docker, Galaxy, Reproducibility, scRNA-Seq

1 Introduction

1.1 Single-Cell RNA-Seq and Galaxy-rCASC

Single-cell sequencing is seen nowadays as an essential tool to investigate cellular heterogeneity and identify cell subpopulation-specific signatures [1]. Several bioinformatics approaches are now available to researchers as software packages for the analysis of datasets produced using techniques such as single-cell RNA-seq [2] (scRNA-Seq), ATAC-seq [3, 4] (scATAC-Seq), and others (e.g., scChIP-Seq [5], scMethyl-HiC [6]). However, relatively few computational workflows offer analysis flexibility and ensure long-term computational reproducibility through a user-friendly environment. One of those is rCASC [7–9], which has been

Raffaele A. Calogero and Vladimir Benes (eds.), *Single Cell Transcriptomics: Methods and Protocols*,
Methods in Molecular Biology, vol. 2584, https://doi.org/10.1007/978-1-0716-2756-3_16,

implemented as a modular workflow that aims to simplify and automate scRNA-Seq data analysis for biologists. While rCASC can be utilized through a Java interface, this frontend is not available when using the workflow on a remote server, a pretty common setup due to the significant computational resources required to run a scRNA-Seq analysis. Therefore, integrating rCASC into the Galaxy [10] workflow manager by developing Galaxy-rCASC has been a straightforward solution to increase its accessibility and improve usability. In fact, Galaxy provides a user-friendly and intuitive web-based GUI and, at the same time, makes available several convenient functionalities that further enhance the reproducibility-oriented features of rCASC. Galaxy-rCASC implements a scRNA-Seq analysis workflow that, starting from the sequencing *fastq* files, leads to cell clustering and feature selection, allowing the identification of (sub)populations and their signature key genes. Finally, an additional advantage provided by Galaxy-rCASC is the direct access to our computational resources, adequate to handle several users simultaneously.

In summary, Galaxy-rCASC allows performing an exploratory but informative and reproducible scRNA-Seq analysis on 10× Genomics data without requiring any coding or system administration skills, furthermore offloading the computational load to our dedicated compute infrastructure.

2 Materials

The public Galaxy instance Galaxy-rCASC can be accessed at aisc.hpc4ai.unito.it. The instance provides rCASC tools for the analysis of 10× Genomics Chromium scRNA-Seq data going from the generation of the count matrix to filtering, clustering, and feature selection. Since rCASC, and scRNA-Seq data analysis in general, requires a significant amount of computing resources to run properly, Galaxy-rCASC has been deployed on an Ubuntu 20.04 server, featuring 256 GB RAM and 24 CPU cores, that has been made publicly available by the University of Turin as an ELIXIR-IT service. In order to access the service, the user needs to register with the instance and then log in.

2.1 Example Dataset

The example dataset used for this tutorial comes from the 3′ sequencing of a sample composed of human acute myeloid leukemia (AML) and retina cell lines. Sequencing was performed using a NextSeq500 in the EMBL genomics-Core lab (Heidelberg). We use data from just one of the four lanes from the original sequencing to shorten the overall computing time needed for this analysis. Since the two cell lines in this example dataset are very different, we may expect two neatly distinct clusters, with AML potentially further subdividing into sub-clusters reflecting different subpopulations in the sample.

The archive containing the dataset can be downloaded at https://zenodo.org/record/5513620/files/counting_s1_fastq_small.tar.gz. The dataset comprises three files: *sample1_S1_L001_I1_001.fastq*, *sample1_S1_L001_R1_001.fastq*, and *sample1_S1_L001_R2_001.fastq*. The first file contains the sample index and can be safely ignored (info on the meaning of the sample index can be found at https://support.10xgenomics.com/single-cell-gene-expression/sequencing/doc/specifications-sample-index-sets-for-single-cell-3), while, of the other two files, one contains the reads with cell barcodes and UMIs (Unique Molecular Identifiers) and the other the sequenced cDNA. The *CellRanger-Count* function of rCASC will use those files to compute the count matrix (*see* Subheading 3).

3 Methods

3.1 The rCASC Workflow

The rCASC workflow is organized in several steps, each handled by one or more rCASC functions. Although for the sake of brevity, we report here a brief description of each step, a comprehensive account of the method can be found in the original rCASC publication [7].

1. **Count matrix generation**: an rCASC analysis starts from the *fastq* files obtained from the samples demultiplexing, usually provided by the sequencing facility. The count matrix is computed by mapping the reads on a reference transcriptome, and then by counting how many reads (with different UMIs) in each cell of a sample overlap each gene. The dimension of the resulting matrix will be *nG* x *nC*, where *nG* represents the number of genes (rows of the matrix) and *nC* the number of cells (columns of the matrix). Galaxy-rCASC employs the 10× Genomics CellRanger 5.0 software for this purpose [11].
2. **Count matrix quality control**: single-cell assays usually include a fraction of low-quality cells (e.g., broken or dead cells) that can be identified by checking the expression of mitochondrial and ribosomal genes. This step identifies and excludes low-quality cells (columns) from the count matrix that might confound the rest of the analysis [12].
3. **Count matrix annotation**: the count matrix created in the first step employs Ensembl gene ID annotations. In this step, gene IDs are converted to gene symbols [13] to improve the readability of the results.
4. **Count matrix filtering**: Single-cell RNA-seq tends to generate sparse matrices, that is, matrices containing many zero values. This outcome is due to several reasons, both technical and

biological. Since it is generally not possible to know if a gene with zero expression across all the cells of a sample is not expressed or if this is due to some technical artefact, genes with very low expression (supported by less than four UMIs) across all cells are filtered out. There is no general rule about how many genes should be considered in the following clustering analysis. In our experience, working with ~1000 up to ~4000 expressed genes provides reasonably robust results.

5. **Cell classification by clustering**: scRNA-Seq aims to find and characterize the cell (sub)populations present in the sample, grouping cells according to the expression of their genes. rCASC provides several clustering algorithms (i.e., SIMLR, griph, kmeans+Tsne, Seurat), each with different characteristics. The choice of which one to employ depends on the sample since one might perform better on a specific dataset and worse on another. That is why rCASC associates a *stability score* to the cells in a cluster. In general, higher average stability scores correspond to higher confidence clustering. The stability score ranges from 0 to 1, with lower values corresponding to lower confidence. Indeed, using this score, it is possible to perform the feature selection step only for those clusters that can be considered bona fide according to this measure.
6. **Feature selection:** this analysis aims to single out the key genes whose expression can be considered specific for each cluster identified in the previous step. The user can choose between two methods for this analysis. The first one is COMET [14], an algorithm that evaluates the most important genes in each cluster. COMET employs a derivative of the hypergeometric test to identify a single gene or combination of genes, which provide an optimal representation of each cluster. The second method [15], based on an ANOVA-like approach, compares each cluster with a reference cluster selected by the user.

3.2 Using Galaxy-rCASC

In this section, we will guide the reader through the analysis of the scRNA-Seq sample introduced in 2.*1*, detailing the use of the Galaxy-rCASC tools that perform each step in the workflow, their input and output, and how to configure and launch them through the Galaxy interface. In the tools description, the following style will be used to present tools options: **Name of the option:** *tutorial value*, description.

Galaxy-rCASC has been initially developed exploiting the ELIXIR-Italy Laniakea@ReCaS [16] service resources by creating a development Galaxy server that was used to port the rCASC tools on Galaxy, writing and testing the tool wrappers (https://github.com/pmandreoli/rCASC_wrappers) and the Galaxy-rCASC server-specific configuration. When ready, a dedicated server based on this work was deployed on the University of Turin cloud infrastructure, using the very same configuration tools, e.g., the

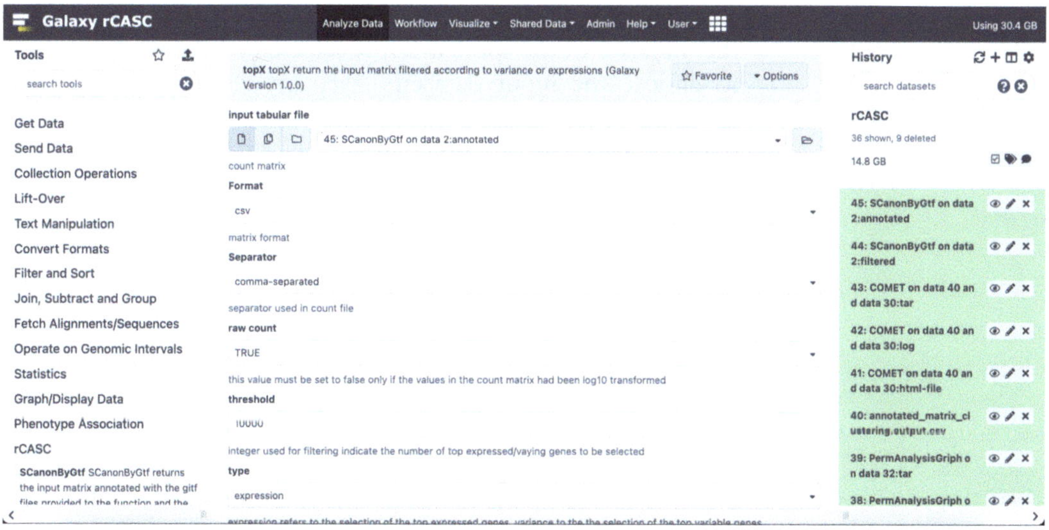

Fig. 1 The Galaxy primary user interface comprises the three main panels: tools panel (left), a central panel (displaying the interface of a tool in the figure), and the history panel (right). Finally, the navigation bar is on the top of the GUI

Ansible (www.ansible.com) roles developed in the context of the Laniakea [17] project, thus achieving seamless transition from the development environment to the current production public server.

The public “Galaxy-rCASC” instance is reachable at aisc.hpc4ai.unito.it. In order to have access to the instance and start using the service, follow the brief procedure described at https://github.com/pmandreoli/rCASC_wrappers/blob/v2/README.md.

3.2.1 Using Galaxy

We point the inexpert reader to the very comprehensive general guides and training materials on using Galaxy and all of its features and functions that can be found at https://training.galaxyproject.org/. For brevity, we will describe here only the main components of the interface needed to operate Galaxy-rCASC. In brief, the Galaxy workspace encompasses four main areas: a navigation bar, a tool panel, a central panel, and a history panel (Fig. 1).

- The navigation bar at the top of the interface provides menus and links to most of Galaxy’s functionalities, including the analysis workspace, workflows, the help service, the sharing data service, and user preferences.
- The tool panel provides access to the utility functions (e.g., files uploading) and all the bioinformatics tools available to the user. Usually, they come organized in categories to facilitate browsing. A *search tools* bar is also available in the topmost part of the panel.

- The central panel has two main functions. The first is to display the interface of the tool currently selected by the user, allowing to choose the input dataset(s), set any required or optional parameter, and finally run the tool on the selected input dataset(s). The second function of the panel is to display the content of a dataset.
- The history panel provides access to all the datasets available in the current history. The input data files uploaded by the user and the output of any bioinformatics tool employed on them are considered datasets by Galaxy and displayed in the same fashion since, usually, the output of a tool can be used as the input for another tool along a multi-step data analysis workflow. We recommend creating a new history for each distinct set of samples or analysis to organize the workspace better. Histories can be created and switched using the buttons at the panel's top.

For every step of the analysis, we will describe the employed tool of Galaxy-rCASC, how to properly set it up, and briefly describe its output. The dataset nomenclature of this walkthrough assumes the user is starting the analysis with an empty history.

3.2.2 Loading Data

The initial input of an rCASC workflow is a *tar.gz* archive file containing the reads, from a scRNA-Seq experiment carried out through a 10× platform, as a series of *fastq* files. To upload the archive on Galaxy-rCASC, the user can use the Galaxy paste and fetch function by selecting **Get Data → Upload File → Paste/ Fetch data**, pasting the dataset link into the box, and finally launching the upload using the **start** button.

In this way, the dataset will be automatically imported into Galaxy and become available in the Galaxy history in a matter of a few minutes. The new history item can then be renamed by selecting the "pencil" button that provides access and editing options for the metadata information attached to any dataset.

This is the only dataset that the user needs to upload to perform the analysis since the reference data required by rCASC tools (e.g., the genomic indexes needed by the reads alignment tool) have already been integrated into the platform and are available to be selected when using the workflow tools.

If needed, data can also be uploaded from a local computer by using **Get Data → Upload File → Choose local file**. For datasets exceeding 2GBs, we suggest uploading the files using an FTP client (e.g., Filezilla) and then selecting **Get Data → Upload File → Choose FTP file** (more info on how to proceed are available at https://galaxyproject.org/ftp-upload/).

3.2.3 Generating the Count Matrix

The bioinformatics scRNA-Seq data analysis workflow starts by computing the count matrix from the *fastq* files. In the process, each read in the "R2" *fastq* file is assigned to the cell to which it belongs using the cell barcodes found in the "R1" *fastq* file, mapped on the transcriptome and thus assigned to the gene that produced it (provided no other read with the same UMI has already been assigned to the same gene). The output of this step is a matrix with dimensions *nG* x *nC*, where *nG* represents the number of genes (rows of the matrix), and *nC* the number of cells (columns of the matrix).

The Galaxy-rCASC tool used to generate the count matrix is called *CellRangerCount.* This rCASC function runs the homonymous 10XGenomic software for the analysis of Chromium single-cell data. CellRangerCount incorporates the Illumina function *bcl2fastq*, adding specific features for 10X libraries and employing a simplified sample sheet format. For more information, please refer to https://kendomaniac.github.io/rCASC/reference/cellrangerCount.html.

Hands-On: *CellRangercount*

fastq_collection: counting_s1_fastq_small.tar.gz, *the fastq files from the 10XGenomics experiment, containing the reads associated with each sample. They need to be submitted in the form of a tar.gz archive.*

Cores: 8, *number of processor cores to use. Using the maximum value will not necessarily guarantee access to that amount of resources since cores will be reserved according to the current server load.*

RAM: 12, *gigabytes of RAM to be used for memory allocation. Using the maximum value will not guarantee access to that amount of resources since RAM will be reserved according to the current server load.*

Using Indexed reference transcriptome: hg38-index, *select the reference index corresponding to the genome assembly of the organism used from the drop-down menu. Human (hg38) and mouse (mm10) are the indexes currently available on Galaxy-rCASC.*

The output count matrix will appear in the History panel as a *CSV* file named *CellRangerCount on data 1: matrix.* The tool execution time should be ~70 min using the suggested parameters (may vary based on server loads).

3.2.4 Count Matrix Quality Control

This operation inspects the count matrix to exclude "bad quality" cells from the following analysis steps, including non-viable, dead, or cells that have been damaged during library preparation and that could confound the clustering step. The distinction between "good" and "bad" quality cells is performed by analyzing the expression of mitochondrial and ribosomal genes for every cell in the count matrix. High mitochondrial gene expression is considered a marker for stressed cells, while low expression of ribosomal genes points to a dead or otherwise inactive cell [12].

To perform this step, the user can employ the Galaxy-rCASC tool *MitoriboUmi*.

This tool draws a scatterplot of the cells according to the relative expression of mitochondrial and ribosomal genes with respect to the total gene expression of a class of genes, that is, a biotype according to Ensembl classification (https://m.ensembl.org/info/genome/genebuild/biotypes.html). The default biotype that should be used is *protein_coding* for standard scRNA-Seq samples, but it is possible to choose also other options for particular experimental setups. Using the graph produced by this tool, it will be possible to define a threshold value for the subsequent filtering step to exclude from the matrix any bad quality cell.

Hands-On: *MitoriboUmi*

Input matrix file: CellRangerCount on data 1, *that is the count matrix output of the previous tool.*

Input gtf file: Built-in gtf hg38-gtf, *the* gtf *file containing the gene annotation. It is also possible to upload and use a customised* gtf *file, but we recommend using those provided by Galaxy-rCASC. Currently, we support human (hg38) and mouse (mm10) Ensembl annotations.*

Format: csv, *the format of the input matrix. If the input matrix is the output from* cellRanger *then, the format to choose is* csv. *However, the user can also upload a* txt *or* csv *file containing a count matrix obtained externally using other tools.*

Separator: comma-separated, *the field separator used in the input matrix file. If the input matrix file is the output of* cellRanger, *the separator to select is* comma-separated; *otherwise, it is possible to provide tab-separated files selecting the corresponding* tab-separated *option.*

umiXgene: 3, *the minimum number of mapping UMIs necessary for a gene to be considered expressed in a cell.*

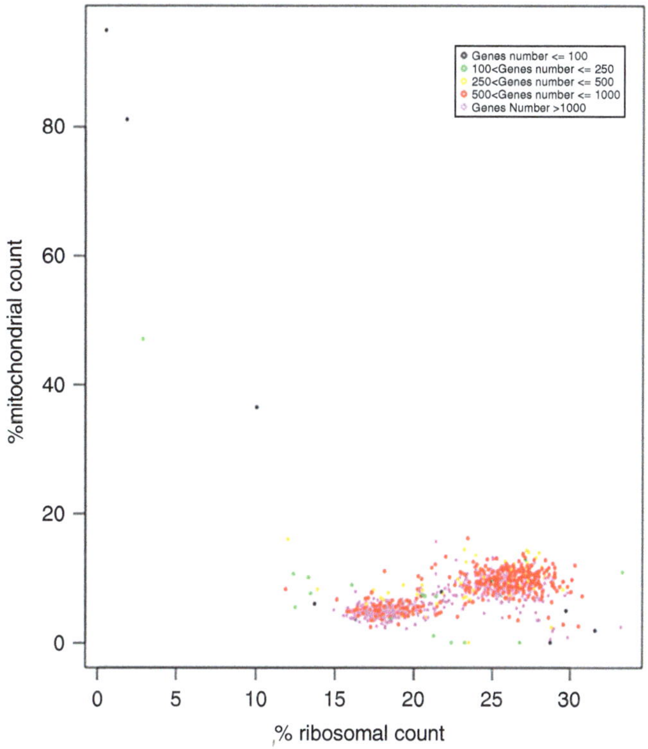

Fig. 2 The MitoriboUmi tool evaluates for each cell (dots in the plot) the expression of ribosomal (x-axis) and mitochondrial genes (y-axis) relative to the selected biotype. Each dot is colored according to the number of genes with more than 3 UMIs, that is, the genes whose expression can be detected. The plot shows how virtually all the cells in the sample are of acceptable quality since they have low mitochondrial gene expression and sufficient ribosomal gene expression

Biotype: protein_coding, *Ensembl biotype of reference, default for standard scRNA-seq is protein_coding. This is the gene class whose expression (total number of mapping UMIs) the tool employs as a reference to compare with the expression of ribosomal and mitochondrial genes.*

In the resulting graph (Fig. 2), each dot represents a cell in the matrix. The x-axis represents the percentage of UMIs assigned to ribosomal genes with respect to the total number assigned to the genes in the chosen biotype, while the y-axis reports the same measure but for mitochondrial genes. According to the figure, almost all the cells in our sample are located in the bottom right corner of the plot and therefore can be considered of overall satisfactory quality, with 15–30% of the UMIs falling on ribosomal genes and less than 15–20% on mitochondrial genes. The color of the dots provides an eye-catching representation of the number of protein-coding genes whose expression can be detected in each cell. This is a critical factor that must be considered before proceeding with the analysis since the classification analysis needs an adequately high amount of expressed genes per cell to provide good results. From our experience, at least ~250 expressed genes per cell should be present on average for obtaining robust results in the clustering step.

3.2.5 Count Matrix Annotation and Cell Filtering

With the information acquired in the previous stage, the user can proceed with the count matrix cell filtering and gene annotation steps. Those two operations are performed by the *Scannobygtf* tool. This function first assigns the corresponding gene symbol to each ENSEMBL ID in the matrix, then filters the cells by removing the ones with too high mitochondrial or too low ribosomal gene expression or an insufficient number of expressed genes. Considering the output of *Mitoribo*, which highlighted an overall good quality for the cells in our sample, we will limit the filtering to retaining the cells with at least 250 expressed genes.

Hands-On: *ScannoByGTF*

input .csv or .txt file: CellRangerCount on data 1, *that is the count matrix output of* CellRangerCount.

Select Build-in gtf: Built-in gtf hg38-gtf, *the* gtf *file containing the gene annotation. It is also possible to upload and use a customized* gtf *file, but we recommend using the ones provided by Galaxy-rCASC. Currently, we support human (hg38) and mouse (mm10) Ensembl annotations. In any case, the user should use the same* gtf *annotation file for the whole workflow.*

Format: csv, *the format of the input matrix. If the input matrix is the output from* cellRanger *then, the format to choose is* csv. *However, the user can also upload a* txt *or* csv *file containing a count matrix obtained externally using other tools.*

Biotype: protein_coding, *Ensembl biotype of reference, default for standard scRNA-Seq is protein_coding. This is the gene class whose expression (total number of mapping UMIs) the tool employs as a reference to compare with the expression of ribosomal and mitochondrial genes. Use the same as with* MitoriboUmi.

mt: FALSE, *a Boolean value. Set to true to keep mitochondrial genes in the annotated count matrix. When there is no* *particular reason to consider them important for the analysis (e.g., interest in metabolic changes), we suggest dropping them since their relative abundance could confound the clustering algorithms masking more subtle differences in other genes underlying true biological heterogeneity.*

ribo_proteins: FALSE, *a Boolean value. Set to true to keep the ribosomal protein genes in the annotated count matrix. Also in this case, when there is no particular reason to consider them important in the analysis, we suggest dropping them since they may introduce noise in the subsequent analysis steps.*

riboStart.percentage: 10, *the start value of the range for ribosomal gene expression* (see *Subheading 3.2), cells within the range are kept.*

riboEnd.percentage: 100, *the end value of the range for ribosomal gene expression* (see *Subheading 3.2), cells within the range are kept.*

mitoStart.percentage: 0, *the start value of the range for mitochondrial gene expression* (see *Subheading 3.2), cells within the range are retained*

mitoEnd.percentage: 20, *the end value of the range for mitochondrial gene expression* (see *Subheading 3.2), cells within the range are retained*

umiXgene: 3, *the minimum number of UMIs necessary to define a gene as expressed. We recommend using the same values as in* MitoriboUmi.

thresholdGenes: 250, *parameter to filter cells according to the number of expressed genes. Cells with a number of expressed genes below the threshold will be removed from the matrix.*

In general, keeping mitochondrial and ribosomal genes in the matrix is a choice that depends on the experimental setting and that should be carefully considered by the user depending on the outstanding biological question(s). However, since our example aims to find genes that characterize and distinguish AML and retina cells, mitochondrial and ribosomal protein genes can be safely discarded.

This tool will provide two output datasets:

- *SCanonByGtf on data 3 and data 2: annotated*
- *SCanonByGtf on data 3 and data 2: filtered*

The first one is the original count matrix annotated using gene symbols. The second one is the count matrix that, in addition to being annotated with gene symbols, has been filtered according to the user-defined parameters.

3.2.6 Count Matrix Gene Filtering

Following cell quality filtering, another critical analysis step is gene filtering. In fact, the count matrices generated from the analysis of scRNA-Seq experiments are usually sparse. That is, most of the elements have a value of zero. When there are no or very few UMIs mapping on a gene across all the cells of a sample, the reason may be technical noise or the fact that the gene is actually not expressed or expressed at very low levels in the sample. Since obtaining a dataset with a low amount of confounding factors is desirable for the clustering algorithm to better classify and group the cells according to gene expression and variance similarities, filtering the count matrix by removing non-informative genes is a simple but effective method to improve cluster quality.

Galaxy-rCASC provides two ways to filter genes:

- Retaining the top *N* most expressed genes
- Retaining the top *N* genes with higher expression variance

Retaining the most expressed genes, that is, the ones with the highest average expression across all the cells, leads to removing all the genes expressed at low levels, which may introduce confounding factors during the clusterization. On the other hand, keeping the genes with the most variable expression across the cells may help the clustering algorithm in the classification since those genes should be the ones that better differentiate the cell subpopulations. We recommend that the user try both strategies and choose the one that leads to better clustering in the downstream analysis. However, the second one should be employed only when the coverage (the total number of mappable UMIs) is reasonably high to provide a meaningful estimation of the variability of gene expression levels.

The rCASC function performing this step is called *TopX*. In our example, the user can filter genes based on their expression.

Hands-On: *TopX*

input csv or txt file: SCanonByGtf on data 3 and data 2: filtered, *the filtered count matrix generated by SCannoByGtf.*

Format: csv, *the format of the input matrix. If the input matrix is the output from* cellRanger *then, the format to choose is* csv. *However, the user can also upload a* txt *or* csv *file containing a count matrix obtained externally using other tools.*

Separator: comma-separated, *the field separator used in the input matrix file. If the input matrix file is the output of* cellRanger, *the separator to select is* comma-separated; *otherwise, it is possible to provide tab-separated files selecting the corresponding* tab-separated *option.*

raw count: true, *this value must be set to false only if the values in the count matrix had been* log_{10} *transformed.*

Threshold: 10000, *genes will be sorted by descending average or variance of UMI counts, and only the top-ranking* N, *with* N *equal to this threshold value, will be kept. For our example, we keep the top 10,000 genes sorted by expression.*

type: *By expression, Galaxy-rCASC provides two methods for filtering genes:*

By expression: keep the genes with the highest mean UMI counts (a good proxy for expression [18]).

By variance: select the genes whose expression is most variable across the cells in the sample.

The tool outputs two datasets:

- *topX on data 5: a pdf file*, containing two histograms that report the genes expression along the dataset before (Fig. 3, red graph) and after filtering (Fig. 3, blue graph). Figure 4 reports the same plots but obtained using variance filtering instead.

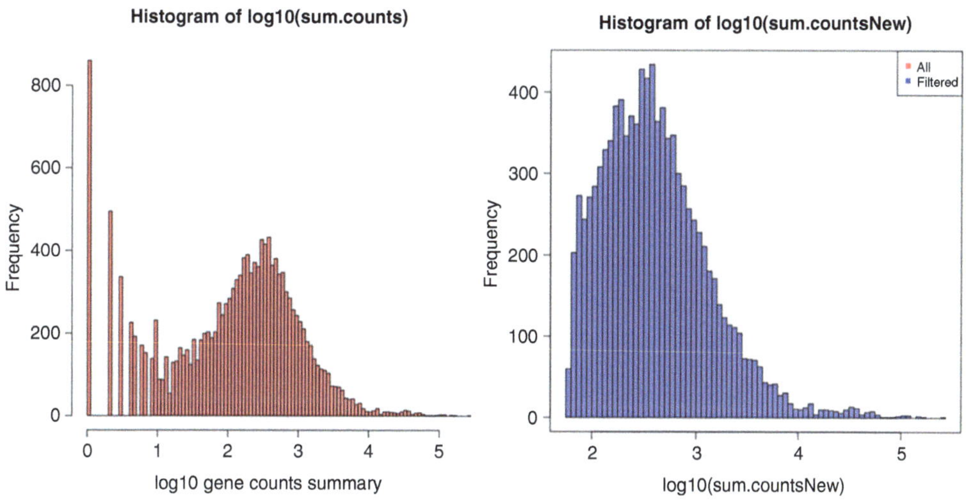

Fig. 3 Histograms of the distribution of genes according to their mean expression before (red graph on the left) and after (blue graph on the right) the topX average expression filtering. In the blue plot, the low expression genes tail found on the left of the red plot has disappeared

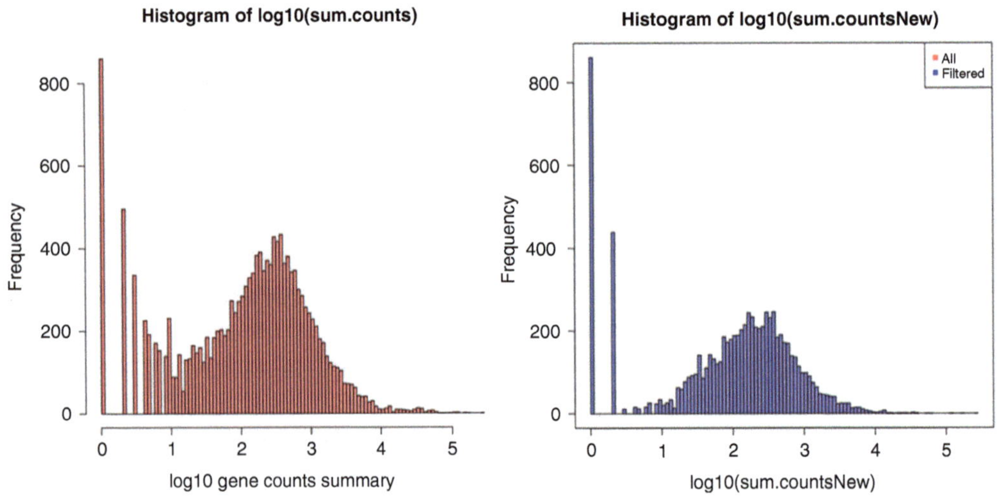

Fig. 4 Histograms of the distribution of genes according to their expression before (red graph on the left) and after (blue graph on the right) the topX variance filtering. While keeping high variability genes should lead to a more fine-grained clustering, this method tends to keep in the matrix a large fraction of genes with low expression and is recommended only for high-coverage samples

- *topX on data 5: filtered*, the count matrix now gene-filtered according to the selected threshold and method.

From the plots in Fig. 3, we can observe that the filtering provided the expected outcome, removing low expression genes from the count matrix and consequently avoiding a potentially confounding factor in the clustering step. Fig. 4 reports the same plots as obtained using the variance filtering.

3.2.7 Clustering

Following the count matrix cleanup and annotation, the user can proceed to the clustering analysis step. The choice of the clustering algorithm is often critical to achieving better results in identifying the subpopulations present in the sample. Galaxy-rCASC provides several clustering algorithms, each of which may be more suitable for one dataset. rCASC calculates a Jaccard index–based score to help the user understand which algorithm provides the best clustering performances for the current dataset: the *cell stability score (CSS)*. This score measures the stability of a cell in a cluster as a consequence of a perturbation induced by removing a random set of cells from the cell population (see rCASC [7] for a more detailed description of CSS). CSS is evaluated for each cell and goes from 0, which can be interpreted as very low confidence that the cell has been assigned to the correct cluster, to 1, which means the cell belongs to the correct cluster with high confidence.

In Galaxy-rCASC, the clustering and the perturbation analysis needed to calculate the CSS for each cell are performed in two steps implemented in the following four tools:

- *permutationClustering* and *PermutationClusteringSeurat*: perform the clustering *N* times perturbing the count matrix, that is, removing a random subset of cells, at each iteration to calculate the probability for each cell to be assigned to each cluster. The user can choose among three clustering algorithms: Kmeans +tsne, griph [19], and SIMLR [19]. In addition, the *PermutationClusteringSeurat* tool has been implemented to support specifically the Seurat [20] clustering algorithm.
- *permutationAnalysis* and *permAnalysisSIMLR*: analyze the output of *PermutationClustering* and *PermutationClusteringSeurat* to calculate the CSS. Once the clusters have been defined and perturbed using *permutationClustering*, with *permutationAnalysis*, it is possible to evaluate the stability score associated with each cell. *permAnalysisGriph* is used for hierarchical clustering methods such as Seurat and Griph, while *permAnalysisSIMLR* is used for non-hierarchical clustering algorithms such as SIMLR and kmeans+tSne.

The clustering algorithms made available by Galaxy-rCASC in the *permutationClustering* tool are:

1. *Kmeans + tSne*: requires in input the number of expected clusters, so the user needs to know (or guess) how many subpopulations are found in the sample.
2. *Griph*: detects and evaluates the optimal number of clusters by itself, creating a dendrogram and selecting the best distance to use in order to generate the clusters.
3. *SIMLR:* uses a machine learning approach to identify the clusters. It is non hierarchical, so, like Kmeans+tSne, it requires an expected cluster number in input.
4. *Seurat:* is considered the current gold standard for hierarchical clustering of scRNA-seq data [21] and is based on the Louvain algorithm [22] applied to K-nearest neighbor (KNN) graphs. It does not require knowing the expected number of clusters a priori.

Having no prior knowledge about the expected number of sub-populations in our sample, the analysis can continue by choosing an algorithm between griph and Seurat. Seurat is among the most used methods for scRNA-Seq; thus, we will adopt its clustering approach in this tutorial. However, even if Seurat employs a widely accepted clustering algorithm, please note that it might not work for every dataset. That is why we implemented in Galaxy-rCASC several alternative algorithms. In this way, users are free to explore and choose the one that provides the most stable, statistically relevant, cluster configuration for each dataset.

Hands-On: *Run permutationClustering/permutationClustering Seurat*

input count matrix: topX on data 5: filtered, *the filtered count matrix obtained from TopX.*

nPerm: 40, *number of perturbation iterations. In order to evaluate a p-value for each cell cluster assignment, the entire dataset is perturbed* N *times, where* N *corresponds to the nPerm value. From our experience, 40 is a pretty good value since fewer permutations are not enough to produce a significant statistic, and more than 40 permutations might be just too time-consuming.*

permAtTime: 2, *since the perturbation step can require a considerable amount of time, the process can be parallelized to process more than one iteration at a time. Up to a certain point, a higher value means greater parallelization, but the overall time is still dependent on the load of the Galaxy-rCASC host.*

percent: 10, *the percentage of randomly removed cells from the dataset at each perturbation. From our experience, 10 is a pretty reasonable value to use. Less than 10% might be insufficient to generate a significant perturbation, while more than 20% may result in some clusters disappearing altogether, leading to noisy results.*

format: csv, *the format of the input matrix. If the input matrix is the output from* cellRanger, *then the format to choose is* csv. *However, the user can also upload a* txt *or* csv *file containing a count matrix obtained externally using other tools.*

separator: comma-separated, *the field separator used in the input matrix file. If the input matrix file is the output of* cellRanger, *the separator to select is* comma-separated; *otherwise, it is possible to provide tab-separated files selecting the corresponding* tab-separated *option.*

raw count: true, *this value must be set to false only if the values in the count matrix had been* log_{10} *transformed.*

Clustering: *the clustering algorithm to use (griph,SIMLR, kmeans+tsne).*

perplexity: 10, *This value is used only by Kmeans+Tsne. Lowering the value will result in smaller clusters. Ignored by other clustering algorithms.*

seed: 1111, *this is a number used to initialize a pseudorandom number generator. Since clustering algorithms employ* *pseudorandom numbers, they do not always end up with the same results in different runs on the same input. Providing the same seed number as input, we guarantee that the same clustering algorithm, with the same parameters, and run on the same dataset and machine will provide the same result.*

Range1 and Range2: *Those parameters are only required when choosing SIMLR or kmeans+tsne as clustering algorithms. They define the number of clusters to divide the dataset into. All integer numbers between Range1 and Range2 will be used progressively. E.g., setting Range1 at three and Range2 at six, the clustering algorithm will be performed four times, each time passing to the algorithm the required number of clusters from three to six.*

Similarly, the tool *PermutationClusteringSeurat* could perform the clustering and perturbation analysis using Seurat. While the parameters are mostly the same as for *PermutationClustering*, there are two additional parameters:

pcaDimensions: *15*, Seurat performs a dimensionality reduction based on the data's PCA (principal component analysis). This value specifies how many dimensions will be taken into account by the Seurat clustering algorithm.

sparse: *FALSE*, Set true to use specific hierarchical data format hdf5 (https://www.hdfgroup.org/solutions/hdf5/), count matrices, false for default .csv and tabular format.

The permutation clustering step produces several output files that can be directly inspected from within Galaxy by selecting the "*PermutationClusteringSeurat on data 7:html-file*" dataset or saved on your local computer by downloading the corresponding *tar* output in "*PermutationClusteringSeurat on data 7.*"

The output consists of four files:

- *Annotated_matrix_clustering.output.csv*: the output of the clustering analysis. For each cell (row), it reports the cluster to which it has been assigned; its spatial coordinates, that is, the 2D projection of each cell performed by the clustering algorithm; and the silhouette value, a measure of how similar a cell is to its cluster.

- *annotated_matrix_13_clusterP.csv*: output of the perturbation analysis, reports for every cell the cluster to which it was assigned at each permutation.
- *annotated_matrix_13_killedCell.csv*: supplementary output of the perturbation analysis. Each column reports the cells that have been excluded from the cluster at each iteration.
- *hist.pdf*: a histogram plot reporting the frequency of the number of clusters obtained in all the permutations. Low variation in the number of obtained clusters points to greater clustering stability.
- *matrixName_vioplot.pdf*: a violin plot of the silhouette score distribution [23]. The silhouette score is assigned to each cell and ranges from −1 to 1, with higher values indicating a cell well matches the cluster it belongs to. When too many cells have low or negative values, the clustering may have too many or too few clusters.

The report produced by the *PemutationClusteringSeurat* tool is preparatory, and the analysis needs to be completed by processing the output with the corresponding permutation analysis tool.

Since we used Seurat as a clustering algorithm, we will use permAnalysisGriph (that can process both the output of GRIPH and Seurat) for the next step. With this tool, we will assign a CSS to each cell and produce plots of cell clusters and their stability.

PermAnalysisGriph is used to analyze *PermutationClustering* results performed using hierarchical clustering algorithms like Griph and Seurat. Instead, use the *PermAnalysisSIMLR* tool to analyze the clustering output of not hierarchical ones, such as SIMLR and Tsne.

Hands-On: *PermAnalysisGriph* and *PermAnalysisSIMLR*

Tar_Clustering: PermutationClusteringSeurat on data 6, *tar output of* PermutationClustering.

Format and separator: *same as with the previous tools.*

sp: 0.8, *the minimum fraction of cells in common between two permutations for two clusters to be considered as the same cluster. This value can be set between 0.51 and 0.9. The higher the value, the more stringent the analysis result will be. We strongly suggest keeping the* sp *threshold at 0.8. For a more detailed explanation of this threshold, check the Material and Methods section of the original rCASC publication [7].*

The PermAnalysisSIMLR *tool requires the same parameters as* PermAnalysisGriph, *plus two additional ones:*

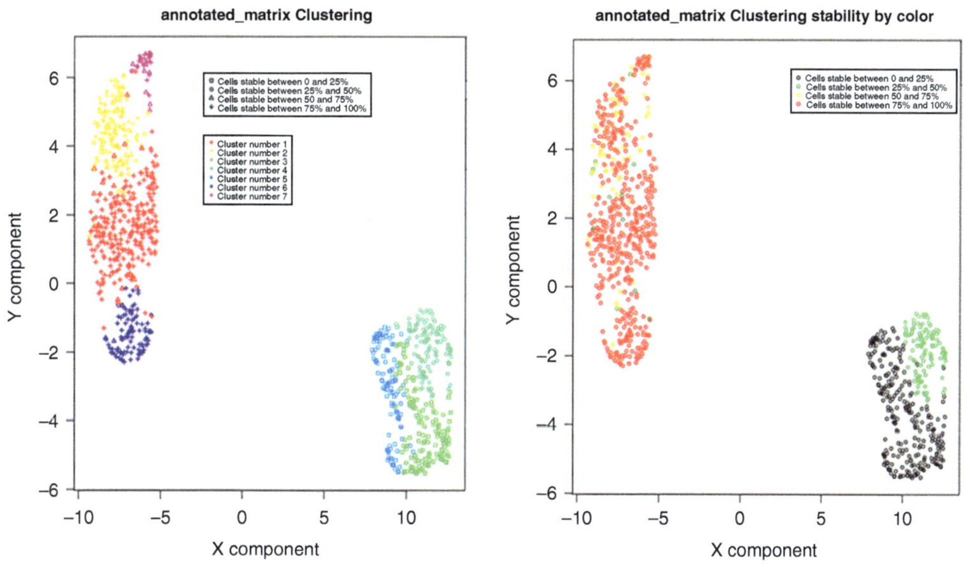

Fig. 5 Clustering stability plots. On the left, each cell is represented by a dot whose shape changes with respect to the stability score, while the color refers to the cluster of assignment. On the right, instead, the colors depend on the stability scores. Since we have data from two very different cell types (AML and retina) in our sample, the clusters identified by the algorithm form two higher-order clusters, or superclusters. The one on the left corresponds to AML cells; the high stability score obtained by those cells can be explained by the existence of well-defined subpopulations (identified by clusters 1, 2, 6, 7) within this cell type. On the other hand, retina cells, corresponding to the supercluster on the bottom right, obtained lower CSSs. This behaviour can be interpreted as a general higher homogeneity of the cells in this supercluster, leading to more blurred differences among the three clusters (3, 4, 5) that form the retina cells supercluster

> ***Range1 and Range2:*** *Those parameters are only required when choosing SIMLR or kmeans+tsne as clustering algorithms (see the hands-on* permutationClustering*). If used in the previous step, set the same values here.*

The two tools produce several outputs, among which two files are particularly interesting to continue our analysis: *clustering.output.csv* and *Stability_plot.pdf*.

1. *Clustering.output.csv*: contains a breakdown of the analysis for each cell—the cluster of assignment, the stability score, intraCluster and extraCluster quality scores, and the silhouette value. They are evaluated using the silhouette algorithm [23], checking the average distance of each cell with respect to the cells that belong to the same cluster (intraCluster) and the other clusters (extraCluster).
2. *Stability_plot.pdf* (Fig. 5): a scatterplot of the cells highlighting clusters and their stability. Cells are plotted as dots whose shape is defined by their CSS, while the dot's color marks the cluster they are assigned. The 2D plotting coordinates are the ones

evaluated by the clustering algorithm. The average stability score depends on the performance of the clustering algorithm and the consistency of the data. The higher the CSS, the less the perturbations influenced the clustering.

As we can see from the stability plot (Fig. 5), Seurat identified two superclusters composed of a total of seven clusters. The two superclusters correspond to AML (formed by clusters 1, 2, 6, 7) and retina cells (clusters 3, 4, 5). The clusters can be interpreted as subpopulations within the two main cell types.

Due to the high stability of the AML clusters obtained by *PermAnalysis*, we will focus the feature selection step on clusters 1, 2, 6 and 7. On the other hand, the instability of clusters 3, 4 and 5 can be explained by the higher overall homogeneity of the retina cells that tend to confound the clustering algorithm and would make the identification of cluster-characterizing genes in the next step difficult. In similar cases, a way forward can be to perform the feature selection step considering only the stable clusters and then re-run the clustering step using another clustering algorithm or the same with different parameters. In our case, a good option would be to use SIMLR with *range1* = *2* and *range2* = *2* to force the algorithm to merge the highly similar cells of clusters 3, 4 and 5 in a single and broader cluster and then follow with a second feature selection analysis considering only that cluster.

3.2.8 Feature Selection

The final step of the workflow extracts the most representative genes for each cluster, which are the genes whose expression patterns can be considered specific for each cluster. In this step, the user needs to consider the information about cluster stability acquired in the previous analysis, as using unstable clusters could lead to misleading results.

The tool used by Galaxy-rCASC to perform this analysis is COMET, which runs the homonymous COMET tool [14] that allows the identification of gene markers for cell (sub)populations in scRNA-Seq data. Galaxy-rCASC runs COMET using the whole human (or mouse) gene annotation. For more information on COMET, please refer to the documentation at https://hgmd.readthedocs.io/en/latest/.

Hands-On: *COMET*

input file: topX on data 5: filtered, *the TopX filtered count matrix.*

input clustering: PermutationClusteringSeurat on data 6 (annotated_matrix_clustering.output), *generated by the* PermutationClusteringSeurat *tool.*

X: 0.15, *X-param of the XL-mHG test, a threshold value used by COMET. We strongly suggest not changing this value or checking COMET documentation before doing that.*

number of gene combinations: 2, *COMET will evaluate from two to four combinations of genes to detect the best combination representing each cluster (refer to the COMET documentation for more details). Increasing this number will significantly increase the required computational time. We suggest using 2.*

raw count: true, *this value must be set to false only if the values in the count matrix had been log_{10} transformed.*

skip visualization: false, *if set to true, COMET skips the generation of the plot.*

Separator: comma separated, *same as with the previous tools.*

COMET generates several output files that can be accessed directly through Galaxy by selecting the "*COMET html:file*" dataset or downloaded by saving the "*COMET tar output*" dataset.

Among the files produced by COMET, we focus here on the ones with the prefix "singleton" for each cluster, which are the files containing possible single-gene markers. The files using the "pair" prefix contain gene pairs, useful for identifying surface proteins to build multi-gene marker panels (refer to COMET's publication [14] and documentation https://hgmd.readthedocs.io/en/latest/Output.html for a more detailed view on output files and how to interpret them).

In particular, for the reader's convenience, we report here from the COMET's manual the content of the singleton files and directories:

- Singleton-all_ranked.csv: a csv file with the single gene marker list containing the overall rankings based on the xl-mHG p-value and the absolute fold change. Includes genes with True Positive value greater than 0.15 and negation genes [authors' note: genes whose expression appears to be avoided in a cluster] if they have a True Negative value of greater than 0.5.
- singleton_positive_markers_ranked: Single gene marker list as above, except without the negation genes. Useful if the focus is on positive markers.
- Singleton-full_unranked.csv: csv file ALL single genes tested, with no rank. Useful for finding genes that did not pass the thresholds in singleton_all_ranked.
- Folder singleton: plots that show the discrete and continuous version of the top-performing single genes in Singleton-all_ranked. (Fig. 6).

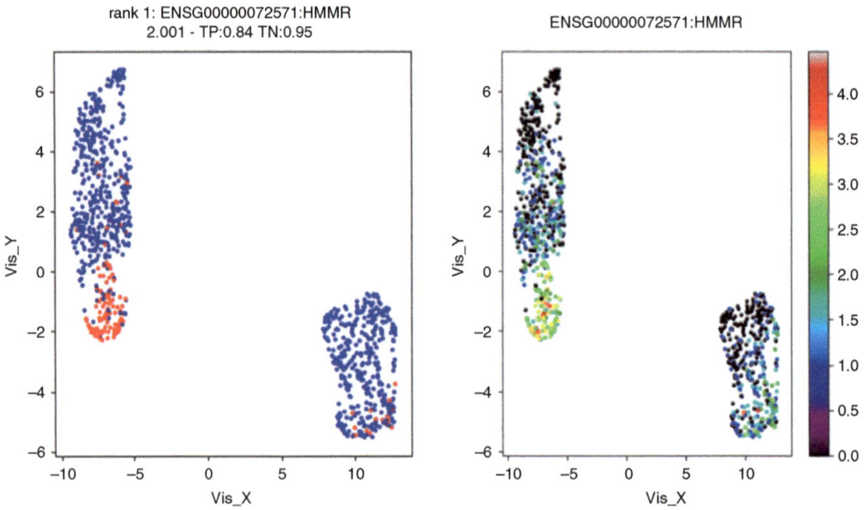

Fig. 6 On the left a COMET plot of the discrete distribution of the gene HMMR (identified by COMET as a singleton in cluster 6). Red dots represent cells where the gene is expressed, while blue dots represent cells with no or very low HMMR expression. The plot on the right shows the distribution for the same gene using continuous values. HMMR appears to be mostly expressed in cluster 6 in a presumptive subpopulation of AML cells

Using the output files of COMET, we can get insights into the cell sub-populations present in each cluster. For example, the genes present in the *Singletons_all_ranked* COMET files can be used to perform downstream gene ontology and functional enrichment analyses to characterize, e.g., cell types, functions, differentiation stage.

Galaxy rCASC also supports the *Anova*-like edgeR [15] tool for feature selection.

This tool evaluates the differential expression of the genes among all the groups against a reference one. A typical use case for *Anova-like* is when there are different conditions for the same cell type in a sample, for example, activated and naive T cells as in [24].

Hands-On: *Run Anova-Like*

count matrix: *the filtered count matrix obtained from TopX.*

Format and separator: *same as with the previous tools.*

cluster-file: *the output generated from permutationClustering.*

reference cluster: *the cluster that will be used as reference. All the other clusters will be evaluated against this one. That means that the genes found in this way are differentially expressed genes with respect to the selected one.*

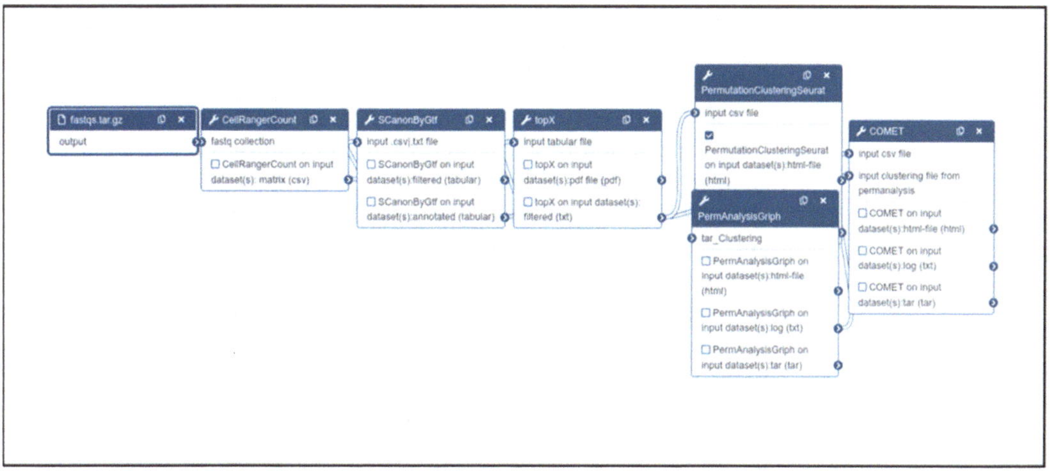

Fig. 7 The Galaxy-rCASC workflow implementing the analysis herein described

logCPM_threshold: *expression value threshold (expressed in* log_2 *of CPM [12]) for the genes to be considered in the analysis.*

FDR_threshold: *0.05, the false discovery rate threshold, usually set at 0.05 or 0.01.*

LogFC_threshold: *1, absolute* log_2 *fold change threshold (logFC).*

The output of the *anovaLike* tool is a list of differentially expressed genes, with respect to the reference cluster, for each of the other clusters, filtered using the user-defined thresholds. The complete list of genes ranked by logFC and p-value is also provided.

3.2.9 Using the rCASC Workflow

All the steps described up to COMET have been integrated into the rCASC workflow (Fig. 7) found in the published workflow section of Galaxy-rCASC. The workflow automatically runs all the Galaxy-rCASC tools described in this guide up to COMET, presetting the parameters to the ones we recommended for the example dataset.

The workflow can also be simply customized to be tailored for other datasets. For more information on creating and using Galaxy workflows, follow the tutorial at https://training.galaxyproject.org/training-material/topics/galaxy-interface/tutorials/workflow-editor/tutorial.html.

4 Final Considerations

We illustrated how to use Galaxy-rCASC to perform an exploratory and reproducible analysis workflow on scRNA-Seq data, including the generation of the count matrix, quality control and filtering, identification of cell types and subtypes using different clustering algorithms, and finally, the identification of gene expression signatures specific for each cell sub-population.

Performing the equivalent rCASC analysis workflow using the command-line interface could be rather complex for users without adequate system administration and coding skills and requires adequate local computational resources. Additionally, thanks to Galaxy, both basic and advanced users can leverage the convenient features for reproducibility, workflow automation and data sharing offered by this platform, and the presence in the same work environment of integrated basic utilities and data retrieving functionalities.

Finally, the Galaxy-rCASC implementation allows to easily exploit significant computational resources, sharing them among multiple users through a user-friendly graphical interface, which is not possible using the original desktop rCASC Java application.

References

1. Choi YH, Kim JK (2019) Dissecting cellular heterogeneity using single-cell RNA sequencing. Mol Cells 42(3):189–199. https://doi.org/10.14348/molcells.2019.2446
2. Kharchenko PV (2021) The triumphs and limitations of computational methods for scRNA-seq. Nat Methods 18:723–732. https://doi.org/10.1038/s41592-021-01171-x
3. Buenrostro JD, Wu B, Chang HY, Greenleaf WJ (2015) ATAC-seq: a method for assaying chromatin accessibility genome-wide. Curr Protoc Mol Biol 109:21 29 21–21 29 29. https://doi.org/10.1002/0471142727.mb2129s109
4. Yan F, Powell DR, Curtis DJ, Wong NC (2020) From reads to insight: a hitchhiker's guide to ATAC-seq data analysis. Genome Biol 21(1):22. https://doi.org/10.1186/s13059-020-1929-3
5. Rotem A, Ram O, Shoresh N, Sperling RA, Goren A, Weitz DA, Bernstein BE (2015) Single-cell ChIP-seq reveals cell subpopulations defined by chromatin state. Nat Biotechnol 33(11):1165–1172. https://doi.org/10.1038/nbt.3383
6. Li G, Liu Y, Zhang Y, Kubo N, Yu M, Fang R, Kellis M, Ren B (2019) Joint profiling of DNA methylation and chromatin architecture in single cells. Nat Methods 16(10):991–993. https://doi.org/10.1038/s41592-019-0502-z
7. Alessandri L, Cordero F, Beccuti M, Arigoni M, Olivero M, Romano G, Rabellino S, Licheri N, De Libero G, Pace L, Calogero RA (2019) rCASC: reproducible classification analysis of single-cell sequencing data. Gigascience 8(9). https://doi.org/10.1093/gigascience/giz105
8. Alessandri L, Cordero F, Beccuti M, Licheri N, Arigoni M, Olivero M, Di Renzo MF, Sapino A, Calogero R (2021) Sparsely-connected autoencoder (SCA) for single cell RNAseq data mining. NPJ Syst Biol Appl 7(1):1. https://doi.org/10.1038/s41540-020-00162-6
9. Alessandri L, Ratto ML, Contaldo SG, Beccuti M, Cordero F, Arigoni M, Calogero RA (2021) Sparsely connected autoencoders: a multi-purpose tool for single cell omics analysis. Int J Mol Sci 22(23). https://doi.org/10.3390/ijms222312755
10. Afgan E, Nekrutenko A, Grüning BA, Blankenberg D, Goecks J, Schatz MC, Ostrovsky AE, Mahmoud A, Lonie AJ, Syme A, Fouilloux A, Bretaudeau A, Nekrutenko A, Kumar A, Eschenlauer AC, DeSanto AD, Guerler A, Serrano-Solano B, Batut B, Grüning BA, Langhorst BW, Carr B, Raubenolt BA, Hyde CJ, Bromhead CJ, Barnett CB, Royaux C, Gallardo C, Blankenberg D, Fornika DJ, Baker D, Bouvier D, Clements D, de Lima Morais DA, Tabernero DL, Lariviere D, Nasr E, Afgan E, Zambelli F, Heyl F, Psomopoulos F, Coppens F, Price GR, Cuccuru G, Corguillé GL, Von Kuster G, Akbulut GG, Rasche H, Hotz H-R, Eguinoa I, Makunin I, Ranawaka IJ, Taylor JP, Joshi J, Hillman-Jackson J, Goecks J, Chilton JM, Kamali K,

Suderman K, Poterlowicz K, Yvan LB, Lopez-Delisle L, Sargent L, Bassetti ME, Tangaro MA, van den Beek M, C ˇech M, Bernt M, Fahrner M, Tekman M, Föll MC, Schatz MC, Crusoe MR, Roncoroni M, Kucher N, Coraor N, Stoler N, Rhodes N, Soranzo N, Pinter N, Goonasekera NA, Moreno PA, Videm P, Melanie P, Mandreoli P, Jagtap PD, Gu Q, Weber RJM, Lazarus R, Vorderman RHP, Hiltemann S, Golitsynskiy S, Garg S, Bray SA, Gladman SL, Leo S, Mehta SP, Griffin TJ, Jalili V, Yves V, Wen V, Nagampalli VK, Bacon WA, de Koning W, Maier W, Briggs PJ (2022) The Galaxy platform for accessible reproducible and collaborative biomedical analyses: 2022 update. Nucleic Acids Res 50(W1): W345–W351. https://doi.org/10.1093/nar/gkac247

11. Zheng GXY, Terry JM, Belgrader P, Ryvkin P, Bent ZW, Wilson R, Ziraldo SB, Wheeler TD, McDermott GP, Zhu J, Gregory MT, Shuga J, Montesclaros L, Underwood JG, Masquelier DA, Nishimura SY, Schnall-Levin M, Wyatt PW, Hindson CM, Bharadwaj R, Wong A, Ness KD, Beppu LW, Deeg HJ, McFarland C, Loeb KR, Valente WJ, Ericson NG, Stevens EA, Radich JP, Mikkelsen TS, Hindson BJ, Bielas JH (2017) Massively parallel digital transcriptional profiling of single cells. Nat Commun 8(1):14049. https://doi.org/10.1038/ncomms14049
12. Ordonez-Rueda D, Baying B, Pavlinic D, Alessandri L, Yeboah Y, Landry JJM, Calogero R, Benes V, Paulsen M (2020) Apoptotic cell exclusion and bias-free single-cell selection are important quality control requirements for successful single-cell sequencing applications. Cytometry A 97(2):156–167. https://doi.org/10.1002/cyto.a.23898
13. Tweedie S, Braschi B, Gray K, Jones TEM, Seal RL, Yates B, Bruford EA (2021) Genenames. org: the HGNC and VGNC resources in 2021. Nucleic Acids Res 49(D1):D939–D946. https://doi.org/10.1093/nar/gkaa980
14. Delaney C, Schnell A, Cammarata LV, Yao-Smith A, Regev A, Kuchroo VK, Singer M (2019) Combinatorial prediction of marker panels from single-cell transcriptomic data. Mol Syst Biol 15(10):e9005. https://doi.org/10.15252/msb.20199005
15. Robinson MD, McCarthy DJ, Smyth GK (2010) edgeR: a Bioconductor package for differential expression analysis of digital gene expression data. Bioinformatics 26(1): 139–140. https://doi.org/10.1093/bioinformatics/btp616
16. Tangaro MA, Mandreoli P, Chiara M, Donvito G, Antonacci M, Parisi A, Bianco A, Romano A, Bianchi DM, Cangelosi D, Uva P, Molineris I, Nosi V, Calogero RA, Alessandri L, Pedrini E, Mordenti M, Bonetti E, Sangiorgi L, Pesole G, Zambelli F (2021) Laniakea@ReCaS: exploring the potential of customisable Galaxy on-demand instances as a cloud-based service. BMC Bioinform 22(Suppl 15):544. https://doi.org/10.1186/s12859-021-04401-3
17. Tangaro MA, Donvito G, Antonacci M, Chiara M, Mandreoli P, Pesole G, Zambelli F (2020) Laniakea: an open solution to provide Galaxy "on-demand" instances over heterogeneous cloud infrastructures. Gigascience 9(4). https://doi.org/10.1093/gigascience/giaa033
18. Phipson B, Zappia L, Oshlack A (2017) Gene length and detection bias in single cell RNA sequencing protocols. F1000Res 6:595. https://doi.org/10.12688/f1000research.11290.1
19. Zheng GX, Terry JM, Belgrader P, Ryvkin P, Bent ZW, Wilson R, Ziraldo SB, Wheeler TD, McDermott GP, Zhu J, Gregory MT, Shuga J, Montesclaros L, Underwood JG, Masquelier DA, Nishimura SY, Schnall-Levin M, Wyatt PW, Hindson CM, Bharadwaj R, Wong A, Ness KD, Beppu LW, Deeg HJ, McFarland C, Loeb KR, Valente WJ, Ericson NG, Stevens EA, Radich JP, Mikkelsen TS, Hindson BJ, Bielas JH (2017) Massively parallel digital transcriptional profiling of single cells. Nat Commun 8:14049. https://doi.org/10.1038/ncomms14049
20. Satija R, Farrell JA, Gennert D, Schier AF, Regev A (2015) Spatial reconstruction of single-cell gene expression data. Nat Biotechnol 33(5):495–502. https://doi.org/10.1038/nbt.3192
21. Luecken MD, Theis FJ (2019) Current best practices in single-cell RNA-seq analysis: a tutorial. Mol Syst Biol 15(6):e8746. https://doi.org/10.15252/msb.20188746
22. Blondel DB, Guillaume J-L, Lambiotte R, Lefebvre E (2008) Fast unfolding of communities in large networks. J Stat Mech Theory Exp 10:P10008
23. Rousseeuw PJ (1987) Silhouettes: A graphical aid to the interpretation and validation of cluster analysis. J Comput Appl Math. 2053–65. 0377042787901257. https://doi.org/10.10381016/0377-0427(87)90125-7
24. Pace L, Goudot C, Zueva E, Gueguen P, Burgdorf N, Waterfall JJ, Quivy JP, Almouzni G, Amigorena S (2018) The epigenetic control of stemness in CD8(+) T cell fate commitment. Science 359(6372):177–186. https://doi.org/10.1126/science.aah6499

Chapter 17

Bringing Cell Subpopulation Discovery on a Cloud-HPC Using rCASC and StreamFlow

Sandro G. Contaldo, Luca Alessandri, Iacopo Colonnelli, Marco Beccuti, and Marco Aldinucci

Abstract

The idea behind novel single-cell RNA sequencing (scRNA-seq) pipelines is to isolate single cells through microfluidic approaches and generate sequencing libraries in which the transcripts are tagged to track their cell of origin. Modern scRNA-seq platforms are capable of analyzing up to many thousands of cells in each run. Then, combined with massive high-throughput sequencing producing billions of reads, scRNA-seq allows the assessment of fundamental biological properties of cell populations and biological systems at unprecedented resolution.

In this chapter, we describe how cell subpopulation discovery algorithms, integrated into rCASC, could be efficiently executed on cloud-HPC infrastructure. To achieve this task, we focus on the StreamFlow framework which provides container-native runtime support for scientific workflows in cloud/HPC environments.

Key words Single-cell RNA sequencing, Cloud computing, HPC environment, StreamFlow

1 Introduction

RNA sequencing (RNA-seq) became a routine method in biomedical research for discovering gene expression patterns across millions of cells.

However, RNA-seq is typically performed in bulk, which does not allow to retain the relevant biological differences among cells. To deal with this crucial aspect, single-cell RNA-seq (scRNA-seq) analysis was developed. In scRNA-seq, each sequencing library represents a single cell. Therefore, scRNA-seq allows assessing the fundamental biological properties of cells in more detail.

Among the specific bioinformatics tools proposed in the literature for scRNA-seq, rCASC (https://github.com/kendomaniac/rCASC) [1–3] is specifically designed to provide an integrated analysis environment for cell subpopulation discovery providing a

Raffaele A. Calogero and Vladimir Benes (eds.), *Single Cell Transcriptomics: Methods and Protocols*, Methods in Molecular Biology, vol. 2584, https://doi.org/10.1007/978-1-0716-2756-3_17,

high level of flexibility and enabling computation reproducibility as part of the data analysis. rCASC provides, in a single computational environment, (i) raw data preprocessing, (ii) subpopulation discovery via different clustering approaches, and (iii) cluster-specific gene signature detection. These three steps require different computational resources and computing models, leading to a workflow which requires a heterogeneous computing environment, e.g., CPU and GPU architectures, multi-cores hardware, and multi-server environment. In particular, clustering is the most computationally demanding activity.

In this chapter, we describe how the cell subpopulation discovery functions implemented in rCASC can be brought on a cloud-HPC infrastructure using StreamFlow (https://streamflow.di.unito.it) [4]. In such context, StreamFlow has been leveraged to execute the workflow on top of a hybrid cloud-HPC environment.

2 Materials

2.1 Minimal Hardware/Software Requirements

The analyses described in this chapter perform best if implemented in a HPC cloud environment, e.g., https://hpc4ai.it/.

Docker installation (https://docs.docker.com/get-docker/) and R installation are required (https://cran.r-project.org/). Specifically, the installation of *devtools* library and the rCASC github package (https://github.com/kendomaniac/rCASC) is required.

2.2 Exemplary Dataset

As exemplary dataset we used was RNA5c, which is available at the GitHub repository https://github.com/kendomaniac/single_cell_transcriptomics/tian_et_al_2018.git. This dataset includes the human lung adenocarcinoma cell lines H2228, H1975, A549, H838, and HCC827 mixed at the same ratio. RNA-5c includes 3904 cells: 1242 cells of A549, 436 cells of H1975, 749 cells of H2228, 879 cells of H838, and 598 cells of HCC827 cells. The raw data set is available at GEO repository GSM3618014 [5]. The count table in GitHub has the cell types associated with each cell name. Analysis results are available at https://github.com/Gepiro/rCASC_StreamFlow/tree/main/Results.

2.3 Used Hardware

For the experiments reported in this chapter, we used eight virtual machines: each of them with 8 cores, 32 GB RAM, Ubuntu 20.04 LTS, and Docker 20.10.6. These machines were provided by the High-Performance Computing for Artificial Intelligence (HPCAI) Center at the University of (https://hpc4ai.unito.it/).

2.4 StreamFlow

The *StreamFlow framework* [4] was created as container-native runtime support for scientific workflows on cloud/HPC environments. In particular, StreamFlow automatically manages the life cycle of complex, multi-container environments, the remote execution of potentially distributed tasks (e.g., MPI and MapReduce), and the data movements between subsequent steps.

StreamFlow was designed to seamlessly integrate with external coordination semantics, allowing users to execute existing workflows on distributed infrastructures without modifying the business logic. In particular, it is fully compliant with the *Common Workflow Language (CWL)* [6] open standard. The same design concept applies to most supported execution environments, described in an external, well-known format whenever such format exists (e.g., Helm charts for Kubernetes deployments or Slurm scripts for HPC workloads).

StreamFlow relies on the *hybrid workflows* paradigm, in which the workflow steps, the topology of deployment locations in charge of their execution, and the mapping relations among steps and locations are included in the same model. This approach fosters portability and reproducibility, directly including the entire execution environment in the workflow specification. Specifically, users rely on a StreamFlow file, conventionally called `streamflow.yml`, to link each step in a workflow with the service that should execute it with a declarative YAML syntax.

2.4.1 Architecture

StreamFlow input is composed of three main elements:

- A workflow description
- One or more deployment descriptions
- A StreamFlow file to bind each step of the workflow with the most suitable execution environment

2.4.2 How to Install

It is possible to install StreamFlow as a Python package with pip command:

```
pip install streamflow
```

StreamFlow requires Python >= 3.8 to be installed on the system. Then the workflow can be executed through the StreamFlow CLI:

```
streamflow run /path/to/streamflow.yml
```

Moreover, StreamFlow can be installed by Docker. To download the latest StreamFlow image, the following command can be used:

```
docker pull alphaunito/streamflow:latest
```

2.5 Clustering Algorithms

There are two main types of clustering algorithms in rCASC implemented in StreamFlow:

- Hierarchical clustering algorithm
- Partitional clustering algorithm

The former type provides a hierarchical decomposition of the set of data using some criteria. The latter one constructs various partitions and then evaluates them by some criteria. Essentially the key differences among these clustering types are related to running time and partitioning assumptions.

Hierarchical clustering algorithms are typically slower than other clustering methods but require fewer assumptions providing more meaningful and subjective division of clusters. Indeed, the hierarchical clustering algorithm requires as input only a similarity measure, while partial clustering requires stronger assumptions such as the number of clusters and the initial centers.

2.5.1 Hierarchical Clustering

Griph

Griph [7] is an R package for the analysis of single-cell RNA-sequencing data. It provides a graph-based approach-based Louvain modularity. Thus, its clustering approach is like agglomerative clustering methods, where every node is initially assigned to its own community, and communities are subsequently built by iterative merging.

2.5.2 Partitional Clustering

tSne + K-means

tSne [8] + K-means [9] is often used for the analysis of single-cell RNA-seq data. Indeed, firstly data reduction is performed using the tSne method, and then cells are clustered by the k-means method, which aims partitioning the points into k groups, such that the sum of squares from points to the assigned cluster centers is minimized.

SIMLR

SIMLR [10], single-cell interpretation via multi-kernel learning, is a framework for learning similarity measures from single-cell RNA-seq data in order to perform dimensionality reduction, clustering, and visualization. In particular, the clustering task is provided through k-mean clustering.

3 Methods

3.1 Executing Command

To run the workflow, users must execute the following command:

```
bash testTime.sh
```

The script below is present in testTime.sh file. The first part of the script, i.e., lines 1–6, checks if the output directory already exists, otherwise the script creates it. Then the second part implements the StreamFlow workflow execution.

```
DIR="ResultsWorkflow"

if [ -d "$DIR" ]; then
  echo "$DIR exists"
else
  mkdir ResultsWorkflow
fi

now=$(date)
echo "Start time : $now" >| time.txt
streamflow run path/to/testTime.yml -outdir ResultsWork-
flow
now=$(date)
echo "End time : $now" >> time.txt
```

3.2 StreamFlow's File

The testTime.yml reported in hereafter contains the description of three elements required by StreamFlow as described in Subheading 2.4.

```
#!/usr/bin/env streamflow
version: v1.0
workflows:
  master:
    type: cwl
    config:
      file: /path/to/testTime.cwl
      settings: path/to/config.yml
    bindings:
      - step: /
        target:
          model: ssh-model
models:
  ssh-model:
    type: ssh
    config:
      nodes:
        - worker1
        - worker2
        - worker3
        - worker4
        - worker5
        - worker6
        - worker7
      username: ubuntu
      sshKey: /home/ubuntu/.ssh/id_rsa
      maxConcurrentSessions: 8
```

The above description of the testTime.yml file includes a link to other two files reporting the workflow description (i.e., testTime.cwl) and the deployment descriptions, for instance, R scripts or count Matrix (i.e., config.yml).

testTime.cwl file

```
#!usr/bin/env cwl-runner
cwlVersion: v1.0
class: Workflow
requirements:
  ScatterFeatureRequirement: [11]
inputs:
  fileRPermutationP: File
  skeletonPermutationP: File
  index_prova: string[]
  matrix: File
  nCluster: File
outputs:
  mtxKilledCell:
    type:
      type: array
      items: File
    outputSource: permutationP/mtxKilledCell
  mtxPermutationP:
    type:
      type: array
      items: File
    outputSource: permutationP/mtxPermutationP

steps:
  permutationP:
    run: permutationClusteringP.cwl
    scatter: index
    in:
      fileRPermutationP: fileRPermutationP
      mtxTopX: matrix
      skeleton: skeletonPermutationP
      index: index_prova
      nCluster: nCluster
    out: [mtxKilledCell, mtxPermutationP]
```

The script above is divided into the following sections:

- inputs → the inputs necessary to each workflow step (e.g., R scripts and the additional parameters)
- outputs → files or directories saved inside the output folder.
- steps → the description of the tasks executed in step.

config.yml file

```
skeletonPermutationZero:
  class: File
  path: ../Zero/permutationClusteringSW.R
skeletonPermutationP:
  class: File
  path: ../RFiles/skeleton_PermutationClusteringP.R
matrix:
  class: File
  path: ../dataset/RNA-5c.csv
```

The script above contains a list of the deployment description. For each item is specified:

- class → the item's class (e.g., File, Directory)
- path → the item's relative path

Finally, the execution environment is specified in the last part of testTime.yml, where ssh protocol is the execution environment for parallelization of the workflow on 8 nodes.

In Fig. 1, the implemented workflow schema with its inputs and outputs is shown.

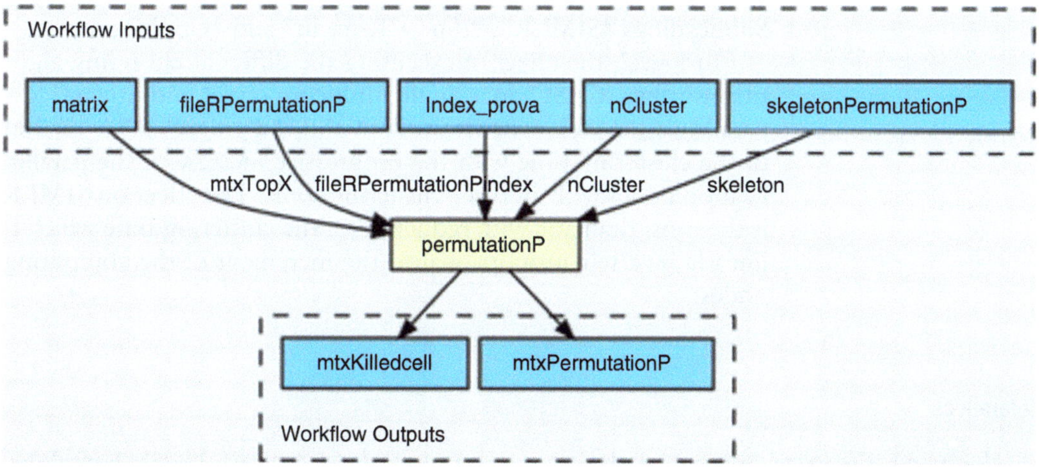

Fig. 1 Implemented workflow schema: The parallelized clustering algorithm takes in input: count matrix file; the permutation file, which is the script required to execute the parallelization; the permutation index; the number of clusters to be uses for the partitioning; the total number of permutations to be performed. For each permutation that runs independently, the following files are generated: mtxkilledCell, where all the cells were removed as a consequence of the stored bootstrap algorithm; mtxPermutationP, where all the remaining cells are labeled according to the belonging cluster

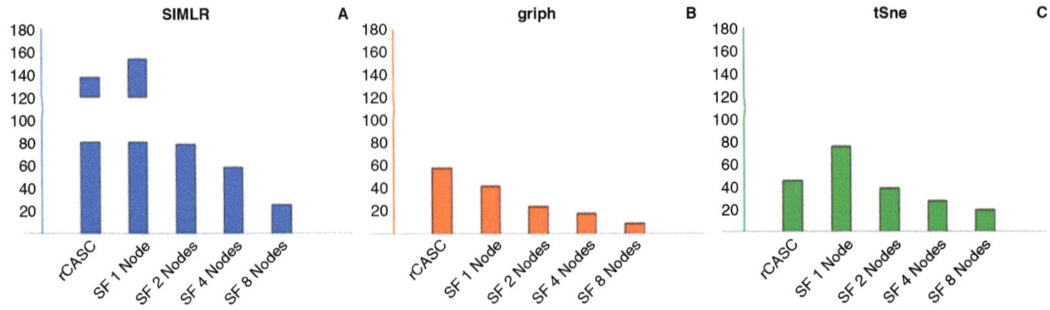

Fig. 2 Time required to cluster the 3904 cells of the RNA5c dataset, using 80 bootstraps respectively on a multi-core server with rCASC and using StreamFlow on 1, 2, 4, and 8 nodes. (**a**) SIMLR, (**b**) K-means + tsne, and (**c**) Griph

3.3 Analysis

We evaluated the speed-up achievable in clustering using StreamFlow with respect to a single multi-core server. rCASC implements a metrics focused at evaluating the cluster stability: cell stability score (CSS) [1]. CSS requires the perturbation of the initial dataset, i.e., removal of a random subset of cells, and the clustering of the perturbed dataset. Perturbation/clustering are repeated multiple times to evaluate the overall stability of cells partitioning, measuring how many times each cell remains in a specific cluster independently by the effect of the perturbations disturbing the overall dataset structure.

We tested on the RNA-5c (3904 cells), *see* exemplary dataset Subheadings "SIMLR", "tSne + K-mean" and "Griph".

In Fig. 2, the effect of executing the different clustering algorithms using rCASC on a single multicore server and StreamFlow parallelization is shown. StreamFlow shows a progressive reduction of the clustering time with the progressive increase of the parallelization nodes with all three clustering tools. The effect on SIMLR is the most dramatic with reduction of the clustering time which is linearly inversely proportional to the increment of the computing nodes.

References

1. Alessandri L, Cordero F, Beccuti M, Licheri N, Arigoni M, Olivero M, Di Renzo MF, Sapino A, Calogero R (2021) Sparsely-connected autoencoder (SCA) for single cell RNAseq data mining. NPJ Syst Biol Appl 7(1):1. https://doi.org/10.1038/s41540-020-00162-6
2. Alessandri L, Ratto ML, Contaldo SG, Beccuti M, Cordero F, Arigoni M, Calogero RA (2021) Sparsely connected autoencoders: a multi-purpose tool for single cell omics analysis. Int J Mol Sci 22(23). https://doi.org/10.3390/ijms222312755
3. Alessandri L, Cordero F, Beccuti M, Arigoni M, Olivero M, Romano G, Rabellino S, Licheri N, De Libero G, Pace L, Calogero RA (2019) rCASC: reproducible classification analysis of single-cell sequencing data. Gigascience 8(9). https://doi.org/10.1093/gigascience/giz105
4. Colonnelli I, Cantalupo B, Merelli I, Aldinucci M (2021) StreamFlow: cross-breeding cloud

with HPC. IEEE Trans Emerg Topics Comput 9(4). https://doi.org/10.1109/TETC.2020.3019202

5. Tian L, Dong X, Freytag S, Le Cao KA, Su S, JalalAbadi A, Amann-Zalcenstein D, Weber TS, Seidi A, Jabbari JS, Naik SH, Ritchie ME (2019) Benchmarking single cell RNA-sequencing analysis pipelines using mixture control experiments. Nat Methods 16(6): 479–487. https://doi.org/10.1038/s41592-019-0425-8
6. Crusoe M, Abeln S, Iosup A, Amstutz P, Chilton J, Tijanić N, Ménager H, Soiland-Reyes S, Goble C (2022) Methods included: standardizing computational reuse and portability with the common workflow language. Commun ACM 65:54–63. https://doi.org/10.48550/arXiv.2105.07028
7. Serra D, Mayr U, Boni A, Lukonin I, Rempfler M, Challet Meylan L, Stadler MB, Strnad P, Papasaikas P, Vischi D, Waldt A, Roma G, Liberali P (2019) Self-organization and symmetry breaking in intestinal organoid development. Nature 569(7754):66–72. https://doi.org/10.1038/s41586-019-1146-y
8. van der Maate L, Hilton G (2008) Visualizing Data using t-SNE. J Mach Learn Res 9
9. Tavazoie S, Hughes JD, Campbell MJ, Cho RJ, Church GM (1999) Systematic determination of genetic network architecture. Nat Genet 22(3):281–285. https://doi.org/10.1038/10343
10. Wang B, Zhu J, Pierson E, Ramazzotti D, Batzoglou S (2017) Visualization and analysis of single-cell RNA-seq data by kernel-based similarity learning. Nat Methods 14(4):414–416. https://doi.org/10.1038/nmeth.4207
11. Zheng GX, Terry JM, Belgrader P, Ryvkin P, Bent ZW, Wilson R, Ziraldo SB, Wheeler TD, McDermott GP, Zhu J, Gregory MT, Shuga J, Montesclaros L, Underwood JG, Masquelier DA, Nishimura SY, Schnall-Levin M, Wyatt PW, Hindson CM, Bharadwaj R, Wong A, Ness KD, Beppu LW, Deeg HJ, McFarland C, Loeb KR, Valente WJ, Ericson NG, Stevens EA, Radich JP, Mikkelsen TS, Hindson BJ, Bielas JH (2017) Massively parallel digital transcriptional profiling of single cells. Nat Commun 8:14049. https://doi.org/10.1038/ncomms14049

Chapter 18

Profiling RNA Editing in Single Cells

Adriano Fonzino, Graziano Pesole, and Ernesto Picardi

Abstract

RNA editing is a widespread molecular phenomenon occurring in a variety of organisms. In humans, it mainly involves the deamination of adenosine to inosine (A-to-I) in double-stranded RNAs by ADAR enzymes. A-to-I RNA editing has been investigated in different tissues as well as in diverse experimental and pathological conditions. By contrast, its biological role in single cells has not been explored in depth. Recent methodologies for cell sorting in combination with deep sequencing technologies have enabled the study of eukaryotic transcriptomes at single cell resolution, paving the way to the profiling of their epitranscriptomic dynamics.

Here we describe a step-by-step protocol to detect and characterize A-to-I events occurring in publicly available single-cell RNAseq experiments from human alpha and beta pancreatic cells.

Key words RNA editing, scRNAseq, Single cell

1 Introduction

RNA editing is a relevant epitranscriptomic molecular phenomenon occurring in a variety of organisms through mechanistically and evolutionary unrelated machineries [1]. In humans, as well as in vertebrates, the most common RNA editing event involves the deamination of adenosine (A) in inosine (I) by ADAR enzymes acting on double RNA strands [2]. The majority of human A-to-I changes reside in non-coding regions, generally introns or 3'UTRs containing repetitive elements in opposite orientation, and tend to be involved in the innate immune response preventing the interferon activation by MDA5 [2]. Only a small fraction of RNA editing events occurs in coding protein genes and can change the codon identity (recoding events) leading to novel protein isoforms [2, 3]. RNA editing deregulation at these sites has been linked to several human diseases, including neurological disorders and cancer [4, 5].

Raffaele A. Calogero and Vladimir Benes (eds.), *Single Cell Transcriptomics: Methods and Protocols*, Methods in Molecular Biology, vol. 2584, https://doi.org/10.1007/978-1-0716-2756-3_18,

A-to-I RNA editing can be detected through massive transcriptome experiments and ad hoc computational workflows in which raw RNAseq data are quality checked, mapped onto the reference genome, and resulting multiple alignments are traversed to look at A-to-G changes (I is commonly recognized as G by cellular machineries and sequencing enzymes) [6]. Large-scale genomic investigations have revealed that A-to-I RNA editing is pervasive in the human transcriptome and occurs in a tissue-specific manner [7, 8]. Millions of known events are currently collected in the specialized REDIportal database [9], a unique resource for studying RNA editing in its biological and genomic context.

Recently, REDIportal has been used in combination with REDItools [10], a widespread package for RNA editing detection in RNAseq data, to profile known A-to-I events in single cells RNAseq experiments from human brains, proving that RNA editing is also a cell type specific mechanism [11]. Nonetheless, many aspects of A-to-I RNA editing in single cells have not yet been explored in detail. For example, editing levels in single cells show a peculiar bimodal distribution with values close to 0 or 1, suggesting that RNA editing in single cells could be an all or nothing phenomenon [11, 12].

To facilitate the profiling of A-to-I RNA editing in single-cell RNAseq experiments, here we describe a simple and reproducible protocol using publicly available datasets from human alpha and beta pancreatic cells [13].

2 Materials

The following protocol is based on the combined use of REDItools [10] and REDIportal [9]. While REDItools are Python scripts to call RNA editing in RNAseq data [10], REDIportal is a comprehensive database of A-to-I RNA editing events in human and mouse [9]. REDItools are currently used in Linux and Mac OS X equipped machines. They could not run on Windows-based computers due to the Pysam module dependency that is not natively available for this environment (*see* **Note 1**). REDItools require aligned RNAseq data from individual cells. Although this step can be accomplished using different splice-aware programs such as GSNAP [14] or HISAT [15], we suggest STAR [16] because of its accuracy and speed.

Although there are no specific restrictions in the use of the present protocol for calling RNA editing in bulk tissues, its profiling in single-cell RNAseq data is effective only for reads generated according to the SMART-seq2 sequencing protocol because it enables the full-length sequencing of transcripts [17]. Our procedure could still be applied to scRNAseq data produced by alternative protocols oriented to the sequencing of the 3′ ends (such as DROPseq, CELseq, InDROP, or Chromium 10x) [18]. However, results could not provide a complete overview of the A-to-I repertoire.

2.1 Prerequisites

2.1.1 Hardware

64-bit computer running Linux, Mac OS, or other Unix-based operating system with at least 32GB of RAM (for genome mapping using STAR) and at least 500 GB of free disk space (*see* **Note 2**).

2.1.2 Software

The following programs are required to correctly run the entire procedure:

- REDItools version 1.2 or superior [10] available at https://github.com/BioinfoUNIBA/REDItools for calling RNA editing sites. REDItools require the Python interpreter (at least version 2.7) and the external pysam module.
- SAMtools version 1.9 or superior [19], including bgf and tabix executables from the HTSlib (http://www.htslib.org/download/), for dealing with BAM files and indexed textual tables.
- STAR version 2.5 or superior [16] (https://github.com/alexdobin/STAR/releases) to align RNAseq onto a reference genome.
- Fastp version 0.20 or superior [20] (https://github.com/OpenGene/fastp) to pre-process RNAseq read removing low-quality regions.
- FeatureCounts version 1.6.0 or superior [21] (http://subread.sourceforge.net/) for counting reads to known human genes.
- RseqQC version 2.6 or superior [22] (http://rseqc.sourceforge.net/) for detecting the strand orientation of aligned reads (through the infer_experiment.py script).
- SRAtoolkit version 2.10.5 or superior (https://github.com/ncbi/sra-tools) to download scRNAseq experiments from SRA by means of the fasterqdum executable.
- RNAEditingIndexer [23] (https://github.com/a2iEditing/RNAEditingIndexer) to calculate the Alu editing index (AEI) from BAM files.
- QEdit to quantify RNA editing in deep transcriptome datasets [24], available at https://github.com/BioinfoUNIBA/QEdit.

2.1.3 Annotations

The human genome assembly hg38 (in fasta format) and gene annotations (in gtf format) are available at the Illumina iGenomes repository (https://emea.support.illumina.com/sequencing/sequencing_software/igenome.html). Known RNA editing events, compliant with the hg38 genome version, can be downloaded from the REDIportal database (http://srv00.recas.ba.infn.it/atlas/download.html) (*see* **Note 3**).

2.1.4 RNAseq Data

Single-cell RNAseq data used in the following procedure are freely available through the SRA database under the BioProject accession PRJNA322355 [13]. The analysis has been restricted to 30 alpha and beta cells with SRA run accessions reported in Table 1.

3 Methods

3.1 Downloading scRNAseq Data

1. Create a main folder named "scRNA" and two subfolders called "alpha" and "beta" for storing alpha and beta reads, respectively:

```
$ mkdir scRNA
$ mkdir scRNA/alpha
$ mkdir scRNA/beta
```

2. Download scRNAseq data in fastq format using the SRA run accessions indicated in Table 1 and the fasterqdum program from the SRAtoolkit package. Repeat the following command line for each experiment, replacing "mySRRsample" with the accession number and "myfolder" with the corresponding cell-line folder:

```
$ fasterq-dump -3 -O myfolder mySRRsample
```

3. Organize fastq files in subfolders named according to the SRA run accession using the following python script to save time:

```
import sys, os, glob

folder=sys.argv[1]
os.chdir(folder)
fq=glob.glob('*.fastq')

d={}
for i in fq:
base=i.split('_')[0]
if base not in d: d[base]=[i]
 else: d[base].append(i)
for i in d:
os.mkdir(i)
files=d[i]
for j in files: os.system('mv %s %s' %(j,i))
```

Table 1
Single-cell RNAseq experiments used in this protocol. For each cell type, we report the corresponding SRA run accession

Cell type	Accession
Alpha	SRR3564196
Alpha	SRR3564197
Alpha	SRR3564200
Alpha	SRR3564201
Alpha	SRR3564202
Alpha	SRR3564203
Alpha	SRR3564204
Alpha	SRR3564205
Alpha	SRR3564206
Alpha	SRR3564207
Alpha	SRR3564209
Alpha	SRR3564211
Alpha	SRR3564212
Alpha	SRR3564213
Alpha	SRR3564214
Alpha	SRR3564217
Alpha	SRR3564219
Alpha	SRR3564220
Alpha	SRR3564222
Alpha	SRR3564223
Alpha	SRR3564224
Alpha	SRR3564226
Alpha	SRR3564230
Alpha	SRR3564232
Alpha	SRR3564233
Alpha	SRR3564235
Alpha	SRR3564236
Alpha	SRR3564240
Alpha	SRR3564241
Alpha	SRR3564242
Beta	SRR3564208

(continued)

Table 1
(continued)

Cell type	Accession
Beta	SRR3564225
Beta	SRR3564228
Beta	SRR3564231
Beta	SRR3564237
Beta	SRR3564238
Beta	SRR3564247
Beta	SRR3564248
Beta	SRR3564252
Beta	SRR3564258
Beta	SRR3564261
Beta	SRR3564266
Beta	SRR3564269
Beta	SRR3564271
Beta	SRR3564279
Beta	SRR3564281
Beta	SRR3564290
Beta	SRR3564302
Beta	SRR3564327
Beta	SRR3564333
Beta	SRR3564337
Beta	SRR3564361
Beta	SRR3564384
Beta	SRR3564387
Beta	SRR3564388
Beta	SRR3564399
Beta	SRR3564407
Beta	SRR3564408
Beta	SRR3564415
Beta	SRR3564430

Save the script in the file makeFolders.py and run it passing the cell folders, one at a time:

```
$ python makeFolders.py alpha
$ python makeFolders.py beta
```

3.2 Preprocessing and Alignment of Reads

1. Download and uncompress into the scRNA folder the hg38 human genome in FASTA format as well as gene annotations from the Illumina iGenomes repository:

```
$ wget http://igenomes.illumina.com.s3-website-us-
east-1.amazonaws.com/Homo_sapiens/UCSC/hg38/Homo_sapien
s_UCSC_hg38.tar.gz
$ tar -xvf Homo_sapiens_UCSC_hg38.tar.gz
```

The whole genome will be located in Homo_sapiens/UCSC/hg38/Sequence/WholeGenomeFasta/genome.fa, while gene annotations in gtf format will be present in Homo_-sapiens/UCSC/hg38/Annotation/Genes/genes.gtf.

2. Build a reference genome index for STAR into a novel folder named stardb:

```
$ mkdir stardb
$ STAR --runMode genomeGenerate --genomeDir stardb --gen-
omeFastaFiles Homo_sapiens/UCSC/hg38/Sequence/WholeGeno-
meFasta/genome.fa   --sjdbGTFfile   20   Homo_sapiens/UCSC/
hg38/Annotation/Genes/genes.gtf --sjdbOverhang 75
```

where --**runMode** is the type of the run ("genomeGenerate" specifies the genome indices generation job), --**genomeDir** is the path to the reference genome to be indexed, --**sjdbGTFfile** indicates the location of the file with annotated transcripts in the standard GTF format (STAR will extract splice junctions from this file and use them to improve the mapping accuracy), and --**sjdbOverhang** is the length of the genomic sequence surrounding the annotated junction to be used in constructing the splice junctions database.

3. Trim scRNAseq by fastp. Enter into each alpha and beta subfolder and run the following command, replacing "mySRRsample" with the accession number and "myfolder" with the corresponding cell-line folder.

```
$ cd myfolder
$ cd mySRRsample
$ fastp -i mySRRsample_1.fastq -I mySRRsample_2.fastq -o
mySRRsample_trim_1.fastq -O mySRRsample_trim_2.fastq --
detect_adapter_for_pe -q 25 -n 2 -l 50 -y -x -w 10
$ cd ../../
```

where **-i** [read1 input file name], **-I** [read2 input file name], **-o** [read1 output file name], **-O** [read2 output file name], **-q** [INT] "base phred quality" (e.g., 25), **-n** [INT] "discard reads with more than a fixed number of N bases," **-l** [INT] "reads length cutoff, reads shorter than this value will be discarded" (e.g., 50), **-y** "low complexity filter, defined as the percentage of base that is different from its next base (base[i]!= base [i+1])," **-x** "activate polyX trimming in 3' ends," **-w** "cpu threads number," and **--detect_adapter_for_pe** "adapter detection for paired end reads".

4. Align scRNAseq reads to the reference genome with STAR. Enter into each alpha and beta subfolder and run the following command, replacing "mySRRsample" with the accession number and "myfolder" with the corresponding cell-line folder.

```
$ cd myfolder
$ cd mySRRsample
$ STAR --twopassMode Basic --runThreadN 8 --genomeDir
../../stardb --genomeLoad NoSharedMemory --outFileName-
Prefix mySRRsample_ --outReadsUnmapped Fastx --outSAMtype
BAM SortedByCoordinate --outSAMstrandField intronMotif --
outSAMattributes All --outFilterType BySJout --outFilter-
MultimapNmax 1 --alignSJoverhangMin 8 --alignSJDBover-
hangMin 1 --outFilterMismatchNmax 999 --
outFilterMismatchNoverLmax 0.04 --alignIntronMin 20 --
alignIntronMax 1000000 --alignMatesGapMax 1000000 --read-
FilesIn mySRRsample_trim_1.fastq mySRRsample_trim_2.
fastq.gz
$ cd ../../
```

where:

--**twopassMode** enables the two passes algorithm.

--**runThreadN [INT]** is the number of cpu threads on the user's machine (e.g., 4).

--**genomeDir** is the folder containing the indexed genome.

--**genomeLoad** NoSharedMemory indicates the program to store distinct copies of reference in the computer's RAM for each alignment job.

--**outFileNamePrefix** is a string to be included as output filename.

--**outReadsUnmapped** indicates the program to store unmapped and partially mapped reads (mate1/2) in separate fasta/fastq files.

--**outSAMtype** SortedByCoordinate specifies the alignment output format (e.g., BAM) that will be sorted according to a selected criterion (e.g., coordinates).

--**outSAMstrandField** intronmotif tells the program to discard reads with inconsistent and/or non-canonical introns.

--**outSAMattributes** describes which SAM flags *(16)* will be present in the output (e.g., ALL means NH HI AS nM NM MD jM j).

--**readFilesCommand** specifies a command line that will be executed for each of the input files (e.g., zcat, since the input reads are compressed with gzip).

--**outFilterType** indicates the program to keep only those reads that contain junctions that passed filtering into SJ.out.tab.

--**outFilterMultimapNmax** is the maximum number of loci a read is allowed to map to; if set to 1 only unique reads are shown in the output.

--**alignSJoverhangMin** is minimum spliced reads overhang (e.g., 8 nucleotides).

alignSJDBoverhangMin is the minimum spliced reads overhang for annotated splicing sites (splice junctions db inferred by STAR from reference genome and annotations).

--**outFilterMismatchNmax** is the maximum number of mismatches per reads pair, large numbers for this parameter (e.g., 999) switch off the filter.

--**outFilterMismatchNoverLmax** is the maximum number of tolerated mismatches relative to read length (e.g., 0.04, for 2x100b max number of mismatches is 0.04*200 = 8b for the paired reads).

--**alignIntronMin** indicates STAR the minimum intron length to consider during the alignment process (e.g., 20).

--**alignIntronMax** is the maximum intron length (e.g., 1,000,000).

--**alignMatesGapMax** is the maximum genomic distance between mates (e.g., 1,000,000).

--**readFilesIn** is the path to input read1 (and, if present, read2).

5. Index aligned scRNAseq reads with SAMtools. Enter into each alpha and beta subfolder and run the following command, replacing "mySRRsample" with the accession number and "myfolder" with the corresponding cell-line folder.

```
$ cd myfolder
$ cd mySRRsample
$ mv mySRRsample_Aligned.sortedByCoord.out.bam mySRRsam-
ple.bam
$ samtools index mySRRsample.bam
$ cd ../../
```

6. Infer the orientation of scRNAseq reads using the infer_experiment.py script from the RSeQC package. It derives by inference the orientation of reads from the orientation of known transcript annotations. Enter into each alpha and beta subfolder and run the following command, replacing "mySRRsample" with the accession number and "myfolder" with the corresponding cell-line folder.

```
$ wget --no-check-certificate
https://sourceforge.net/projects/rseqc/files/BED/Human_-
Homo_sapiens/ hg38_RefSeq.bed.gz
$ gunzip hg38_RefSeq.bed.gz

$ cd myfolder
$ cd mySRRsample
$ infer_experiment.py -r ../../hg38_RefSeq.bed -s 2000000
-i mySRRsample.bam
```

where:

-r [REFGENE_BED] is the gene annotations file (in BED format) for the reference human genome assembly hg38.

-s [SAMPLE_SIZE] is the number of reads sampled from SAM/-BAM in order to infer the strand orientation.

-i [INPUT_FILE] is the input alignment file in SAM or BAM format.

```
$ cd ../../
```

3.3 Calling Known RNA Editing Candidates

1. Download and uncompress hg38 REDIportal annotations into the scRNA folder:

```
$ wget http://srv00.recas.ba.infn.it/webshare/ATLAS/
donwload/TABLE1_hg38.txt.gz
$ gunzip TABLE1_hg38.txt.gz
```

2. Convert hg38 REDIportal annotations in gtf using the following python script:

```
import sys, os

f=open(sys.argv[1])
for i in f:
 if i.startswith('Region'): continue
 l=i.strip().split('\t')
 if not l[0].startswith('chr'): continue
 gid='gene_id "%s:%s"; transcript_id "%s:%s";' %(l[0],l
[1],l[0],l[1])
 line=[l[0],'hg38','atlas',l[1],l[1],".",l[4],'.',gid]
 print('\t'.join(line))
f.close()
```

Save the script in the file convertAtlasToGtf.py and run it with the command:

```
$ python convertAtlasToGtf.py TABLE1_hg38.txt > atlas.
gtf
```

3. Sort and index hg38 REDIportal annotations:

```
$ sort -k1,1 -k4,4n atlas.gtf > atlas.sorted.gtf
$ bgzip -@ 10 atlas.sorted.gtf
$ tabix -p gff atlas.sorted.gtf.gz
```

4. Call RNA editing using REDItools and known REDIportal annotations. Enter into each alpha and beta subfolder and run the following command, replacing "mySRRsample" with the accession number and "myfolder" with the corresponding cell-line folder.

```
$ cd myfolder
$ cd mySRRsample
$ REDItoolDnaRna.py -t 10 -i mySRRsample_hg38.bam -f
```

```
Homo_sapiens/UCSC/hg38/Sequence/WholeGenomeFasta/genome.
fa -c 1,1 -T ../../atlas.sorted.gtf.gz -o editing -p -u
-v 0 -n 0 -G ../../atlas.sorted.gtf.gz -e -m 20,20 -q
30,30
```

where:

-**f** is the path to reference genome in fasta file format.

-**i** refers to scRNAseq in BAM format.

-**o** is the output folder in which all results will be stored.

-**t** is the number of threads for job (e.g., 10). It indicates how many processes should be launched and assigned to different threads. Each thread will work on an individual chromosome/region.

-**m** is the minimum mapping quality score for WGS and RNAseq reads, respectively (e.g.,30,255); it should be tuned depending on the alignment program. Default values are (25,25).

-**v** is the minimum number of RNAseq reads supporting the variation (e.g., 0). Default value is 3.

-**q** is the minimum quality score for WGS and RNAseq reads, respectively (e.g., 30,30). Default values are (25,25).

-**c** is the minimum coverage for WGS and RNAseq reads, respectively (e.g., 1,1). Default values are [10].

-**e** indicates the script to exclude multiple hits in RNAseq.

-**n** is the minimum RNA editing frequency to be considered (e.g., 0).

-**u** tells the script to consider mapping quality set with option -m.

-**p** tells the script to use paired and concordant RNAseq reads only.

-**T** tells the script to work only on positions specified in a given gtf file (here it is the gtf file with REDIportal annotations).

-**G** tells the script to infer the strand orientation using annotations provided by an external gtf file (here it is the gtf file with REDIportal annotations).

Each execution of REDItoolDnaRna.py will create an output folder named "editing" containing a subfolder in which the script will store results. The main output is a simple textual table called outTable_XXXX, where XXXX is a random number generated at each run. The output folder includes also a parameters.txt file containing all selected options as well as the main command line.

5. Collect RNA editing events in a textual table for downstream analyses using the following python code. Only sites supported by at least 10 reads will be selected.

```
import sys, os
import glob, gzip

folders=['alpha','beta']

isgzip=0
tabgz=glob.glob('alpha/SRR*/editing/*/outTable*.gz')
if len(tabgz)>0: isgzip=1

if isgzip:
 a=["alpha_"+os.path.basename(x) for x in glob.glob('al-
pha/SRR*/editing/*/outTable*.gz')]
 b=["beta_"+os.path.basename(x) for x in glob.glob('beta/
SRR*/editing/*/outTable*.gz')]
else:
 a=["alpha_"+os.path.basename(x) for x in glob.glob('al-
pha/SRR*/editing/*/outTable*')]
 b=["beta_"+os.path.basename(x) for x in glob.glob('beta/
SRR*/editing/*/outTable*')]
ab=a+b
ids={}
for i in range(len(ab)):
 ids[ab[i]]=i

d={}
for folder in folders:
 os.chdir(folder)
 for i in os.listdir('.'):
 if i.startswith('SRR'):
 os.chdir(i)
 if gzip:
 tab=glob.glob('editing/*/outTable*.gz')[0]
 else:
 tab=glob.glob('editing/*/outTable*')[0]
 tabname=folder+'_'+os.path.basename(tab)
 id=ids[tabname]
 #print(tab)
 if gzip: f=gzip.open(tab)
 else: f=open(tab)
 xx,xxx=0,0
 for j in f:
 if gzip: line=j.decode('utf-8')
 else: line=j
 if line.startswith('Region'):continue
 l=line.strip().split('\t')
 #print(l)
 if l[7]!='AG': continue
```

```
xxx+=1
if int(l[4])<10: continue
pos=l[0]+':'+l[1]
xx+=1
if pos not in d:
d[pos]=['-' for x in range(len(ab))]
d[pos][id]=l[8]
else:
d[pos][id]=l[8]
f.close()
print('%s\t%i\t%i' %(tabname,xxx,xx))
os.chdir('..')
os.chdir('..')

f=open('editing.txt','w')
f.write('Position\t'+'\t'.join(ab)+'\n')
for i in d:
 p=d[i]
 f.write(i+'\t'+'\t'.join(p)+'\n')
f.close()
```

Copy the above code in a file called CollectEditing.py and run it with the command:

```
$ python CollectEditing.py
```

The script will generate a table named editing.txt in which each row will be an editing site while each column will be a cell. Further filters could also be applied to restrict the number of sites to those edited in a consistent fraction of cells (for example 40 or 50% of all cells).

3.4 Calculating the Alu Editing Index

The global RNA editing activity of a human cell can be measured through the Alu Editing Index (AEI), the weighted average of editing events occurring in all adenosines within Alu elements [23]. The AEI can be calculated using the aligned BAM files and the RNAEditingIndexer program.

1. Create two folders (into the mail scRNA folder) for storing BAM files:

```
$ mkdir bamalpha
$ mkdir bambeta
$ cp alpha/*/*.bam* bamalpha
$ cp beta/*/*.bam* bambeta
```

2. Run the RNAEditingIndexer program in each cell folder:

```
$ RNAEditingIndex -d bamalpha -o bamalpha/indexer -os
bamalpha --genome hg38 --paired_end -f .bam
$ RNAEditingIndex -d bambeta -o bambeta/indexer -os bam-
beta --genome hg38 --paired_end -f .bam
```

Where:

-**d** specifies the input directory

-**o** is the output directory

-**os** is the directory for the summary output

--**genome** specifies the built-in genome version to use

--**paired_end** treats the data as paired end

-**f** specifies a suffix of the BAM files to run on

At the end of each run, the program will create an output file named EditingIndex.csv containing AEI values per single scRNA experiment.

3. Collect AEI indices using the following python code and assuming that the outputs of the RNAEditingIndexer program are bamalpha/EditingIndex.csv and bambeta/EditingIndex.csv.

```
import sys, os

print('AEI\tCell_Type')
f=open('bamalpha/EditingIndex.csv')
for i in f:
 if i.startswith('StrandDecidingMethod'): continue
 l=i.strip().split(',')
 print('%s\t%s' %(l[5],'Alpha'))
f.close()
f=open('bambeta/EditingIndex.csv')
for i in f:
 if i.startswith('StrandDecidingMethod'): continue
 l=i.strip().split(',')
 print('%s\t%s' %(l[5],'Beta'))
f.close()
```

Copy the above code in a file named CollectAEI.py and run it with the command:

```
$ python CollectAEI.py > aei.txt
```

The script will direct the output to the aei.txt file and AEI values per cell will be printed out in individual rows.

4. Explore AEI results by creating a box plot and performing a statistical test to assess the significance of the RNA editing activity between alpha and beta cells. Use the following python code and the aei.txt file.

```
import sys, os
import seaborn as sns
import matplotlib.pyplot as plt
import pandas as pd
from scipy import stats

sns.set_theme(style="whitegrid")

data='aei.txt'
df = pd.read_table(data, sep='\t')
ax = sns.boxplot(x="Cell_Type", y="AEI", data=df)
plt.savefig('aei.png')
plt.close()
a=df.loc[df['Cell_Type'] == "Alpha"]['AEI']
b=df.loc[df['Cell_Type'] == "Beta"]['AEI']
mann=stats.mannwhitneyu(a,b,alternative='two-sided')
t_test=stats.ttest_ind(a,b)
print(mann)
print(t_test)
```

Copy the above code in a file named ReadAEI.py and run it with the command:

```
$ python ReadAEI.py
MannwhitneyuResult(statistic=250.0,                pva-
lue=0.0031829591616507357)
Ttest_indResult
```

```
(statistic=-2.9365876772717168pvalue=0.004752039009296634)
```

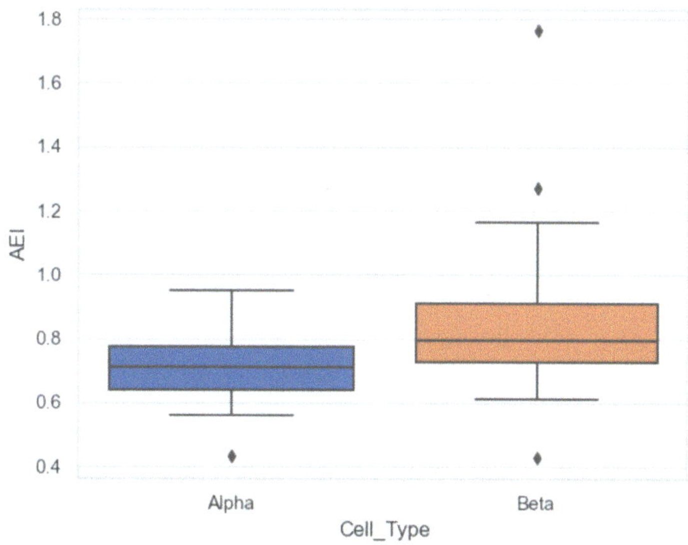

Fig. 1 Box plot of AEI values. AEI was calculated per each scRNAseq using the RNAEditingIndexer program. The distribution of AEI values indicates that the global RNA editing activity is higher in beta cells than in alpha cells, confirming the cell-type specificity of RNA editing

The execution will print out the results of two statistical tests, the Mann-Whitney test and t-test, indicating that the RNA editing activity is significantly different (p-value <0.05) between alpha and beta cells. Additionally, the script will generate the box plot named aei.png reported in Fig. 1.

3.5 Calculating the Recoding Index

Similarly to the AEI, recoding activity due to RNA editing can be quantified calculating the weighted average over all known recoding sites residing in coding protein genes, referred to as the recoding editing index (REI) [5, 24].

1. Select recoding RNA editing events from the REDIportal database. Here, each RNA editing event is annotated according to three main resources such as Gencode, RefSeq, and UCSC. To select a reliable set of recoding sites, only positions falling in non-repetitive genomic regions, annotated as non-synonymous and supported by at least 100 independent RNAseq samples, will be taken into account. Recoding sites can be extracted using the following python code:

```
import sys, os

f=open(sys.argv[1])

for i in f:
 if i.startswith('Region'): continue
```

```
 l=i.strip().split('\t')
 #print(l[11])
 #print(l[6])
 #print(l[22])
 if l[6]=='NONREP' and l[11].startswith('nonsynonymous')
and l[15].startswith('nonsynonymous') and l[19].starts-
with('nonsynonymous'
) and int(l[22])>=100:
 print (i.strip())
f.close()
```

Copy the above code in a file named CollectRecoding.py and run it with the command:

```
$ python CollectRecoding.py TABLE1_hg38.txt > recoding.
txt
```

2. Calculate the REI index through the GetREI.py script of the QEdit repository. For each alpha and beta scRNAseq run the following command:

```
$ python QEdit/scripts/GetREI.py -i myREDItable -r
recoding.txt -o myREDItable.rec
```

where myREDItable is the REDItools output table (obtained at Subheading 3.3, **step 4**), sorted and indexed by tabix:

```
$ bgzip outTable_XXXX
$ tabix -s1 -b2 -e2 -cR outTable_XXXX.gz
```

3. Collect REI values using the commands:

```
$ grep ">" */SRR*/editing/DnaRna_*/outTable_*.gz.rec |
awk 'BEGIN { FS = "/" }; {print $1,$5}' | awk 'BEGIN { FS
= " " }; {print $1"\t"$4}' > rei
$ sed '1s/^/Cell_Type\tREI\n/' rei > rei.txt
$ rm rei
```

4. Explore REI results by creating a box plot and performing a statistical test to assess the significance of the RNA editing activity between alpha and beta cells at recoding sites. Use the following python code and the rei.txt file.

```
import sys, os
import seaborn as sns
import matplotlib.pyplot as plt
import pandas as pd
from scipy import stats

sns.set_theme(style="whitegrid")

data='rei.txt'
df = pd.read_table(data, sep='\t')
ax = sns.boxplot(x="Cell_Type", y="REI", data=df)
plt.savefig('rei.png')
plt.close()
a=df.loc[df['Cell_Type'] == "Alpha"]['REI']
b=df.loc[df['Cell_Type'] == "Beta"]['REI']
mann=stats.mannwhitneyu(a,b,alternative='two-sided')
t_test=stats.ttest_ind(a,b)
print(mann)
print(t_test)
```

Copy the above code in a file named ReadREI.py and run it with the command:

```
$ python ReadREI.py
MannwhitneyuResult(statistic=647.0,                    pva-
lue=0.0036704200322962635)
Ttest_indResult
```

(statistic=2.4781375589153627pvalue=0.016139298289795285)

The execution will print out the results of two statistical tests, the Mann-Whitney test and t-test, indicating that the RNA editing activity at recoding sites is significantly different (p-value <0.05) between alpha and beta cells. Additionally, the script will generate the box plot named rei.png reported in Fig. 2.

3.6 Expression of ADAR Genes

It could be useful to have a look at the expression of ADAR genes in order to correlate their values to the activity of RNA editing. To this aim, FeatureCounts will be used for counting reads to known human genes.

1. Run FeatureCounts replacing "myBAMs" with the list of BAM files (separated by a space):

```
$ featureCounts -a Homo_sapiens/UCS
C/hg38/Annotation/Genes/genes.gtf  -o  counts_all  --pri-
mary -p -B -C -T 8 myBAMs
```

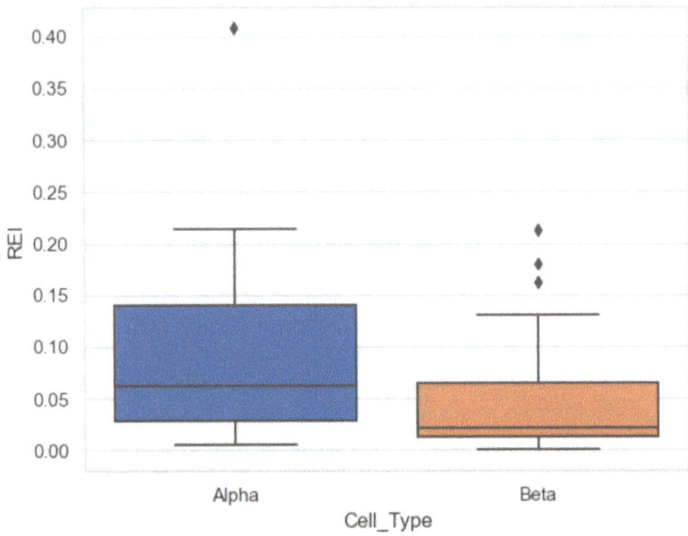

Fig. 2 Box plot of AEI values. REI was calculated per each scRNA experiment taking into account the known fraction of recoding events from REDIportal. The plot indicates that, in contrast with AEI, the RNA editing activity at recoding sites is higher in alpha cells than in beta cells

2. Convert raw counts to TPM (transcripts per million) using the following R code:

```
## Adapted from https://gist.github.com/slowkow/
c6ab0348747f86e2748b#file-counts_to_tpm-r-L44

tpm <- function(counts, lengths) {
 rate <- counts / lengths
 rate / sum(rate) * 1e6
 }

## read table from featureCounts output
args <- commandArgs(T)
tag <- tools::file_path_sans_ext(args[1])
ftr.cnt <- read.table(args[1], sep="\t", stringsAsFac-
tors=FALSE, header=TRUE)
library(dplyr)
library(tidyr)

ftr.tpm <- ftr.cnt %>%
 gather(sample, cnt, 7:ncol(ftr.cnt)) %>%
 group_by(sample) %>%
 mutate(tpm=tpm(cnt, Length)) %>%
 select(-cnt) %>%
 spread(sample, tpm)
write.table(ftr.tpm, file=paste0(tag, "_tpm.txt"),
sep="\t", row.names=FALSE, quote=FALSE)
```

Save the above code to a file named CountsToTPM.R and run it with the command:

```
$ Rscript CountsToTPM.R counts_all
```

3. Collect ADAR gene expression values using the following python code:

```
import sys, os

print('Cell_Type\tGene\tTPM')
a,b=[],[]
f=open('counts_all_tpm.txt')
for i in f:
 if i.startswith('Geneid'):
 h=i.strip().split('\t')
 for j in range(len(h)):
 if h[j].startswith('alpha'): a.append(j)
 if h[j].startswith('beta'): b.append(j)
 continue
 l=i.strip().split('\t')
 if l[0]=="ADAR":
 for j in range(len(l)):
 if j in a: print('%s\t%s\t%s' %('Alpha','ADAR',l[j]))
 if j in b: print('%s\t%s\t%s' %('Beta','ADAR',l[j]))
 if l[0]=="ADARB1":
 for j in range(len(l)):
 if j in a: print('%s\t%s\t%s' %('Alpha','ADARB1',l[j]))
 if j in b: print('%s\t%s\t%s' %('Beta','ADARB1',l[j]))
f.close()
```

Save the above code to a file named CollectADAR.py and run it with the command:

```
$ python CollectADAR.py > adar.txt
```

4. Create a box plot to visualize gene expression values using the following python code:

```
import sys, os
import seaborn as sns
import matplotlib.pyplot as plt
import pandas as pd
from scipy import stats
```

```
sns.set_theme(style="whitegrid")

data='adar.txt'
df = pd.read_table(data, sep='\t')
df2 = df.loc[df['Gene'] == "ADAR"]
df3 = df2.loc[df2['TPM'] > 0]
ax = sns.boxplot(x="Cell_Type", y="TPM", data=df3)
plt.savefig('adar.png')
plt.close()
a=df3.loc[df3['Cell_Type'] == "Alpha"]['TPM']
b=df3.loc[df3['Cell_Type'] == "Beta"]['TPM']
```

Save the above code to a file named ADARplot.py and run it with the command:

```
$ python ADARplot.py
```

It will generate the box plot depicted in Fig. 3. To visualize a plot for the ADARB1 gene expression, replace "ADAR" with "ADARB1" in the above python code.

4 Notes

1. Windows users can import a docker containerized version available at http://srv00.recas.ba.infn.it/webshare/reditools_docker.tar.gz. To complete the import, put the tar file under the user folder (e.g., C:\Users\user folder\reditools_docker.tar.gz).

   In the Windows command line, run the docker load command:

```
C:\Users\user folder> docker load -i reditools_docker.
tar.gz
```

The file will be loaded as a docker image and available through the command:

```
C:\Users\user folder> docker images
```

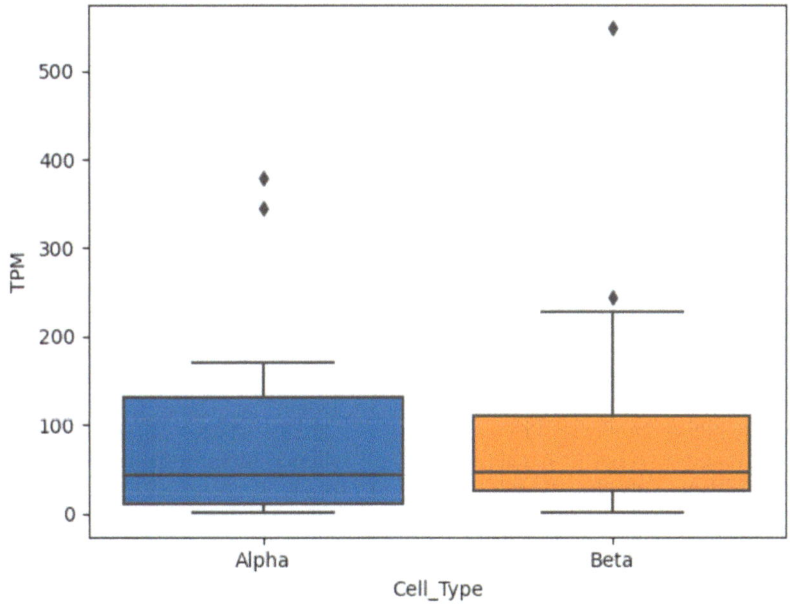

Fig. 3 Box plot of ADAR expression. *TPM values for the ADAR gene were calculated across both cell populations*

Finally run a container from the reditools_docker image using the command:

```
C:\Users\user folder> docker run -i -t reditools_docker /
bin/bash
```

2. The entire protocol has been run on a Linux (CentOS 7) machine equipped with 40 cores, 256 GB of RAM, and several TB of disk space. Although the procedure is not compute intensive, the analyses can be speeded up by using multiple cores (at least 4 threads).
3. To get REDIportal annotations, open the link http://srv00.recas.ba.infn.it/atlas/download.html select "*Homo sapiens*" as Organism Name and "hg38" as Genome Version. Then click on the Get Link button and a link will appear in the text box for the direct download or through wget (or curl) in a terminal.

Acknowledgments

We kindly thank the Italian node of ELIXIR (Elixir-IIB) and the Bari ReCaS DataCenter for computing support.

References

1. Gott JM, Emeson RB (2000) Functions and mechanisms of RNA editing. Annu Rev Genet 34:499–531
2. Eisenberg E, Levanon EY (2018) A-to-I RNA editing – immune protector and transcriptome diversifier. Nat Rev Genet 19:473–490
3. Pinto Y, Cohen HY, Levanon EY (2014) Mammalian conserved ADAR targets comprise only a small fragment of the human editosome. Genome Biol 15:R5
4. Gallo A, Locatelli F (2012) ADARs: allies or enemies? The importance of A-to-I RNA editing in human disease: from cancer to HIV-1. Biol Rev Camb Philos Soc 87:95–110
5. Silvestris DA, Picardi E, Cesarini V et al (2019) Dynamic inosinome profiles reveal novel patient stratification and gender-specific differences in glioblastoma. Genome Biol 20:33
6. Lo Giudice C, Tangaro MA, Pesole G et al (2020) Investigating RNA editing in deep transcriptome datasets with REDItools and REDIportal. Nat Protoc 15:1098–1131
7. Picardi E, Manzari C, Mastropasqua F et al (2015) Profiling RNA editing in human tissues: towards the inosinome Atlas. Sci Rep 5: 14941
8. Tan MH, Li Q, Shanmugam R et al (2017) Dynamic landscape and regulation of RNA editing in mammals. Nature 550:249–254
9. Mansi L, Tangaro MA, Lo Giudice C et al (2021) REDIportal: millions of novel A-to-I RNA editing events from thousands of RNA-seq experiments. Nucleic Acids Res 49: D1012–D1019
10. Picardi E, Pesole G (2013) REDItools: high-throughput RNA editing detection made easy. Bioinformatics 29:1813–1814
11. Picardi E, Horner DS, Pesole G (2017) Single cell transcriptomics reveals specific RNA editing signatures in the human brain. RNA 23(6): 860–865
12. Ansell BRE, Thomas SN, Bonelli R et al (2021) A survey of RNA editing at single-cell resolution links interneurons to schizophrenia and autism. RNA 27:1482–1496
13. Enge M, Arda HE, Mignardi M et al (2017) Single-cell analysis of human pancreas reveals transcriptional signatures of aging and somatic mutation patterns. Cell 171:321–330.e14
14. Wu TD, Nacu S (2011) Fast and SNP-tolerant detection of complex variants and splicing in short reads. Bioinformatics 26:873–881
15. Kim D, Langmead B, Salzberg SL (2015) HISAT: a fast spliced aligner with low memory requirements. Nat Methods 12:357–360
16. Dobin A, Davis CA, Schlesinger F et al (2013) STAR: ultrafast universal RNA-seq aligner. Bioinformatics 29:15–21
17. Picelli S, Faridani OR, Björklund AK et al (2014) Full-length RNA-seq from single cells using Smart-seq2. Nat Protoc 9:171–181
18. Haque A, Engel J, Teichmann SA et al (2017) A practical guide to single-cell RNA-sequencing for biomedical research and clinical applications. Genome Med 9:75
19. Li H, Handsaker B, Wysoker A et al (2009) The sequence alignment/map format and SAMtools. Bioinformatics 25:2078–2079
20. Chen SZY (2018) fastp: an ultra-fast all-in-one FASTQ preprocessor. Bioinformatics 34(6)
21. Liao Y, Smyth GK, Shi W (2014) featureCounts: an efficient general purpose program for assigning sequence reads to genomic features. Bioinformatics 30:923–930
22. Wang L, Wang S, Li W (2012) RSeQC: quality control of RNA-seq experiments. Bioinformatics 28:2184–2185
23. Roth SH, Levanon EY, Eisenberg E (2019) Genome-wide quantification of ADAR adenosine-to-inosine RNA editing activity. Nat Methods 16:1131–1138
24. Lo Giudice C, Silvestris DA, Roth SH et al (2020) Quantifying RNA editing in deep transcriptome datasets. Front Genet:11

Chapter 19

Practical Considerations for Complex Tissue Dissociation for Single-Cell Transcriptomics

Stephanie T. Pohl, Maria Llamazares Prada, Elisa Espinet, and Renata Jurkowska

Abstract

Single-cell and single-nucleus RNA sequencing have revolutionized biomedical research, allowing analysis of complex tissues, identification of novel cell types, and mapping of development as well as disease states. Successful application of this technology critically relies on the dissociation of solid organs and tissues into high-quality single-cell (or nuclei) suspensions.

In this chapter, we examine several key aspects of the tissue handling workflow that need to be considered when establishing an efficient tissue processing protocol for single-cell RNA sequencing (scRNA-seq). These include tissue collection, transport, and storage, as well as the choice of the dissociation conditions. We emphasize the importance of the tissue quality check and discuss the advantages (and potential limitations) of tissue cryopreservation. We provide practical tips and considerations on each of the steps of the processing workflow, and comment on how to maximize cell viability and integrity, which are critical for obtaining high-quality single-cell transcriptomic data.

Key words Tissue dissociation, Cryopreservation, Enzymatic digestion, Single-cell suspension, Tissue viability, Lung tissue, Single-cell transcriptomics, Single-cell RNA-seq

1 Introduction

Recent advances and widespread implementation of the single-cell technologies have transformed basic and biomedical research, allowing unprecedented insights into complex tissues [1–3], identification of previously unknown cell types [4, 5], reconstitution of developmental trajectories [6, 7], and mapping of disease states and drivers [8–12]. As we are moving into the era of personalized medicine, single-cell RNA sequencing (scRNA-seq), single-nucleus RNA-seq (snRNA-seq), and single-cell multiomics sequencing (sc multiomics-seq) hold great promise for biomarker discovery for disease diagnosis, prognosis, and response to treatment [13, 14].

Raffaele A. Calogero and Vladimir Benes (eds.), *Single Cell Transcriptomics: Methods and Protocols*, Methods in Molecular Biology, vol. 2584, https://doi.org/10.1007/978-1-0716-2756-3_19,

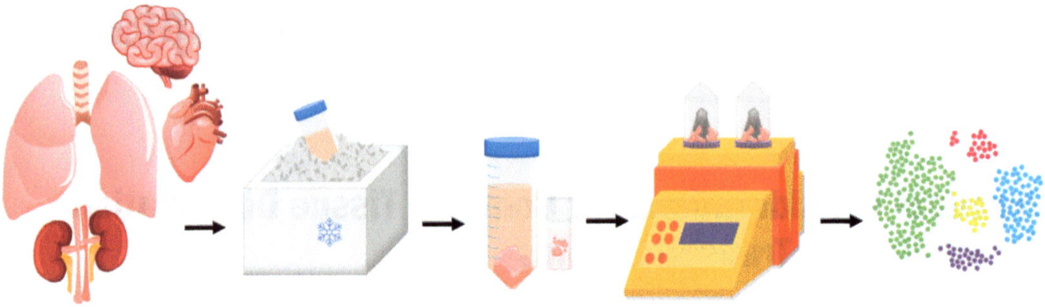

Fig. 1 Schematic representation of the main key steps of tissue processing workflow that need to be considered for single cell transcriptomics, including tissue collection and transport, tissue storage, and tissue dissociation

For single-cell transcriptomics, solid organs and tissues need to be dissociated to release individual cells from their extracellular matrix. When tissue complexity or storage format (e.g., fixed or snap frozen samples) preclude intact cell isolation, suspensions of single nuclei can be prepared [15, 16]. With the development of large collaborative single-cell initiatives, like Human Cell Atlas Project [2, 17, 18], numerous human tissue processing protocols compatible with the scRNA-seq workflow have been developed and benchmarked. Importantly, preparation of a high-quality single-cell suspension is a key prerequisite for obtaining high-quality single-cell transcriptomics data [19]. Therefore, each parameter of the tissue processing protocol needs to be carefully tested and optimized. Independent of the exact platform used, the ideal dissociation protocol should preserve cellular composition of the original tissue, yield single-cell suspension with high viability, and maintain high RNA integrity. Moreover, some tissues are more difficult to work with than others, and their processing may require extensive optimization.

When developing a tissue processing workflow for single-cell RNA-seq several steps need to be carefully considered and optimized (Fig. 1), including the following:

1. Tissue collection (patient inclusion criteria, sampling method, required amount, transport logistics).
2. Tissue storage before dissociation (storage buffer, time and temperature, possibility of cryopreservation/freezing).
3. Tissue quality check and histological characterization.
4. Tissue dissociation into single-cell suspension (dissociation temperature, selection of the enzyme mix, and dissociation procedure).

In the following text, we discuss the importance of each of these steps and provide practical considerations on how to maximize cell viability and integrity, which are critical for obtaining high-quality single-cell RNA-seq data. Our workflow, described in detail in Subheading 3, was originally developed for processing human lung tissue, but similar considerations should be given for other complex tissues and organs.

Tissue Collection

Patient Inclusion Criteria

As single cell -omics technologies remain expensive, usually relatively small samples sizes are analyzed [20]. Hence, for disease studies, it is important to establish strict patient inclusion criteria for prospective tissue collection to ensure the best possible sample characterization and matching of the control and disease groups. Selection criteria depend on the exact experimental question and need to be set up specifically for each project. We usually match our lung donors in terms of age, gender, body mass index, smoking status, and smoking history. In addition, whenever possible, we also recommend collecting patient medical history and tests results for the best possible characterization of the included samples and disease classification.

Tissue Sampling

In addition to patient inclusion criteria, the choice of tissue collection method needs to be considered depending on availability, feasibility, required amount, and suitability to the research question. Tissue can be obtained via different sampling methods (e.g., from surgical resections or biopsy extraction, from life or deceased donors). The choice of starting material, location of the collected tissue piece within the organ, and the sampling method are important parameters as they may yield different cellular representation in scRNA-seq analysis.

Tissue Transport and Storage (Before Dissociation)

The tissue piece is typically collected in transport medium from the surgery theater and sent on ice to the processing laboratory via an organized transport route (e.g., courier). The transport medium should allow an optimized storage of fresh tissue and maintain high viability. There must be sufficient buffering capacity and osmolarity. A variety of organ and tissue transport solutions have been developed [21] and can be tested when optimizing a new workflow. For human tissue transport, country-specific human tissue regulations apply and require special packaging, labeling, and tracking of the material. Typically, a double-sealed container system is required. During tissue procurement and transport, care should be taken to prevent contamination, cross-contamination between samples, mislabeling, and deterioration of the tissue. To obtain the highest possible tissue and cell viability, it is essential to keep the overall time from tissue collection to final processing as short as possible. Hence, the logistics of tissue collection, transport, and processing needs to be well organized and optimized.

Standard tissue storage formats, like formalin-fixed paraffin-embedded (FFPE) or flash-frozen, do not yield viable cell suspensions and are therefore not compatible with scRNA-seq analysis (can be used for snRNA-seq ans sc multiomics). Hence, most single-cell transcriptomic studies to date have relied on direct processing of fresh tissue received from the hospital. Processing of fresh tissue is attractive, as it does not introduce any further tissue treatments steps apart dissociation, minimizing additional biases [19, 22]. However, fresh tissue processing poses several challenges, especially in clinical settings, including access to specialized equipment and specialist facilities where the material needs to be dissociated immediately and further processed to avoid sample degradation [22]. This restricts complex experimental designs, does not allow histological evaluation before processing, and may introduce technical biases related to processing of fresh samples in multiple batches [22, 23]. Alternative protocols include cold storage of the tissue [24], addition of various fixatives and stabilizers to dissociated cells or minced tissues [25–29], or cryopreservation [30–32].

Tissue Storage

In order to allow sufficient time to assess the quality of the collected tissue and to design and organize the follow-up processing, clinical fresh specimens may be stored by cryopreservation. Cryopreservation entails controlled freezing of the small pieces of tissue in the presence of a cryopreserving agent (e.g., DMSO), protecting tissue, and cell viability [30]. Alternatively, collected samples can also be cryopreserved as dissociated single-cell suspensions for subsequent scRNA-seq analysis [29]. The main advantage of cryopreservation is that it in enables a longer-term storage of the viable tissue. Hence, it allows thorough sample quality check before further processing, permitting exclusion of low-quality samples [30]. In addition, separating location and time of tissue collection from downstream scRNA-seq analysis facilitates complex designs and collaboration between multiple laboratories. Prior to cryopreservation, tissues can be macroscopically separated in areas of interest (e.g., airways from lung parenchyma, or different parts of tumors, distal, central) allowing to expand and define the collection of viable tissue. However, small biases in population proportions might occur in cryopreserved samples [32, 33]. Hence, consistency across samples for a given tissue and experiment is recommended.

Tissue Quality Check Before Downstream Analysis

As mentioned above, tissue cryopreservation enables the pathological evaluation of the sample before downstream processing, ensuring that only high-quality samples are processed and sequenced. For this, representative tissue slides (i.e., adjacent to the tissue piece that is cryopreserved) can be stained with hematoxylin and eosin (H&E) and analyzed by experienced pathologists. If needed, specific informative markers can be used in immunohistochemistry analysis for further evaluation (e.g., cytokeratins for certain tumors). For diseased samples, histopathological evaluation allows better patient characterization based on disease phenotypes (e.g., for lung tissue: presence of emphysema, fibrosis, alveolar thickening, extent and type of immune infiltration, tumor cell content, and necrosis). For non-diseased (control) samples, it confirms the absence of disease phenotypes. Accurate selection of control and diseased samples based on both patient medical history and histological tissue characterization will result in more meaningful data. For example, the presence of tumor cells or extensive inflammation in control samples increase the experimental noise and may mask disease-associated gene expression signatures. Hence, tissue quality check is critical, as it enables exclusion of low-quality samples, which is important for the discovery of disease-relevant changes. Additionally, having access to representative tissue slides of the cryopreserved tissue allows the validation and investigation of the genes identified by scRNA-seq analysis in a spatial approach and at a protein level.

Tissue Dissociation into Single-Cell Suspension

Dissociation of an organ or tissue into a single-cell suspension often requires a lengthy and complex protocol. Several parameters of the dissociation protocol need to be considered and optimized, including dissociation temperature, time, as well as the choice of the enzyme mix and dissociation device.

Digestion temperature Dissociation temperature is an important parameter that needs to be tested and optimized when developing a new tissue processing workflow, as the use of warm or cold dissociation may alter gene expression profiles. Tissue dissociation is commonly conducted using proteolytic enzymes that require incubation at 37 °C for optimal activity. Digestion at 37 °C may however introduce gene expression artifacts in specific cell populations (e.g., induction of heat shock proteins and immediate-early response genes) that correlate with the digestion time since mammalian transcriptional machinery is also active at this temperature [33, 34]. Hence, alternative protocols have been developed based on cold-adapted proteases that show high activity at low temperatures, where mammalian transcription is minimized [35, 36]. Additionally, computational approaches can be used to remove dissociation-affected cells from single-cell data sets in silico [34].

Dissociation enzymes For tissue digestion several enzymes are typically used in various combinations and concentrations (e.g., trypsin, collagenase, liberase, or dispase), often aided with mechanical agitation or dissociation. The choice of the enzymatic mix typically depends on the starting material and needs to be carefully optimized. Notably, when optimizing a new workflow for fresh (or cryopreserved) tissue, different enzymatic combinations should be tested, as they may lead to different representations of cellular populations [37].

(Lung) Tissue Processing Workflow for scRNA-seq

Our complete tissue processing workflow, originally developed for human lung and described in detail in Subheading 3, consist of four main steps: (1) tissue collection, (2) tissue cryopreservation, (3) tissue quality check, and (4) tissue dissociation (Figs. 1 and 2). First, tissue is collected in transport medium from the surgery theater and sent to the processing laboratory. After initial washing, representative tissue pieces are collected for FFPE for subsequent histological analysis. The remaining material is cut into small pieces (approx. 4 × 4 mm) and cryopreserved for long-term storage (Fig. 1 *top row*). The FFPE slides are stained by H&E and evaluated by an experienced pathologist. This tissue quality check is a critical step that allows characterization and classification of the samples, and most importantly, the exclusion of low-quality material. After histopathological evaluation, the cryopreserved tissue is thawed, washed, and dissociated into homogenous single-cell suspension using a combination of mechanical and enzymatic processing. Finally, dissociated single-cell suspension with high viability can be subjected to a scRNA-seq workflow (not covered in detail in this chapter).

2 Materials and Equipment

Required buffers, media, solutions, and specialist equipment are provided separately for the different steps of the workflow.

Tissue Collection and Transport:

1. Transport buffer (CO_2-i +++): CO_2-independent medium as base medium (Thermo Fisher Scientific), 1% BSA, 1% penicillin/ streptomycin, 1% amphotericin B.
2. 50 mL canonical (Falcon) tubes.
3. Resealable plastic bag.
4. Sterile forceps.
5. Transport box filled with ice.

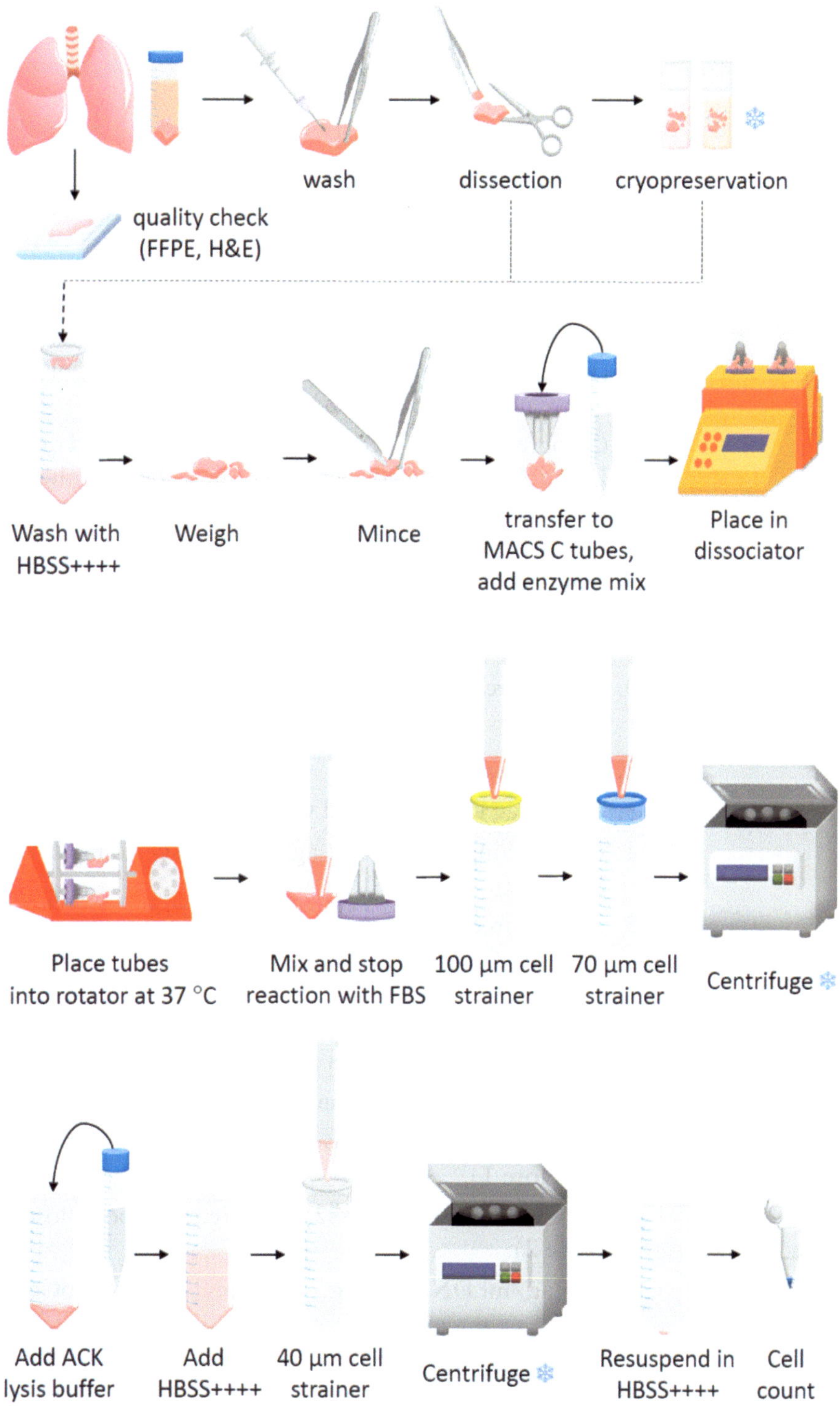

Fig. 2 Schematic representation of the detailed steps of the tissue processing workflow described in Subheading 3 (for details refer to Subheading 3)

Tissue Washing and Cryopreservation:

1. Plastic Petri dishes.
2. 15 mL and 50 mL canonical (Falcon) tubes.
3. Cryogenic tubes, 2.0 mL.
4. Syringes (2.5 mL and 5 mL).
5. Syringe needles (0.5 × 25 mm).
6. 10% neutral buffered formalin (Sigma).
7. Tissue wash buffer (HBSS++++): Hanks balanced salt solution (HBSS) without Ca2+, Mg2+ and phenol red as base buffer (Fisher Scientific), 2 mM EDTA, 1% BSA, 1% penicillin/streptomycin, 1% amphotericin B.
8. Cryopreservation medium for lung tumor and parenchyma: Dulbecco's Modified Eagle Medium (DMEM), high glucose, GlutaMAX™ as base medium (Thermo Fisher Scientific), 20% Fetal Bovine Serum (FBS), 10% Dimethylsulfoxide (DMSO).
9. Optional: Cryopreservation medium for airways and vessels (or dissociated single-cell suspensions): Cryostor (Sigma, C2874).
10. Sterile forceps and dissection scissors.
11. Balance.
12. Rocking table (at 4 °C).
13. Two large ice boxes filled with ice.
14. Cell freezing containers (e.g., Mr. Frosty from Thermo Fisher Scientific).
15. −80 °C freezer.
16. Liquid nitrogen dewar and freezing boxes.

Tissue Dissociation:

1. Gentle MACS™ C tubes (Miltenyi Biotec).
2. Falcon cell strainers 40 μm, 70 μm, 100 μm.
3. Human tumor dissociation kit (Miltenyi Biotec, 130-095-929) (*see* **Note 1**).
4. ROCK inhibitor, Y-27632, 2HCl (e.g., Adooq Bioscience) (*see* **Note 2**).
5. 100 μg/ml DNase I (e.g., ProSpec-Tany TechnoGene).
6. 20% Fetal Bovine Serum (FBS), heat-inactivated (Gibco).
7. ACK lysis buffer (Thermo Fisher Scientific).
8. Trypan Blue solution 0.4%, sterile.
9. Sterile scalpels and forceps.
10. Gentle MACS dissociator (Miltenyi Biotec) (*see* **Note 3**).

11. MACSmix™ tube rotator (Miltenyi Biotec).
12. Hemocytometer Counting Chamber (e.g., Neubauer Improved).
13. Water bath set to 37 °C.
14. Balance.

Optional:

15. FACS analyzer/sorter.
16. SyTOX blue viable dye.

3 Methods

Tissue Collection, Transport, and Initial Processing:

1. Prepare 50 mL conical tubes filled with around 30 mL of CO_2-independent medium (CO_2-i+++) and store them at 4 °C until use (*see* **Note 4**).
2. Collect lung tissue piece from the surgery, place it inside a 50 mL conical tube containing CO_2-independent medium+++, carefully label the tube, put it in inside a plastic bag, seal tightly, and keep on ice during transport in a sealed container. Transfer to the sterile tissue processing area as quickly as possible (*see* **Note 5**).
3. Under a class 2 cell culture hood disinfect forceps and cutting tools with 70% ethanol, put them on a Petri dish, and wait until ethanol has completely evaporated before starting the dissection procedure.
4. Place the tissue piece on a Petri dish, weigh it, and record the tissue mass.
5. Wash the tissue extensively, first with ice-cold PBS and afterward with ice-cold HBSS++++ (*see* **Note 6**).
6. (Optional): Carefully remove the pleura using surgical scissors (*see* **Note 7**).
7. For histological analysis of the tissue, slice representative pieces of the lung tissue/tumor and put in into a 15 mL conical tube containing 10% neutral buffered formalin for fixing. Rock the tube at 4 °C for 24 h. Next day, discard the formalin (requires special waste), wash with PBS, and transfer the tissue to a 15 mL conical tube containing 70% ethanol. Rock for an additional 24 h at 4 °C and keep at 4 °C in 70% ethanol until paraffin embedding (*see* **Note 8**). For next steps on histological processing, refer to the *tissue quality check section* below.
8. Cut the remaining tissue into small pieces (approx. 4 × 4 mm) using surgical scissors (*see* **Note 9**).

9. Proceed with tissue cryopreservation (next section) or *if processing fresh tissue, continue with tissue dissociation from* **step 3** (section after next) (*see* **Note 10**).

Tissue Cryopreservation: (Optional)

1. Transfer the small lung tissue pieces into a 2.0 mL cryotubes so that it each tube is filled no more than the 500 μl mark (equals approx. 10–15 pieces per tube).
2. Add 1–1.4 mL of ice-cold cryopreserving medium to fill the tubes, close the tubes, and flip them to distribute the medium within the tissue pieces. Make sure there are no tissue pieces stuck to the lid after flipping the tube.
3. Incubate the tubes on ice until tissue sinks in the tube (maximum15 min). This enables the cryopreserving medium to penetrate the tissue pieces.
4. Place the tubes in the freezing containers (e.g., Mr. Frosty) and bring them to a −80 °C freezer. In the following days, transfer the tubes containing cryopreserved tissue to a storage box pre-cooled at −80 °C and place it in the liquid nitrogen dewar for long-term storage (*see* **Note 11**).

Tissue Quality Check:

1. Embed the fixed tissue pieces in paraffin blocks. Cut 4 μm slides using a microtome and transfer to adhesive glass slides. Perform H&E and immunohistochemistry (IHC) staining, if needed, for subsequent histological analysis (*see* **Note 12**).
2. H&E/IHC-stained tissue slides should be evaluated by an experienced pathologist to (1) confirm and score disease phenotypes (for diseased samples) and (2) confirm the absence of disease phenotypes in control samples (*see* **Note 13**).

Tissue Dissociation:

1. Thaw cryopreserved lung tissue pieces for 2 min in a water bath set to 37 °C (skip this step if fresh tissue is used) (*see* **Note 14**).
2. Quickly transfer all tissue pieces by flipping the thawed tubes onto a cell strainer (use forceps if same pieces are left in the tube) and wash them twice with HBSS++++ (skip this step if fresh tissue is used).
3. Transfer the tissue to a new Petri dish tube and weigh the tissue pieces on a balance. It is important to determine the exact weight of the used tissue to adjust the amount of enzyme and MACS C dissociation tubes that will be needed (*see* **Note 15**).
4. Under a class 2 cell culture hood, using scalpels or surgical scissors and forceps mince the tissue thoroughly into smaller pieces before transferring to MACS C tubes for mechanical and enzymatic dissociation.

5. Transfer the minced tissue to a MACS C tube containing 4.5 mL of CO_2-i+++, add the enzyme mix from the human tissue dissociation kit, consisting of 200 μL enzyme H, 100 μL enzyme R, 50 μL enzyme A, 10 μM ROCK inhibitor, and 100 μL DNaseI (*see* **Note 16**).
6. Close the MACS C tubes tightly and place them into the MACS dissociator for mechanical dissociation (*see* **Note 17**). Following the recommended program for lung tissue in the following order:

Dissociation step	Dissociator program	Incubation
1	h_tumor_01	15 min at 37 °C on MACSmix rotator
2	h_tumor_01	15 min at 37 °C on a MACSmix rotator
3	h_tumor_02	15 min at 37 °C on a MACSmix rotator
4	h_tumor_02	Transfer cells to class 2 cell culture hood

7. Pipet the dissociated mix up and down using a 10 mL serological pipet to support dissociation, aiming for a smooth mixture without clumps. Avoid bubbles.
8. Stop the dissociation process by adding 2 mL of 20% FBS per MACS C tube (*see* **Note 18**).
9. To remove any leftover aggregates, tissue pieces and undissolved tissue matrix, pass the suspension through cell strainers of descending mesh size. Start with 100 μm, then transfer the flowthrough to a 70 μm cell strainer and collect into a 50 mL conical tube.
10. To maximize cell recovery, wash the tubes and strainers with 10 mL HBSS++++, pass through the strainer, and combine with the cell suspension.
11. Centrifuge the cell suspension at 4 °C, 300 g for 8 min.
12. Resuspend cell pellet in 4 mL ACK lysis buffer and incubate for 4 min at room temperature (*see* **Note 19**).
13. Add 20 mL HBSS++++ and pass the cell suspension through a 40 μm cell strainer.
14. Centrifuge the cell suspension at 4 °C, 1.400 rpm (around 300 g) for 8 min and resuspend the cell pellet in 500 μL HBSS++++.
15. Take an aliquot of the cell suspension and determine cell number and viability. For example, when using trypan blue, mix the

aliquot with trypan blue solution in a 1:1 ratio, load into a counting chamber (e.g., Improved Neubauer Chamber) for cell counting and to determine cell viability (*see* **Note 20**).

16. As a final step, we recommend calculating the number of cells obtained per tissue weight by dividing the counted cell number by the initial mass of tissue used for dissociation (*see* **Note 21**).
17. The cells can be further processed by fluorescence-activated cell (FACS) sorting to enrich specific populations or cryopreserved as single-cell suspension before proceeding with the scRNA-seq workflow (if processed from fresh tissue).

4 Notes

1. Upon arrival, the kit should be stored at 4 °C. After reconstitution (following manufacturer's instructions), the enzymes should be aliquoted and stored at −20 °C.
2. Y-27632 is a selective, cell-permeable inhibitor of Rho-associated kinase (ROCK). It is used to increase the survival of human stem cells after dissociation into single cells. We usually prepare a 10 mM stock in DMSO, and store in small aliquots at −20 °C for 6 months to 1 year. Once thawed, the aliquot should be stored at 4 °C and used within 2 weeks.
3. There are newer MACS dissociator models available with eight cartridges and integrated heaters (e.g., gentleMACS™ Octo Dissociator with heaters; 130-096-427, Miltenyi Biotec), which enable performing the dissociation procedure in one device. When using gentleMACS™ Octo, MACSmix™ tube rotator is not required.
4. We typically obtain 0.5–10 g of human lung tissue from the surgery, but the protocol can be upscaled if needed for larger pieces (use multiple 50 mL conical tubes or larger containers for transport).
5. Care should be taken to carefully label the tubes, especially if multiple samples are being collected (e.g., tumor, parenchyma) to maintain traceability and avoid sample swap. To obtain the highest possible tissue and cell viability, the overall time from tissue collection to final processing should be as short as possible. We therefore usually process the tissue with several people in parallel. It is helpful to bring all the necessary materials to the class 2 hood area in advance of the tissue arrival and prepare and label all the required tubes and Petri dishes. We usually also prepare at least two large ice boxes to maintain all the buffers ice-cold during tissue processing.
6. The washing step is required to remove as much blood as possible from the tissue. Larger tissue pieces may be cut into

smaller parts if washing is not efficient. Using a syringe with a needle, infuse and wash the tissue piece gently, by "inflating it" with ice-cold PBS while holding it with the forceps. Repeat this step at least 4–5 times using ice-cold HBSS++++.

7. The removal of the pleura is optional and depends on whether this part is of interest for analysis. The easiest way to remove the pleura is to lift the tissue with the forceps and cut out the pleura with scissors as close as possible and parallel to the surface.
8. If the tissue piece is heterogenous, we recommend taking representative pieces for FFPE from different parts and compartments, respectively, for a more representative characterization and classification of the sample (e.g., presence of disease phenotypes in different lung compartments, like airways, vessels, or alveoli). Inclusion of this step allows collection of a sample for H&E/IHC staining and pathological evaluation (for more detailed discussion of the utility of that step, refer to **Note 13**). In addition, as the FFPE samples are collected from the same piece of tissue that is subjected to the scRNA-seq, they can be used for subsequent validation of the transcriptomic results (e.g., by immunofluorescence (IF) or IHC). Alternative tissue embedding methods (e.g., OCT) compatible with H&E, IF, and IHC staining can be used to collect tissue for evaluation and subsequent validation. For FFPE samples, it is important not to extend the incubation time in the formalin solution beyond 24 h, as this may lead to over fixation. Once washed and transferred to ethanol, the tissue pieces can be stored at 4 °C for several days before embedding.
9. If an initial pre enrichment of different tissue compartments or parts is desired (e.g., for the lung: distal versus proximal, tumor versus adjacent parenchyma, or large airways versus alveolar parenchyma), the material can be separated prior cutting to small pieces via manual macroscopic dissection and the different fractions cryopreserved separately. We typically first cut the tissue into several smaller pieces and process one piece at the time (with several people in parallel). The lung tissue should be cut into small pieces between 9 mm^2 and 16 mm^2. The uncut lung should be kept in cold HBSS++++ (on ice) while processing the rest.
10. Tissue cryopreservation permits thorough histological characterization of the collected tissue before time-consuming and expensive downstream dissociation and scRNA-seq. Additionally, it also enables more complex experimental design, as after thawing, samples from multiple donors can be processed and dissociated in parallel, technical biases related to multiple batch processing of fresh samples are avoided. However, small biases in the proportions of different cell populations may occur in

cryopreserved samples [27, 33]; therefore, careful consideration of the experimental design and consistency in the processing and storage across samples for a given tissue, samples, and study is strongly recommended.

11. We recommend the use of cell freezing containers (e.g., Mr. Frosty) to ensure a gradual temperature decrease (1 °C/min). For short-term storage (up to 2 weeks), the samples can be stored at −80 °C, but for longer storage the samples should be transferred to liquid nitrogen dewar or a −150 °C freezer to maintain their viability. When transferring, the frozen vials should be transported on dry ice since thawing of the samples will lead to decreased cell viability.
12. Many research institutions offer tissue embedding and staining services. Alternatively, paraffin embedding and H&E staining can also be outsourced (e.g., we successfully used service from Morphisto (Morphisto GmbH, Frankfurt, Germany) in the past). When cutting the slides by yourself, we recommend using adhesive glass slides (e.g., Superfrost Plus, Thermo Fisher). Subsequent drying of the sections O/N in a 40 °C oven removes excess water and enhances adhesion.
13. Careful histological evaluation of the collected tissue is critical, as it enables exclusion of low-quality samples. From our experience with the collection of human lung tissue from control donors with preserved lung function (as assessed by spirometry), only about 35–50% of samples showed normal histological pattern expected for a control lung [30]. Other samples showed presence of moderate to extensive emphysema or fibrosis, alveolar wall thickening, as well as immune infiltration and other alterations. Although the implementation of the strict tissue-quality step results in a significant dropout of samples before sequencing, it results in a controlled setup, where the control and disease groups are divided not only based on clinical parameters, but also on histopathology of the same tissue piece which is used for downstream analysis.
14. Transport the tubes containing cryopreserved tissue on dry ice after their retrieval from the liquid nitrogen dewar. The use of the water bath at 37 °C enables a fast thawing of the tissue samples, which is important to maintain high cell viability. When thawing multiple tubes, use a floating tube rack to hold the tubes immersed in the water bath.
15. The dissociation kit manufacturer recommends using max. 1 g tissue per MACS C tube. If using 2 g of tissue, it is recommended to double the amount of enzyme mix, ROCK inhibitor and DNase I, since manufacturers' protocol indications are for 1 g of tissue.

16. The choice of the best dissociation conditions needs to be considered carefully and each parameter experimentally optimized. This includes choice of the enzyme mix, dissociation temperature, time, and dissociation device. For the dissociation of the lung parenchyma and lung tumors, we obtained very good results using tumor dissociation kit from Miltenyi Biotech and a gentle MACS dissociator. However, the efficiency of the dissociation largely depends on the type of tissue used, sampling method, and requires a careful optimization of the digestion parameters. Refer to Miltenyi website for guidance on available dissociation kits. Alternatively, digestion enzymes can be purchased separately and mixed in different composition during optimization steps. In this case, we would recommend using high purity enzymes to avoid unspecific enzymatic activities. After dissociation, we recommend running a flow cytometry analysis to analyze distribution of different cell populations and their viability. Ideally, the dissociation protocol will lead to an efficient release of all cell types (including rare and fragile cells), while preserving their integrity and variability.
17. Different models of the MACS dissociators are available. Gentle MACS™ dissociator allows processing two samples in parallel and has no automated heating. Hence, the samples need to be transferred between the MACSmix™ tube rotator in the cell culture incubator and the bench-top dissociator. This can be time consuming and not ideal when processing multiple samples, as it may compromise cell viability. There are newer models available with eight cartridges and integrated heaters (e.g., gentleMACS™ Octo Dissociator with heaters; 130-096-427, Miltenyi Biotec), which enable performing the entire procedure in one device, making the dissociation process more efficient, easier, and faster.
18. Addition of FBS inactivates the enzyme mix. Try to avoid foaming when pipetting up and down, since it may lead to decreased yield of single cells after the dissociation procedure. We use heat inactivated FBS to destroy complement that may lead to cell lysis by antibody binding.
19. This step is optional and may not be needed if the sample has low red blood cell content (<20%). It is recommended if the tissue looks "bloody," because if the erythrocytes are not removed, they will add to the total number of loaded cells, increasing the amount of sequencing reads needed for coverage of cells of interest.
20. Trypan blue will stain dead cells dark blue, while leaving live cells unstained. It is compatible with any manual hemocytometer counting chamber and can be also used with many

automated cell counters; however, other dyes can be used. In addition, cell viability of the total suspension, as well as of specific fractions (e.g., epithelial or immune), can also be evaluated by flow cytometry using viability dyes (e.g., SyTOX staining). The correct cell count is key as it ensures appropriate cell loading and droplet formation in the subsequent scRNA-seq workflow.

The viability of the cell suspension should be as high as possible, ideally above 90%, but at least 75%. We typically achieve 85–90% viability for cryopreserved tumors and 78–94% for cryopreserved parenchyma [30]. We successfully dissociated cryopreserved lung tissue from smokers, as well as COPD donors with various stages of COPD, maintaining viability above 80%, indicating that this protocol can be used to profile both normal and COPD tissue. If higher viability is required, enrichment of viable cells by FACS can be performed. This step leads to the removal of debris, damaged and dead cells, as well as cell doublets and aggregates.

21. This step is not strictly required, but it helps to estimate how well the dissociation protocol worked. It is also very helpful when comparing different dissociation conditions and enzyme mixes. The yield varies depending on the type of the sample (e.g., lung tumors are usually easier to dissociate than parenchyma), location of the lung the sample was taken from (high content of vessels or airways may decrease the yield), as well as the donor group (e.g., fibrotic lungs are harder to dissociate than normal lungs).

References

1. Travaglini KJ et al (2020) A molecular cell atlas of the human lung from single-cell RNA sequencing. Nature 587(7835):619–625
2. Regev A et al (2017) The human cell atlas. Elife 6:e27041
3. Litvinukova M et al (2020) Cells of the adult human heart. Nature 588(7838):466–472
4. Montoro DT et al (2018) A revised airway epithelial hierarchy includes CFTR-expressing ionocytes. Nature 560(7718):319–324
5. Plasschaert LW et al (2018) A single-cell atlas of the airway epithelium reveals the CFTR-rich pulmonary ionocyte. Nature 560(7718): 377–381
6. Buenrostro JD et al (2018) Integrated single-cell analysis maps the continuous regulatory landscape of human hematopoietic differentiation. Cell 173(6):1535–1548 e16
7. Tusi BK et al (2018) Population snapshots predict early haematopoietic and erythroid hierarchies. Nature 555(7694):54–60
8. Morse C et al (2019) Proliferating SPP1/MERTK-expressing macrophages in idiopathic pulmonary fibrosis. Eur Respir J 54(2): 1802441
9. Vieira Braga FA et al (2019) A cellular census of human lungs identifies novel cell states in health and in asthma. Nat Med 25(7): 1153–1163
10. Kim N et al (2020) Single-cell RNA sequencing demonstrates the molecular and cellular reprogramming of metastatic lung adenocarcinoma. Nat Commun 11(1):2285
11. Seumois G, Vijayanand P (2019) Single-cell analysis to understand the diversity of immune cell types that drive disease pathogenesis. J Allergy Clin Immunol 144(5):1150–1153

12. Ramachandran P et al (2019) Resolving the fibrotic niche of human liver cirrhosis at single-cell level. Nature 575(7783):512–518
13. Davis-Marcisak EF et al (2021) From bench to bedside: single-cell analysis for cancer immunotherapy. Cancer Cell 39(8):1062–1080
14. Yofe I, Dahan R, Amit I (2020) Single-cell genomic approaches for developing the next generation of immunotherapies. Nat Med 26(2):171–177
15. Habib N et al (2016) Div-Seq: single-nucleus RNA-Seq reveals dynamics of rare adult newborn neurons. Science 353(6302):925–928
16. Lacar B et al (2016) Nuclear RNA-seq of single neurons reveals molecular signatures of activation. Nat Commun 7:11022
17. Rozenblatt-Rosen O et al (2017) The human cell atlas: from vision to reality. Nature 550(7677):451–453
18. Schiller HB et al (2019) The human lung cell atlas: a high-resolution reference map of the human lung in health and disease. Am J Respir Cell Mol Biol 61(1):31–41
19. Haque A et al (2017) A practical guide to single-cell RNA-sequencing for biomedical research and clinical applications. Genome Med 9(1):75
20. Svensson V, Vento-Tormo R, Teichmann SA (2018) Exponential scaling of single-cell RNA-seq in the past decade. Nat Protoc 13(4):599–604
21. Guibert EE et al (2011) Organ preservation: current concepts and new strategies for the next decade. Transfus Med Hemother 38(2): 125–142
22. Lafzi A et al (2018) Tutorial: guidelines for the experimental design of single-cell RNA sequencing studies. Nat Protoc 13(12): 2742–2757
23. Conesa A et al (2016) A survey of best practices for RNA-seq data analysis. Genome Biol 17:13
24. Madissoon E et al (2019) scRNA-seq assessment of the human lung, spleen, and esophagus tissue stability after cold preservation. Genome Biol 21(1):1
25. Alles J et al (2017) Cell fixation and preservation for droplet-based single-cell transcriptomics. BMC Biol 15(1):44
26. Attar M et al (2018) A practical solution for preserving single cells for RNA sequencing. Sci Rep 8(1):2151
27. Ruf-Zamojski F et al (2018) Single-cell stabilization method identifies gonadotrope transcriptional dynamics and pituitary cell type heterogeneity. Nucleic Acids Res 46(21): 11370–11380
28. Thomsen ER et al (2016) Fixed single-cell transcriptomic characterization of human radial glial diversity. Nat Methods 13(1):87–93
29. Wohnhaas CT et al (2019) DMSO cryopreservation is the method of choice to preserve cells for droplet-based single-cell RNA sequencing. Sci Rep 9(1):10699
30. Llamazares-Prada M et al (2021) Versatile workflow for cell type-resolved transcriptional and epigenetic profiles from cryopreserved human lung. JCI Insight 6(6):e140443
31. Guillaumet-Adkins A et al (2017) Single-cell transcriptome conservation in cryopreserved cells and tissues. Genome Biol 18(1):45
32. Krjutskov K et al (2016) Single-cell transcriptome analysis of endometrial tissue. Hum Reprod 31(4):844–853
33. Denisenko E et al (2020) Systematic assessment of tissue dissociation and storage biases in single-cell and single-nucleus RNA-seq workflows. Genome Biol 21(1):130
34. van den Brink SC et al (2017) Single-cell sequencing reveals dissociation-induced gene expression in tissue subpopulations. Nat Methods 14(10):935–936
35. Adam M, Potter AS, Potter SS (2017) Psychrophilic proteases dramatically reduce single-cell RNA-seq artifacts: a molecular atlas of kidney development. Development 144(19): 3625–3632
36. O'Flanagan CH et al (2019) Dissociation of solid tumor tissues with cold active protease for single-cell RNA-seq minimizes conserved collagenase-associated stress responses. Genome Biol 20(1):210
37. Cohen M et al (2018) Lung single-cell signaling interaction map reveals basophil role in macrophage imprinting. Cell 175(4): 1031–1044 e18

Index

Raffaele A. Calogero and Vladimir Benes (eds.), *Single Cell Transcriptomics: Methods and Protocols*,
Methods in Molecular Biology, vol. 2584, https://doi.org/10.1007/978-1-0716-2756-3,

MIX
Papier aus verantwortungsvollen Quellen
Paper from responsible sources
FSC® C105338

If you have any concerns about our products,
you can contact us on
ProductSafety@springernature.com

In case Publisher is established outside the EU,
the EU authorized representative is:
Springer Nature Customer Service Center GmbH
Europaplatz 3, 69115 Heidelberg, Germany

Printed by Libri Plureos GmbH
in Hamburg, Germany